# Enhanced Carbon-Based Materials and Their Applications

*Edited by Poh Choon Ooi, Mengying Xie, and Chang Fu Dee*

WILEY-VCH

**Editors**

***Dr. Poh Choon Ooi***
Institute of Microengineering and Nanoelectronics (IMEN)
Universiti Kebangsaan Malaysia
43600 Bangi, Selangor
Malaysia

***Dr. Mengying Xie***
State Key Laboratory of Precision Measuring Technology and Instrument
College of Precision Instruments and Opto-electronics Engineering
Tianjin University
Tianjin, 300072
China

***Dr. Chang Fu Dee***
Institute of Microengineering and Nanoelectronics (IMEN)
Universiti Kebangsaan Malaysia
43600 Bangi, Selangor
Malaysia

**Cover:** © Andrey Suslov/Shutterstock



**Library of Congress Card No.:** applied for

**British Library Cataloguing-in-Publication Data**
A catalogue record for this book is available from the British Library.

**Bibliographic information published by the Deutsche Nationalbibliothek**
The Deutsche Nationalbibliothek lists this publication in the Deutsche Nationalbibliografie; detailed bibliographic data are available on the Internet at <http://dnb.d-nb.de>.



**Print ISBN:** 978-3-527-34802-2
**ePDF ISBN:** 978-3-527-82966-8
**ePub ISBN:** 978-3-527-82967-5
**oBook ISBN:** 978-3-527-82968-2

**Typesetting** Straive, Chennai, India
**Printing and Binding** CPI Group (UK) Ltd, Croydon, CR0 4YY

C107620_311022

# Contents

# Preface

Carbon has drawn the attention of global researchers because of its dimensional structural flexibility, which can be transformed into different materials according to sp, $sp^2$, and $sp^3$ hybridization. Moreover, its outstanding mechanical, electrical, thermal, optical, and chemical properties have been extensively employed in designing and developing from the conventional macro realm for structural, mechanical, and chemical reinforcement/strengthen applications, down to micro/nanodomain, primarily material doping, conductivity, and mobility modulation until all carbon electronics and integrated circuits. Their applications expand cross-disciplinary into physics, chemistry, and biological subjects. Such diverse carbon material applications are not possible to be covered in a book or even in a book series. Therefore, we focus on specific frontier applications of carbon-based smart electronic materials for certain latest advanced areas. This book is intended to provide information about the introduction to the latest development of carbon-based materials, preparation methods, functionalization, and applications of carbon-based nanomaterials. The discussion extends to the sensor, piezoelectric, thermoelectric, actuator, nonvolatile memory (NVM), optoelectronic, energy harvesting, and display applications.

This book begins with an overview of carbon-based materials and their synthesis methods in Chapters 1–3. Chapter 1 comprises a cursory overview of this book, a brief introduction of carbon-based materials and their applications, followed by the book outline. Chapter 2 discusses the recent development of carbon allotropes. The typical carbon-based nanomaterial fabrication and characterization methods have been provided as well. Chapter 3 highlights the recent advances in chemically modified and functionalized low-dimensional carbon-based nanomaterials, especially fullerenes, carbon nanotubes, and graphene. The enhanced intrinsic properties of these functional nanomaterials in optoelectronics, strain–pressure, and gas-sensing applications are also presented.

Advanced topics of enhanced carbon-based materials and their applications are presented in Chapters 4–9. Chapter 4 delineates the discovery of memristor-based memory, substrate options, and the effect of electrode materials. Then, an explicit discussion on solution-processing techniques and plasma polymerization deposition utilized to construct the metal–insulator–metal structure of carbon-based NVM devices will be presented. Chapter 5 reviews the developments in carbonous materials, including fullerene, carbon nanotubes, carbon black, graphene, and carbon quantum dots, with their wide applications in optoelectronics. Chapter 6 presents the working principle of thermoelectric energy harvesting, including materials, devices, and applications. This chapter primarily focuses on organic and hybrid materials, a promising solution for lightweight and conformal thermoelectric power generation. Chapter 7 describes carbon-enhanced piezoelectric materials and demonstrates their applications as sustainable energy harvesters and biomechanical sensors. The introduction of carbon-based nanomaterials into poly(vinylidene fluoride) and its copolymer matrix to enhance their performance as energy harvesters will be emphasized. Chapter 8 provides an overview and discussion of actuators based on carbon nanostructures, particularly carbon nanotubes, graphene, graphene oxide, and fullerene. Subsequently, the factors that allow carbon materials to be one of the most promising actuator resources will be discussed. Chapter 9 elaborates on the display performance enhancement of carbon nanomaterials, especially carbon dots, carbon nanotubes, and graphene/graphene oxide. The remaining challenges and further improvement aspects are also proposed for future optimization and commercialization of carbon-based displays. Finally, Chapter 10 summarizes the overview of this book and the future prospects and challenges of enhanced carbon-based nanomaterials to be overcome before the large-scale production of carbon-based products.

In a nutshell, we hope that the broad spectrum of readers, particularly the scientific and industrial community, will find this book helpful to gain valuable experience and references related to enhanced carbon-based materials and their applications.

18 April 2022

*Poh Choon Ooi*
Institute of Microengineering and Nanoelectronics (IMEN)
Universiti Kebangsaan Malaysia
43600 Bangi
Selangor
Malaysia

*Mengying Xie*
State Key Laboratory of Precision Measuring Technology and Instrument
College of Precision Instruments and Opto-electronics Engineering
Tianjin University
Tianjin 300072
China

*Chang Fu Dee*
Institute of Microengineering and Nanoelectronics (IMEN)
Universiti Kebangsaan Malaysia
43600 Bangi
Selangor
Malaysia

# 1

# Enhanced Carbon-Based Materials and Their Applications

*Poh Choon Ooi[1], Mengying Xie[2], and Chang Fu Dee[1]*

[1] *Institute of Microengineering and Nanoelectronics (IMEN), Universiti Kebangsaan Malaysia, 43600 Bangi, Selangor, Malaysia*
[2] *State Key Laboratory of Precision Measuring Technology and Instrument, College of Precision Instruments and Opto-electronics Engineering, Tianjin University, Tianjin, 300072, China*

## 1.1 Overview

This book entitled "Enhanced Carbon-Based Materials and Their Applications" intends to deliver the synthesis methods and role of carbon-based materials that can be introduced into the device to enhance the performance of several types of electronic devices. The theme of this book is multidisciplinary and provides the broader scientific and industrial community with a timely and carefully referenced overview that provides a summary of carbon-based materials and applications. The book has been written in style to make it accessible for academics to provide an overview of synthesis methods and characterization techniques of carbon nanomaterials. On the other hand, this book also demonstrates the various reported applications for carbon-based and enhanced materials for professionals.

## 1.2 Glance of Carbon-Based Materials

Carbon is the 4th most abundant element in the universe and the foundation of life, despite the various elements on the Earth. Its existence extended from nonliving substances to living objects. All organic substances that are found in nature consist of carbon. It has unique physical and chemical properties in multiple forms [1]. The carbon material is a big family with many allotropes, and they possess unique structure, texture, and properties. These carbon materials are classified according to their C-to-C bonding, namely, sp, $sp^2$, and $sp^3$ hybrid orbitals of carbon atoms. Carbon structures include three primary allotropes known as diamond, graphite, and fullerene. Diamond and graphite are the natural carbon sources that humankind could discover abundantly on Earth [1].

Diamond shows $sp^3$ hybrid orbitals, consisting of a tetrahedron network of carbon atoms. It can be naturally formed under prolonged high pressure and high

*Enhanced Carbon-Based Materials and Their Applications*, First Edition.
Edited by Poh Choon Ooi, Mengying Xie, and Chang Fu Dee.

temperature. In the sp3 hybridized form, the C atom and adjacent C atom are linked up by covalent bonds. Consequently, it leads to a strong bonding and tetrahedral atomic arrangement that allows the diamond to be the hardest substrate among others. Diamond is a colorless transparent substrate, which is heavy and extremely hard. It does not conduct electricity but has high thermal conductivity and melting point. In addition, diamond has high resistance to chemical corrosion. For more than one century, this special allotrope of carbon has only been used as a precious stone in making jewelry, ornaments, and glass cutters. Nonetheless, recent scientific development has extended its application into medical science, material science, semiconductor, and nanotechnology [2–6].

Graphite shows $sp^2$ orbitals. It is a greyish-black opaque substance with a high-melting point and low coefficient of thermal expansion. In addition, graphite also has high thermal, electrical conductivity, chemical, and corrosion resistance. These properties enable wide applications from pencil lead, high-temperature lubricant, paint, printing, plastic industry, and electrode in battery/cell up to the application as nuclear reactor moderator [7, 8]. A fullerene is the third allotropes of carbon whose molecule consists of carbon atoms connected by covalent bonds to form fused rings of five to seven. Fullerenes are polyhedral carbon cages in which $sp^2$-carbons are directly bonded to three neighbors in an arrangement of five- and six-membered rings [9]. Fullerene (also called "buckyball") can be constructed by 24, 28, 32, 36, 50, 60, 70, 72, etc., number of carbon atoms. Compared to other types of carbon allotrope, it can be dissolved in common solvents, for example, carbon disulfide and toluene, at room temperature. Recent research on fullerene has enabled its applications in catalysis, water purification, antiviral and antioxidant activities, biohazard protection, magnetic resonance imaging (MRI) medium for diagnostic agents, and the latest nanotechnology advancement in drug and gene delivery [9–14].

Graphite, carbon fiber, etc., have been widely used in the manufacturing and industrial field. Modern applications for carbon materials are focused on the nanostructured form of carbon elements. The world demand for the carbon nanomaterials market is expected to reach US$ 21.97 Billion in 2026. A compound annual growth rate (CAGR) of 4.1% was predicted during this forecast period [15]. The end-use area in industries includes aviation, energy, electrical and electronics, medical and healthcare, packaging and backend, transportation, automotive, and consumer goods. Its nontoxic and biocompatible characteristics have greatly enhanced the possibility of its applications to areas from nonliving to living things. Figure 1.1 summarizes the allotropes/form/type of carbon in a concentric circle. Carbon-based materials can be categorized into different types, which are carbon nanotubes (CNTs), layered sheets (graphene/graphite), carbon black, fullerene, and quantum dots.

Most carbon-based materials are naturally stiff, lightweight, and highly conductive [16, 17]. In addition, they can be easily engineered to achieve targeted material properties. For example, CNTs have been assembled into fibers with ultrahigh specific strength and stiffness, which are much higher than commercially available engineering fibers [17]. These lightweight, stiff CNT fibers show their potential in

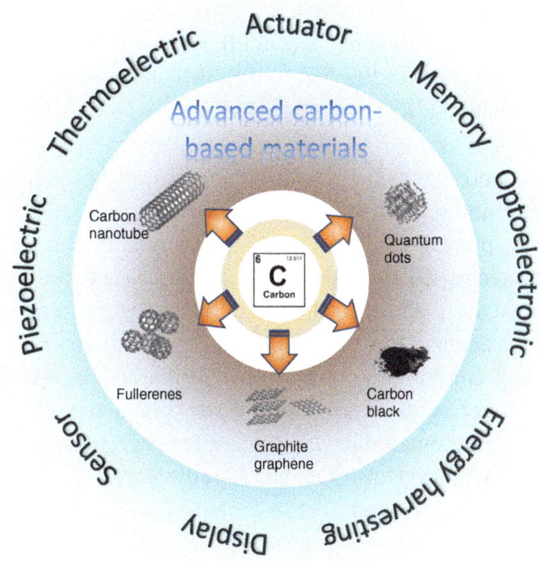


**Figure 1.1** Multiforms of carbon-based materials and their applications in contemporary electronic and optical-related technologies.

many structural applications, such as the construction of energy-efficient airplanes and space structures. Although single-crystal graphene shows excellent properties, single crystals need to be assembled into the macroscopic structure in many practical applications. A single piece of graphene is delicate, but linking them through chemical or physical approaches can form a stronger 3D network [18, 19]. It has been reported that porous structured 3D graphene assembly can be even lighter than air and ten times as strong as mild steel [19]. In addition to the inherent properties of carbon nanomaterials, the high surface area and large pore volume of the 3D porous structure can create more space for the transportation or storage of the electron/ion, gas, and liquid, which enables remarkable sensitivity for sensors, large capacity for supercapacitors, and high-power density batteries [20].

On the other hand, carbon-based materials always face irreversible aggregation and insolubilization issues when preparing aqueous suspensions due to their van der Waals attractions between single crystals and hydrophobicity. Stable and distributed aqueous or organic suspensions are the key factors to enable repeatability and homogeneity, which directly limit their industrial applications. Several functionalization approaches have been utilized to treat carbon-based materials to reduce aggregation and insolubilization [21]. The commonly used method modifier to achieve stabilized aqueous or organic dispersions is an amphiphilic or covalent surfactant. For instance, graphene nanosheet was stably dispersed in water, *N*,*N*-dimethylformamide (DMF), and dimethyl sulfoxide (DMSO) by covalently functionalized polydisperse graphene with amine-terminated ionic liquid [22]. Amphiphilic surfactant poly(sodium 4-styrenesulfonate) (PSS) has been coated on graphitic nanoplatelets to provide stable aqueous dispersion with the aid of ultrasonic treatment. Reduced graphene was stabilized via association with the hydrophobic backbone of PSS, while the hydrophilic sulfonate side groups sustained the graphene-PSS in water [23].

## 1.3 Applications

In this book, we infer "enhanced carbon-based materials" as modified properties of pristine carbon materials derivatives that can be used to enhance the performance of devices. In the examples mentioned above, the formed functional variants in enhanced carbon-based nanomaterials play a crucial role in distinguishing specific applications. Plenty of methods can be used to enhance the carbon materials, and the resulting enhanced carbon-based materials can be categorized as (i) doping, (ii) alloying, and (iii) surface attaching of alien atoms/molecules in order to alter their physical and chemical properties. Those three methods mentioned above achieve the same goal in the nanoscale material form, resulting in surface energy/chemistry change. This is due to its high surface-to-volume ratio of nano-regime properties. The change in the surface energy enhances a better coupling for signal transduction used for different applications with the attachment of dedicated alien substances, especially luminescent molecules/substances for biomarkers, chemical/protein-based drugs/genomic elements for medical applications, moieties for catalysis or plasmonics, and polar/nonpolar materials for modulation on wettability [24–30]. Again, those attachments in different forms enhance the carbon materials and enable the multiple functions used for various applications.

In the trend toward miniaturization, electronics require small, flexible, lightweight, and multifunction devices. These demands have driven the rapid development of nanotechnology. Enhanced carbon-based materials have become the most desirable materials for future electronics. They have driven the development of revolutionary technologies, such as advanced materials, renewable energy, biomedical, and smart electronic devices, due to their superior mechanical and electrical performances. These applications associated with enhanced carbon-based materials have shown low-power consumption, high sensitivity, selectivity, and efficiency. Figure 1.2 shows that the research on carbon-based materials has increased by leaps and bounds in the recent three decades since 1980 [31]. Over the past three decades, the number of published articles with the keywords "carbon" and "carbonuous" has increased tremendously, more than 12 times from 10 690 in 1980 to 137 176 in 2021. Carbon-based materials have shown great versatility since they can be easily functionalized physically and chemically modified with other elements. The excellent and tunable mechanical, electrical, thermal, optical, and chemical properties of carbon-based nanomaterials are desired to be included in electronic gadgets to improve overall performance further [32].

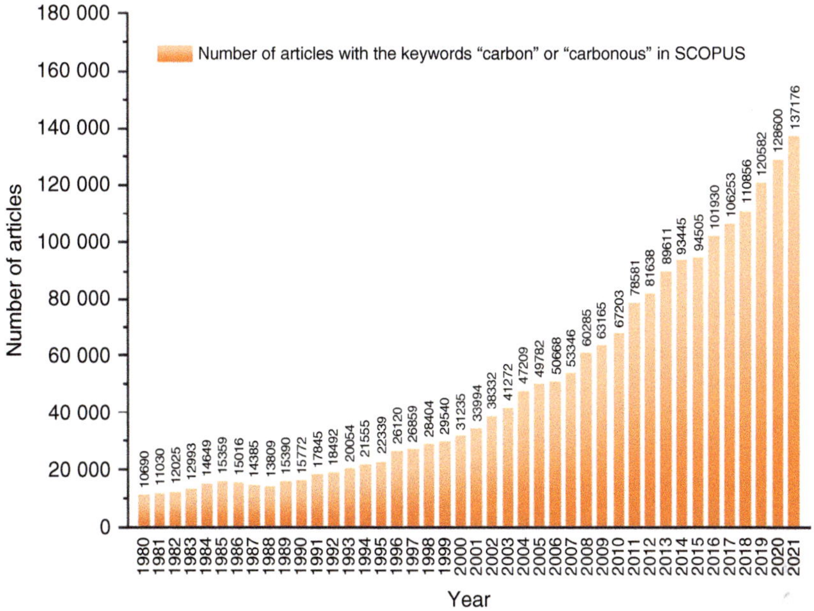


**Figure 1.2** Number of published articles with the keywords "carbon" and "carbonuous" from 1980 to 2021.

## 1.4 Outline of This Book

The book consists of 10 chapters to provide a brief overview of state-of-the-art carbon-based and enhanced nanostructures and their applications. Chapter 2 outlines the fundamental insight into carbon-based materials synthesis and material characterization techniques. Chapter 3 presents the advanced functionalization techniques of carbon-based materials and their applications, such as carbon-based sensors that consist of planar graphene and graphene hybridized material. Their specific material properties, preparation methods, and sensing behaviors are discussed. The following chapters (Chapters 4–9) focus on the overview of carbon-enhanced materials. The integration of these materials is of great importance for the performance enhancement of various future electronics applications, including nonvolatile memories, optoelectronic devices, thermoelectrics, energy harvesters, actuator, and display. In addition, an overview of this book will be summarized in the conclusions, and future directions for the applications of carbon-based and enhanced materials are discussed.

## References

1 Li, Z., Liu, Z., Sun, H., and Gao, C. (2015). Superstructured assembly of nanocarbons: fullerenes, nanotubes, and graphene. *Chemical Reviews* 115 (15): 7046–7117.

**2** Krüger, A., Liang, Y., Jarre, G., and Stegk, J. (2006). Surface functionalisation of detonation diamond suitable for biological applications. *Journal of Materials Chemistry* 16 (24): 2322–2328.

**3** Ando, T. et al. (1996). Direct interaction of elemental fluorine with diamond surfaces. *Diamond and Related Materials* 5 (9): 1021–1025.

**4** Hui, Y.Y., Cheng, C.-L., and Chang, H.-C. (2010). Nanodiamonds for optical bioimaging. *Journal of Physics D: Applied Physics* 43 (37): 374021.

**5** Zhang, Y., Rhee, K.Y., Hui, D., and Park, S.-J. (2018). A critical review of nanodiamond based nanocomposites: synthesis, properties and applications. *Composites Part B: Engineering* 143: 19–27.

**6** Raina, A. and Anand, A. (2018). Effect of nanodiamond on friction and wear behavior of metal dichalcogenides in synthetic oil. *Applied Nanoscience* 8 (4): 581–591.

**7** Pierson, H.O. (2012). *Handbook of Carbon, Graphite, Diamonds and Fullerenes: Processing, Properties and Applications.* Elsevier Science.

**8** Marsden, B.J., Jones, A.N., Hall, G.N. et al. (2017). Graphite as a core material for Generation IV nuclear reactors. In: *Structural Materials for Generation IV Nuclear Reactors* (ed. P. Yvon), 495–532. Woodhead Publishing.

**9** Mateo-Alonso, A., Bonifazi, D., and Prato, M. (2006). Chapter 7 – Functionalization and applications of [60]fullerene. In: *Carbon Nanotechnology* (ed. L. Dai), 155–189. Amsterdam: Elsevier.

**10** Vidal, S., Marco-Martínez, J., Filippone, S., and Martín, N. (2017). Fullerenes for catalysis: metallofullerenes in hydrogen transfer reactions. *Chemical Communications* 53 (35): 4842–4844. https://doi.org/10.1039/C7CC01267E.

**11** Chistyakov, V.A., Smirnova, Y.O., Prazdnova, E.V., and Soldatov, A.V. (2013). Possible mechanisms of fullerene $C_{60}$ antioxidant action. *BioMed Research International* 2013: 821498.

**12** Aoshima, H., Yamana, S., Nakamura, S., and Mashino, T. (2010). Biological safety of water-soluble fullerenes evaluated using tests for genotoxicity, phototoxicity, and pro-oxidant activity. *The Journal of Toxicological Sciences* 35 (3): 401–409.

**13** Montellano, A., Da Ros, T., Bianco, A., and Prato, M. (2011, https://doi.org/10.1039/C1NR10783F). Fullerene C60 as a multifunctional system for drug and gene delivery. *Nanoscale* 3 (10): 4035–4041.

**14** Braun, K. et al. (2010). Gain of a 500-fold sensitivity on an intravital MR contrast agent based on an endohedral gadolinium-cluster-fullerene-conjugate: a new chance in cancer diagnostics. *International Journal of Medical Sciences* 7 (3): 136–146.

**15** Maximize Market Research PVT. Ltd. (2021). *Global Carbon Nanomaterials Market: Industry Analysis and Forecast (2020–2026) by Type, Method, End-Use Industry and Region.* Available: https://www.maximizemarketresearch.com/market-report/global-carbon-nanomaterials-market/104577/ (accessed May 2021).

**16** Geim, A.K. and Novoselov, K.S. (2007). The rise of graphene. *Nature Materials* 6 (3): 183–191.

**17** Zhang, X. et al. Ultrastrong, stiff, and lightweight carbon-nanotube fibers. *Advanced Materials* 19: 4198–4201.

**18** Zeng, J. et al. (2018). 3D graphene fibers grown by thermal chemical vapor deposition. *Advanced materials (Deerfield Beach, Fla.)* 30 (12): e1705380.

**19** Qin, Z., Jung Gang, S., Kang Min, J., and Buehler Markus, J. The mechanics and design of a lightweight three-dimensional graphene assembly. *Science Advances* 3 (1): e1601536.

**20** L. Jiang and Z. Fan, "Design of advanced porous graphene materials: from graphene nanomesh to 3D architectures," *Nanoscale* vol. 6, no. 4, pp. 1922-1945, 2014, https://doi.org/10.1039/C3NR04555B.

**21** Kamran, U., Heo, Y.J., Lee, J.W., and Park, S.J. (2019). Functionalized carbon materials for electronic devices: a review. *Micromachines (Basel)* 10 (4).

**22** H. Yang, C. Shan, F. Li, D. Han, Q. Zhang, and L. Niu, "Covalent functionalization of polydisperse chemically-converted graphene sheets with amine-terminated ionic liquid," *Chemical Communications*no. 26, pp. 3880-3882, 2009, https://doi.org/10.1039/B905085J.

**23** S. Stankovich, R. D. Piner, X. Chen, N. Wu, S. T. Nguyen, and R. S. Ruoff, "Stable aqueous dispersions of graphitic nanoplatelets via the reduction of exfoliated graphite oxide in the presence of poly(sodium 4-styrenesulfonate)," *Journal of Materials Chemistry* vol. 16, no. 2, pp. 155-158, 2006, https://doi.org/10.1039/B512799H.

**24** Norio, T. (2002). Functionalization of carbon material by surface grafting of polymers. *Bulletin of the Chemical Society of Japan* 75 (10): 2115–2136.

**25** Ulman, A. (1996). Wetting studies of molecularly engineered surfaces. *Thin Solid Films* 273 (1): 48–53.

**26** Chan, K.M., Xu, W., Kwon, H. et al. (2017). Luminescent carbon dot mimics assembled on DNA. *Journal of the American Chemical Society* 139 (37): 13147–13155.

**27** Krans, N.A., van der Feltz, E.C., Xie, J. et al. (2018). Attachment of iron oxide nanoparticles to carbon nanotubes and the consequences for catalysis. *ChemCatChem* 10 (16): 3388–3391.

**28** Blanco, M. et al. (2013). Enhanced hydrogen-transfer catalytic activity of iridium *N*-heterocyclic carbenes by covalent attachment on carbon nanotubes. *ACS Catalysis* 3 (6): 1307–1317.

**29** Gupta, B.D., Pathak, A., and Semwal, V. (2019). Carbon-based nanomaterials for plasmonic sensors: a review. *Sensors* 19 (16): 3536.

**30** Awasthi, K., Singh, D., Singh, S.K. et al. (2009). Attachment of biomolecules (protein and DNA) to amino-functionalized carbon nanotubes. *New Carbon Materials* 24 (4): 301–306.

**31** Search for carbonuous and/or carbon from year of 1980 to 2021 [Online]. Available: https://www.scopus.com/search/form.uri?display=basic#basic.

**32** Patel, K.D., Singh, R.K., and Kim, H.-W. (2019). Carbon-based nanomaterials as an emerging platform for theranostics. *Materials Horizons* 6 (3): 434–469.

# 2

# Carbon-Based Nanomaterials: Synthesis and Characterizations

*Yazid Yaakob[1] and Shahira L. Kamis[2]*

[1] *Universiti Putra Malaysia, Department of Physics, Faculty of Science, 43400, Selangor, Malaysia*
[2] *Universiti Teknologi Malaysia, Malaysia-Japan International Institute of Technology (MJIIT), Tribology and Precision Machining (TriPrem) i-Kohza, Department of Mechanical Precision Engineering, 54100, Kuala Lumpur, Malaysia*

## 2.1 Introduction

This era shows rapid progress of technology, especially in electronic fields, along with fast-driven developments of information technology toward industrial revolution 4.0. This trend has been escalated with recent pandemic crisis that urged people around the globe to practice social distance and to adapt "new norms," such as practicing social distance, utilizing contactless gadgets and promoting "work from home" [1]. The rapid advancement now even reached the stage to achieve and control the structural and functional part of a very small component down to the microlevel. These comply with Moore's law, where he predicts that the miniaturization of semiconductor devices (number of transistors per chip) should be doubled every two years [2]. To realize this possibility, the researcher has diligently realigned their research direction toward product integration for high-density, high-speed, high strength-to-weight ratio, and low-energy-consumption devices, but at the same time require low production cost. Nanotechnology, the idea first introduced by Richard P. Feynman in 1959, is the most promising research field that offers the opportunity to exploit novel properties of materials at the nanoscale ($10^{-9}$ m) level, attracting many investors as well as general public attention until now. Among them, carbon nanostructures (CNSs) are the most trending materials within the field of nanoscience and technology owing to their unique properties (high mechanical strength, high electrical and thermal conductivity, excellent optical properties, chemical inertness, and versatility). These superb features have been exploited to develop numerous electronic devices, sensors, batteries, and composites to be employed in the majority of scientific areas.

*Enhanced Carbon-Based Materials and Their Applications*, First Edition.
Edited by Poh Choon Ooi, Mengying Xie, and Chang Fu Dee.

### 2.1.1 Carbon

Carbon is one of the most abundant elements ever known and a major constituent element of all living things, as all organic compounds are comprised of carbon networks. As an element or compound, carbon has been utilized in human civilization and daily life for long centuries. Ever since ancient times, carbon-based materials, such as coal and charcoal, have been used as energy resources. Carbon black and graphite have also been adopted as writing tools. As time progresses, the use of carbon has been creatively manipulated by blending them into other matrix composites to produce enhanced properties of composite materials. For example, "wootz steel" as raw materials to forge Damascus blades that exhibited extraordinary mechanical properties is actually incorporated with carbon nanotubes, which scientists later revealed in the modern day [3]. Graphite and diamonds were recognized during eighteenth century as two existing natural crystalline allotropic forms of carbon structure, yet they exhibit totally different appearances, crystal structures, and properties [4]. Since then, the rapid development of carbon-based materials has taken part in the industrial revolution. As a case in point, the $sp^2$-hybridized graphite is important as raw material for the lead of a pencil, vehicles brake lining, and circuit electrodes, and is further utilized in carbon fiber for sports equipment and aeronautic applications due to their excellent mechanical properties. Diamonds, the $sp^3$-hybridized carbons that possess a sparkle-shiny appearance known as the hardest natural materials to date, have been utilized as jewelry and cutting tools.

### 2.1.2 Allotropes of Carbon

The development of carbon-based nanotechnology has been driven by the refinement of microscopy that opens the path of discovery at the nanoscale. CNS has been identified to consist of a variant of low-dimension allotropes due to its valency, as shown in Figure 2.1 [6]. It was started with the discovery of fullerene in 1985, which triggered the exploration of a new class of low-dimension materials. Then followed by the first observation reports of CNTs in 1991, and later the scotch tape exfoliation of graphene in 2004 [7, 8]. This surprising finding, which is recognized as miracle materials of the century (awarded Nobel Prize in 2010), has triggered even more exploration of other thermodynamically stable two-dimensional (2D) materials (hexagonal boron nitride, molybdenum disulfide, MXenes, etc.). Even though these stories have passed over decades, the exploration of their potential still shows no limit and continues to be a hot topic among researchers and industry players. A brief survey on the Web of Science database for the last five years (2016–2020) with the keywords of the famous "Carbon nanotube" and "Graphene" displayed more than 73 628 and 146 375 articles published, respectively. This outcome indicates that graphene still received more attention than carbon nanotubes and any other carbon allotropes in the scientific community.

Graphene has a fundamental building block of 2D honeycomb lattice, which can be considered the "mother" of other $sp^2$-hybridized building blocks of CNS. This super carbon is already famous for its first-rate electrical, optical, mechanical,

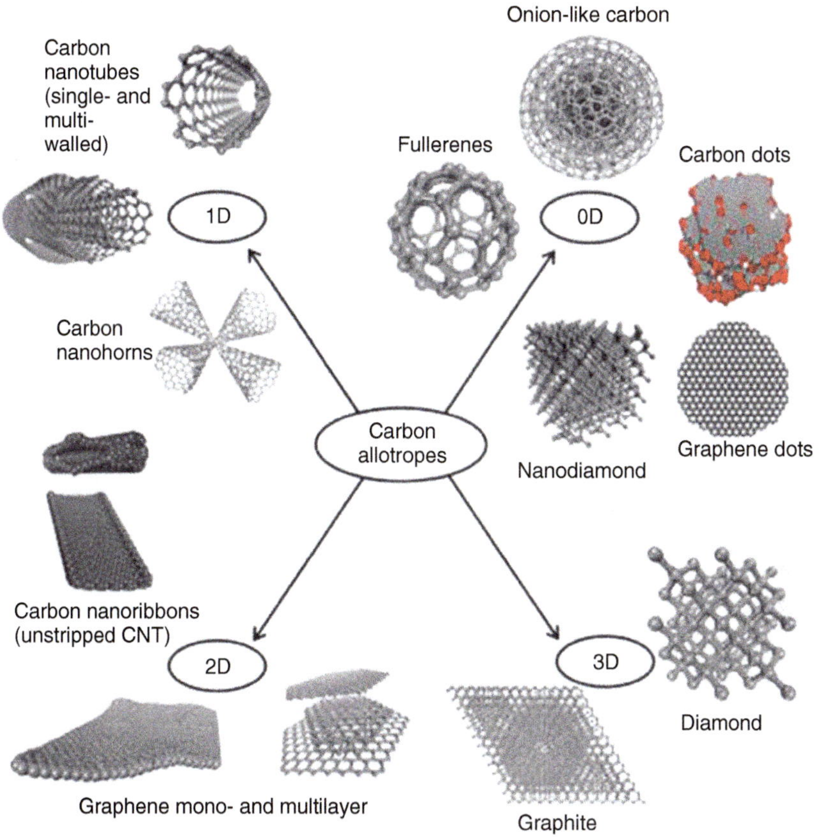


**Figure 2.1** Various carbon allotropes. Source: Reproduced from Savage [5] with the permission of American Chemical Society.

thermal, and chemical properties, making it very flexible to be functionalized and potentially applied in various devices [5]. They even are considered a suitable replacement candidate in semiconductor devices post-silicon. Other than these graphitic ($sp^2$)-dominated forms of CNS, other forms and hybridization, such as amorphous carbon and diamond structure, also exhibit their unique properties. Table 2.1 summarizes some of the CNS's unique properties, which vary depending on their structure forms. For example, three-dimensional diamonds possess high thermal conductivity and inelastic behavior, while zero-dimensional nanodiamonds consisting of $sp^2/sp^3$ hybridization behave more elastic and have lower thermal conductivity [9–12]. Other than basic forms mentioned above, there is also a lot of functionalized and hybrid forms to join the carbon class families, such as carbon dots, graphene quantum dots, graphene oxide, diamond-like carbon and diaphite [14], which have been produced to fill the gap of technology-driven opportunities.

As summarized in Table 2.1, crystalline carbon in different forms and dimensions exhibit contrasting physical and chemical properties. Various physical phenomena

**Table 2.1** Summary of some variation of carbon nanostructures properties.

| Carbon nanostructure material | Dimensions | Hybridization | Hardness | Tenacity | Specific surface area ($m^2\ g^{-1}$) | Thermal conductivity ($W\ m^{-1}\ K^{-1}$) | Electrical conductivity ($S\ cm^{-1}$) |
|---|---|---|---|---|---|---|---|
| Graphite | 3 | $sp^2$ | High | Flexible, nonelastic | ~10–20 | Anisotropic: 1500–2000, 5–10 | Anisotropic: $2–3 \times 10^4$ |
| Graphene | 2 | $sp^2$ | Upper most (for single layer) | Flexible, elastic | ~1500 | 4840–5300 | ~2000 |
| CNTs | 1 | Mostly $sp^2$ | High | Flexible, elastic | ~1300 | 3500 | Structure-dependent |
| Nanodiamonds [9, 10] | 0 | $sp^2/sp^3$ Mostly $sp^3$ | Very high | Elastic [11] | ~300–400 | ~0.3–10 [12] | ~$10^{-11}$ depending on the presence of defects, impurities or dopants |
| Fullerene | 0 | Mostly $sp^2$ | High | Elastic | 80–90 | 0.4 | $10^{-10}$ |

Source: Adapted from Raccichini et al. [13].

and structure-related properties have been explored or discovered. At the same time, the most established CNTs and graphene can be ideal forms of materials to understand the various other forms. Even though specific applications have already made use of CNS, a bunch of fundamental questions still need to be answered before it can totally replace the existing materials and encourage further innovation of new class carbon-related materials. To give some insight on the topic, this chapter aims to provide an overview of the recent development of facile and economical synthesis methods of CNS and their typical characterization techniques. Finally, the potential for detailed characterization of CNS via *in situ* transmission electron microscopy will be introduced and summarized.

## 2.2 Synthesis of Carbon-Based Nanostructures

At the moment, a lot of methods have been developed to synthesize CNS, whether by top-down method or bottom-up method, as illustrated in Figure 2.2 [15]. Both paths have their role, as the top-down approach offers size reduction of existing devices, which is beneficial for architectural structures in connecting macroscopic components. In contrast, the bottom-up methods can build complex devices through strategic atomic arrangement, suitable for nanoscale short-range order of product. These make a perfect combination to produce excellent integration devices based on nanofabrication [16]. CNS fabrication methods have been developed in various ways, either by physical or chemical processes. The common established synthesis methods are exfoliation (physical or chemical), pulsed laser deposition,

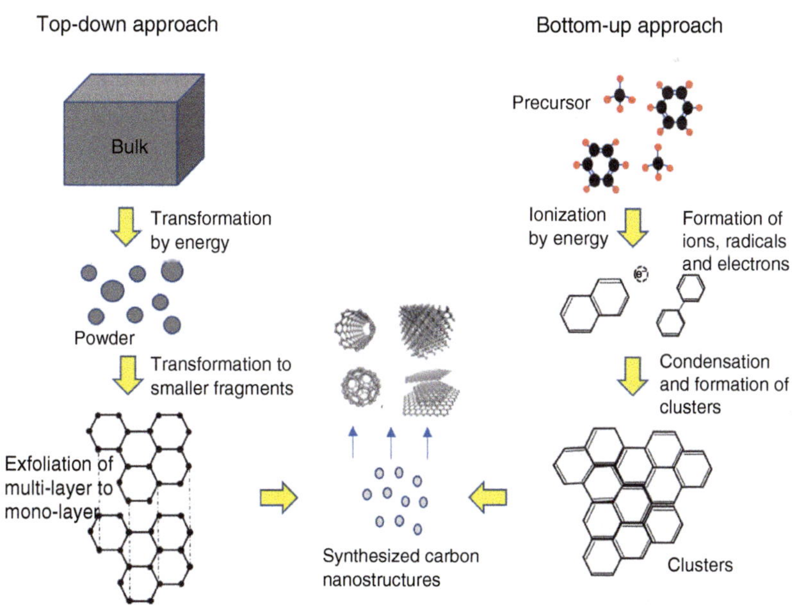


**Figure 2.2** Approaches route for carbon-based nanostructures fabrication. Source: Adapted from Habiba et al. [15].

arc discharge, ion/electron beam irradiation, and physical/chemical vapor deposition. However, most research directions have been moved toward a green and sustainable environment, so low cost, less complexity, and bio/eco-friendly production have become popular. Recently, researchers from Rice University demonstrated a flash Joule heating technique to synthesize turbostratic graphene from plastic waste, which proved to have outstanding properties as filler in polyvinyl alcohol, cement, and concrete composites [17]. Researchers keep coming out with new CNS fabrication techniques and innovations to fulfill industries' desire for efficient and sustainable production. This section will focus on two types of techniques that are considered facile and economical fabrication, namely, chemical vapor deposition and ion irradiation.

### 2.2.1 Chemical Vapor Deposition Technique

The chemical vapor deposition (CVD) method was developed for carbon nanomaterials production in the 1960 and 1970s [18, 19]. High quality and high performance solid coating can be achieved using this method, with great potential for producing large-area deposition. In contrast with other techniques for the growth of carbon nanomaterials, such as discharge using electric arc, chemical or plasma exfoliation, laser ablation, and sputtering, CVD is the most promising technology for producing a high-growth rate, high morphological purity of the grown material. The carbon nanomaterials' synthesis via CVD is achieved through chemical reactions by placing carbon sources into a reactor. Thin films are formed on a heated substrate using thermal or electrical energy sources, such as inductive heating or plasma. Deposition parameters, such as the type of carbon source, catalyst, carrier gas, reaction chamber pressure, and substrate temperature, play a major role in producing thin films composed of carbon nanostructure.

Methane ($CH_4$), ethane ($C_2H_6$), ethylene ($C_2H_4$), and acetylene ($C_2H_2$) are important carbon precursors for the growth of thin film produced by the CVD method [20]. The concentration and purity of carbon source gas can affect the formation of morphology and growth kinetics of thin films [21, 22]. However, these precursors are nonrenewable sources of carbon that originated from fossil fuels. These precursors are also known as inflammable carbonaceous gases, which require safety control and handling precautions. Alternatively, renewable energy sources and nonexplosive carbon sources, especially natural carbon sources and waste materials, are used to reduce fossil fuel dependency and cost. The summary for the use of carbon precursors in the synthesis of the thin films composed of carbon nanomaterials via the CVD method is listed in Table 2.2. A carrier gas, such as hydrogen, nitrogen, or argon, is usually employed to dilute the precursor molecules for chemical reactions.

For efficient carbon nanomaterial production, catalyst particles were introduced into the deposition system. Catalyst plays a key role in the improved quality and controls the size and shape of the carbon nanomaterials produced. Catalysts act as the nucleation site for decomposing the molecules induced carbon deposition on the substrate during the CVD process. Catalysts accelerate reactions by providing the surface area for the adsorption of carbon [36]. Transition metals, such as Cu,

**Table 2.2** Precursor for the synthesis of carbon nanomaterials from different classes of hydrocarbon.

| Carbon nanostructure | Hydrocarbon source | Temperature (°C) | References |
|---|---|---|---|
| *Fossil based* | | | |
| CNFs | Ethylene ($C_2H_4$) | 700 | [23] |
| CNTs | Ethylene($C_2H_4$), Acetylene($C_2H_2$) | 925 | [24] |
| Graphene | Methane ($CH_4$) | 1060 | [25] |
| DLC | Acetylene($C_2H_2$) | 300 | [26] |
| *Natural sources* | | | |
| CNTs | Camphor | 700, 800, and 900 | [27] |
| MWNTs | *Jatropha curcas* | 450–850 | [28] |
| Graphene | Palm oil | 900 and 1000 | [29] |
| *Waste materials* | | | |
| CNTs | Natural rubber glove | 600 and 700 | [30] |
| CNTs | Waste latex | 700 | [31] |
| Graphene | Cooking palm oil | 850–1100 | [32] |
| Graphene | Fruit cover plastic waste and oil palm fiber | 1020 | [33] |
| Graphene | Solid plastic waste | 1020 | [34] |
| Graphene | Chicken fat oil | 1080 | [35] |

Ni, and Pt, are often used as a catalyst to obtain crystalline structure and improve the growth rate of synthesized materials [32, 37, 38].

The nanostructured thin films have a thickness range of a few 100 nm up to a few μm. The CVD of thin films is the result of complicated chemical reactions involving many intermediate steps. As illustrated in Figure 2.3, there are important chemical reactions involved in film growth in CVD, which can be summarized as follows [39]: (a) transport of reactants precursor molecules via gas diffusion at boundary layer region, (b) adsorption and surface diffusion of atoms on that surface, (c) chemical reactions on the substrate, (d) desorption of adsorbed species, and (e) diffusion out of gaseous reaction products through the boundary layer. The boundary layer results from fluid dynamics, which occur when the thin layer of a flowing gas tends to cling to the substrate surface. In this region, the gas stream velocity, concentration of the vapor species, and temperature are not equal to the same parameters in the main gas stream. Therefore, the growth rate of synthesized carbon nanomaterials is generally depending on a wide range of parameters, such as substrate temperature [40], operating pressure [41], carbon precursor/carrier gas ratio [42], and carbon solubility in transition metal catalysts [43].

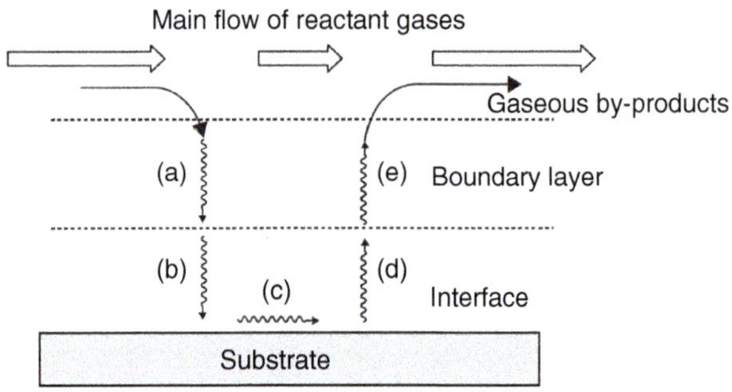


**Figure 2.3** Overview of the chemical interaction process involved in nanostructured thin-film synthesized by CVD. Source: Reproduced from Shang et al. [39] with the permission of Elsevier.

One of the facile CVD techniques is alcoholic catalytic chemical vapor deposition (ACCVD), which is operated up to ~1100 °C to produce CNS using an alcoholic precursor. This method is very simple and flexible in varying CNS growth parameters to fit our resources. They can be operated at atmospheric or low-pressure conditions with a creative heating method (double furnace, spray pyrolysis, heating mantle, etc.) to expand their potential in utilizing various precursor types: solid, liquid, and gas [44]. Figure 2.4 shows an example of a typical CVD procedure to synthesize graphene from solid waste plastic using the double-furnace method.

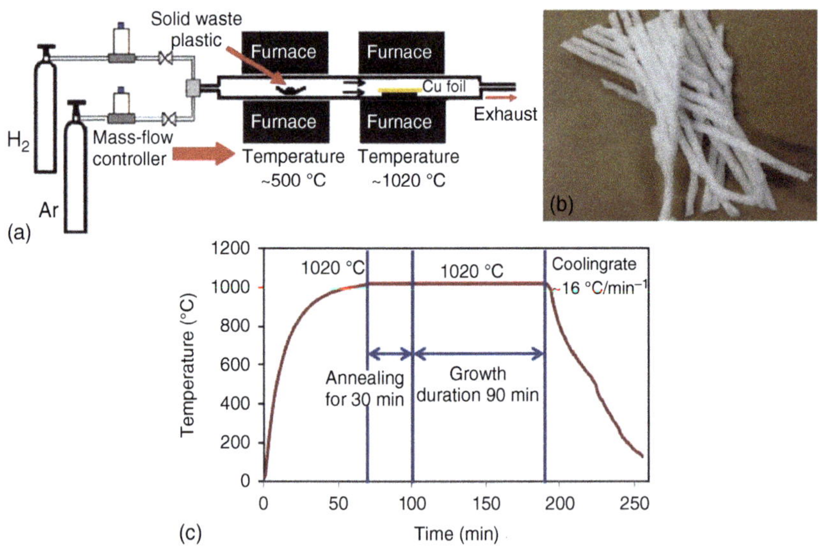


**Figure 2.4** (a) Schematic of the CVD process used for graphene synthesis from solid waste plastic. (b) Photograph of waste plastic used in these experiments. (c) Heating, annealing, growth duration, and cooling rate (~16 °C min$^{-2}$) for the graphene growth process. Source: (a, c) Reproduced from Sharma et al. [34] with the permission of Elsevier, (b) Sharma et al. [34], Reproduced with permission from Elsevier.

Today, various kinds of enhanced CVD processes are also available that can be employed to synthesize CNS, which involve the use of plasmas, ions, photons, lasers, hot filaments, or combustion reactions to increase deposition rates and/or lower deposition temperatures. This method can be categorized based on a heating method, which is discussed in the next subtopic. Out of these techniques, thermal and plasma-enhanced CVD are the two most dominant thin-film deposition methods. Figure 2.5 shows a schematic of thermal and plasma-enhanced CVD reactors.

#### 2.2.1.1 Thermal Chemical Vapor Deposition

Thermal CVD uses higher temperatures (typically between 800 and 2000 °C) to heat hydrocarbon precursors and encourage the chemical reaction in a gas phase that results in thin films grown on the substrate. In 1966, the first thermal CVD on metals was reported to grow highly crystalline graphite films on Ni substrates [18]. Heating processes, such as hot plate heating, resistive heating, induction heating, and radiant heating, are usually used to reach a high temperature. The heat can be delivered externally (hot-wall reactor) or internally (cold-wall reactor). In hot-wall reactors,

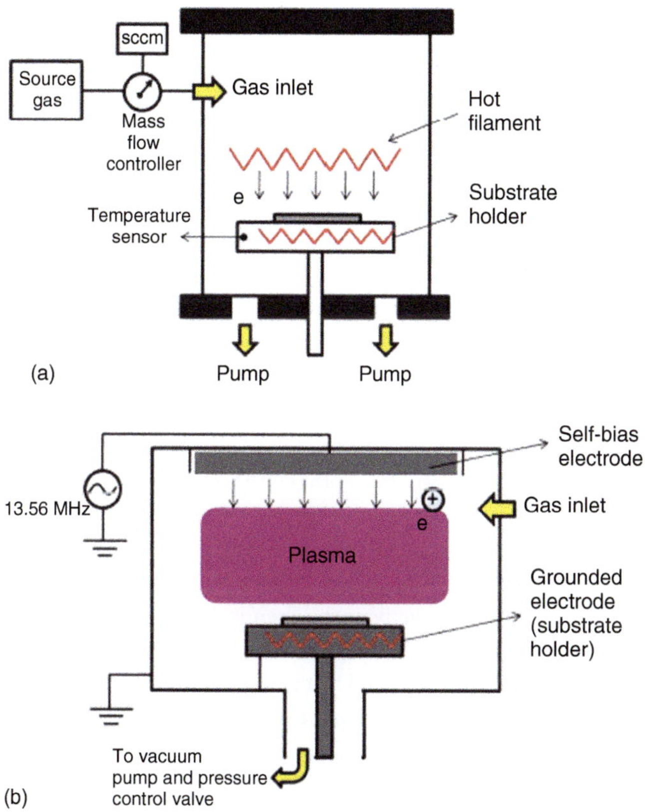


**Figure 2.5** Schematic of (a) thermal and (b) plasma-enhanced CVD reactors. Source: Reproduced from Pessoa et al. [45] with the permission of Elsevier.

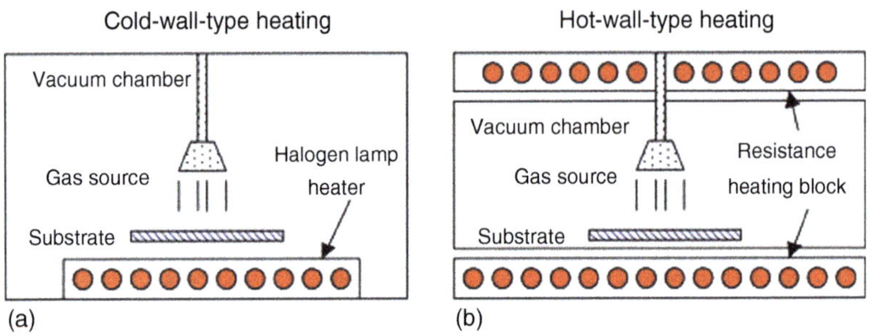


**Figure 2.6** Schematic of typical (a) cold-wall and (b) hot-wall CVD reactors. Source: Reproduced from Kim et al. [46] with the permission of Elsevier.

the whole chamber is heated by an array of heat sources, such as halogen heater or quartz lamps, and the substrate is heated by radiation. In contrast, the substrate is heated by plate-type heaters in cold-wall reactors and the reactor walls are kept cold. Figure 2.6 shows a schematic of typical (a) cold-wall and (b) hot-wall CVD reactors. Besides, research works involving using thermal CVD to synthesize the carbon nanomaterials thin film are summarized in Table 2.3.

Generally, it is desirable to use low-temperature and low-pressure processes in CVD heating systems, which offer significant potential advantages to reduce unwanted gas-phase reactions and enhance the thin-film thickness uniformity. In this respect, the use of low-pressure thermal CVD under 100 Pa with low dissociative temperatures of 800–900 °C has proven capable of producing CNTs by using $C_2H_4$ as carbon precursor [43]. The researchers also proposed using other hydrocarbon precursors, especially $C_2H_2$, as it is easier to decompose at lower temperatures. When using solid or liquid carbon precursors, double thermal CVD is needed, which has two independent heating zones; the first furnace for the precursor furnace and the second furnace for the deposition furnace where the substrate was placed [53].

#### 2.2.1.2 Plasma-Enhanced Chemical Vapor Deposition

Plasma-enhanced CVD (PECVD) is different from other CVD methods in which the chemical reactions on the surface occur after the reacting gases create the plasma. In the PECVD process, a high-frequency voltage is required to ignite plasma by liberating electrons from precursor molecules and atoms for creating thin films. Plasma can ionize and decompose the reactive gas, activate the chemical reaction, and create heat. As a result, the plasma could reduce the deposition temperature (approximately around 200–400 °C) and improve deposition rates. This technique also has the advantage of reducing operating costs involving energy to heat the gas phase and can use for materials and substrates, especially those that are not suitable for higher deposition temperatures. Although PECVD is known as a promising method, the quality of the deposited thin film can be affected by energetic particles (e.g. fast ions and electrons) from the *plasma-surface interaction*, leading to increased surface defect densities.

**Table 2.3** Summary of the related works used thermal CVD for synthesis of carbon nanomaterials thin film.

| Carbon nanostructure material | Carbon precursor | Catalyst | Substrates | Carrier gas and flow rate | Deposition temperature (°C) |
|---|---|---|---|---|---|
| *Cold-wall reactor* | | | | | |
| Graphene [25] | Methane ($CH_4$) | — | Cu foil | Hydrogen (20 sccm) and argon (900 sccm) | 1060 |
| SWCNTs [47] | Methane ($CH_4$) | Al and Fe | Silicon | Hydrogen (20 sccm) | 700–950 |
| SWCNTs and MWCNTs [48] | Ethanol | Co and Mo | Porous alumina | Hydrogen and argon[a)] | 650 and 830 |
| *Hot-wall reactor* | | | | | |
| SWCNTs [49] | Acetylene($C_2H_2$), Methane ($CH_4$) | Ni, Co, and Fe | Thermal Si-oxide | — | 600–950 |
| MWCNTs [50] | Methane ($CH_4$) | Pt–W with MgO | - | Argon (200–500 sccm) | 800 |
| Graphene [51] | Methane ($CH_4$) | Co and Fe | Silicon | Hydrogen (50 sccm) and argon (300 sccm) | 800 |
| *Hybrid cold- and hot-wall reaction chamber* | | | | | |
| Graphene [52] | Methane ($CH_4$) | - | Cu foil | Hydrogen (20 sccm) | 1035 |

a) Flow rates are not mentioned.

The PECVD method can be categorized in different deposition processes depending on the power source that generates plasma. Various types of plasma can be artificially generated by using electric power sources at various plasma excitation frequencies, which could be further divided into three subcategories; audio frequency (10 or 20 kHz), radio frequency (13.56 MHz), and microwave frequency (2.45 GHz) [54]. Among them, radio frequency (RF)-PECVD and microwave (MW)-PECVD are widely used plasma sources for carbon nanostructure thin-film deposition.

According to Wang and Moore [55], CNTs could grow wells by RF-PECVD because there is a higher density of reactive radicals than in dc-PECVD at the same deposition condition. This indicates that RF-PECVD is capable of synthesizing CNT at low temperatures (~180 °C). In another study, it has already been reported that 13.56 MHz of RF-PECVD is capable of growing multilayer graphene on Ni and Co films at a relatively low temperature of 600 °C with 40 W plasma power [56].

In 2011, Qi et al. reported that the thickness of graphene (corresponds to a few graphene layers) prepared by RF-PECVD could be controlled by the ratio of $CH_4$ flow rate in a mixture of reaction gases [57]. Also, RF-PECVD is one of the most popular techniques for the deposition of DLC films because it can improve the quality of the film by producing uniform thickness over a large area.

To date, MW-PECVD has been used extensively to achieve high-deposition rates and/or high fragmentation of precursor material for thin-film deposition. The MW-PECVD process involves applying high electromagnetic radiation (in the GHz range) to generate plasma in the reaction chamber. The plasma is physically transformed into nanoparticles after being heated, diffused, and condensed to a substrate film. For example, Fernandes et al. employed a 2.45 GHz MW-PECVD to obtain growth of the NCD/CNT hybrid films on Si substrate using a $CH_4/H_2$ mixture with Fe catalysts under 700 °C [58]. Another study by Chockalingam et al. successfully synthesized MWCNT graphene-like nanocarbon hybrid film using $CH_4/H_2$ mixture under 600 °C [59]. This hybrid film has the potential to enhance electrical properties by using graphene as connecting particles within the porous CNT network. Cho et al. provided a comparison study between different deposition conditions used in synthesized DLC films; RF-PECVD and MW-PECVD [60]. It was found that both deposition methods produce similar DLC structures, but there are differences in surface roughness and coating thickness.

Recently, there has been an improvement in the CVD method, such as introducing pulsed plasma discharge, in order to reduce the occurrence of gas-phase polymerization caused by continuous-wave discharge [61]. The high-power pulses significantly produced higher plasma density and deposition rates compared to traditional PECVD. In 2018, Mamun et al. proposed that the pulsed dc PECVD system can be considered an alternative to the conventional RF PECVD. It improved film properties and reduced energy consumption to promote the growth of DLC film [26].

### 2.2.2 Ion Irradiation Technique

The irradiation method was developed by utilizing an ion or electron beam to sputter the deposition of samples in the targeted substrate. Normally, the ion/electron beam is known to be destructive and introduces defects toward the target materials' surface. However, these phenomena also will introduce structural change and enhancement of properties of the nanomaterials [62], which lead to a creative and easy way to modify them. For example, Banhart reported that structural transformation of graphitic particles into crystalline diamond occurs when exposed to intense electron beam irradiation of about 10–200 A $cm^{-2}$ current density in transmission electron microscopy (TEM), as shown in Figure 2.12 [63].

Generally, during the ion irradiation process, the bombardment exposure will erode the solid material surface. The erosion rates are characterized by sputtering yield, defined as the mean number of emitted atoms per incident particle, depending on materials. As illustrated in Figure 2.7(a), atoms are ejected from the outer surface layers. Then, the sputtering process will involve the energy transfers and ion incidence angle between many atoms inside the solid [64]. The secondary

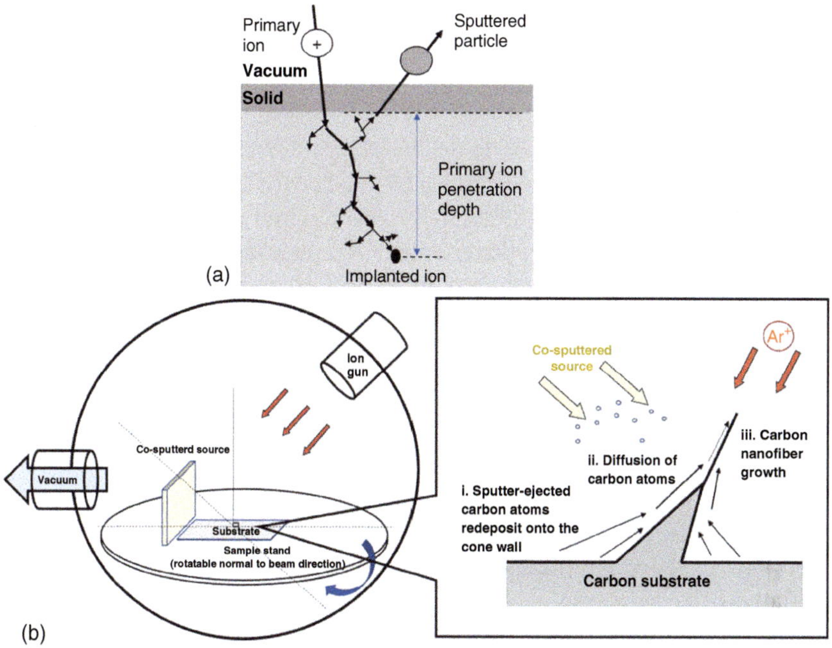


**Figure 2.7** Schematic of (a) ion bombardment process (b) ion irradiation experimental setup and mechanism. Source: Adapted from Refs. [64, 65].

effects induced by ion bombardment, namely, redeposition of sputtered particles, a reflection of primary atoms, and the etching effect, will lead to construction of a cascade structure on a solid surface. By manipulating this process, the development of various types of surface topography occurred (ripple, cones, pyramids, whiskers, etc.) toward beam direction.

Tanemura et al. demonstrated the formation of CNFs from glassy carbon surface by Ar ion irradiation with co-sputtered Mo under ultrahigh vacuum conditions [65]. The CNFs growth were induced by the impurities during the ion bombardment. The growth mechanism can be explained as follow: First, the conical protrusions start to form toward ion irradiation due to surface diffusion, followed by the redeposition of sputter-ejected carbon atoms toward the cone, then, secondary surface diffusion toward the tips. Figure 2.7 illustrates (b) a typical experimental setup for ion irradiation experiment and (c) the CNFs growth mechanism. The construction of various forms of carbon-based nanostructure in the large-scale area is promising by varying the sputtering parameters. This includes composite fabrication of carbon-based nanostructure with other elements, such as metals, semiconductors [66, 67], and even light elements, especially lithium [68]. Ion-induced structural transformation can even promote graphitization at a low temperature of 150 °C by using indium as a catalyst, far lower temperature than the normal graphitization process by other reported CVD methods [69].

## 2.3 Characterization

As CNS has been widely utilized not only among the research community but also in the industrial sector, various methods of characterization have been developed to study and evaluate these materials' quality. Even at fabrication stages, to investigate the morphological and structural characterizations of CNS, several techniques could be used. X-ray diffraction (XRD), X-ray photoelectron spectroscopy (XPS), and infrared and Raman spectroscopy are some of the popular characterization techniques to determine the chemical structure of CNS. However, only a few methods are available to characterize CNS at individual level, for instance, scanning tunneling microscopy (STM) and TEM. This chapter will point out the two most common characterization methods for CNS, namely, Raman spectroscopy and electron microscopy. Both instruments are powerful tools that provide important information about the CNS structure with easy, nondestructive, quick, and high-accuracy measurement directly.

### 2.3.1 Raman Spectroscopic Characterization of Carbon Nanostructure Materials

Raman spectroscopy is the most common nondestructive characterization method for CNS that exhibits structural fingerprint of molecules by utilizing inelastic scattering of monochromatic light from the CNS in vibrational mode. It offers valuable information about the disorder in the $sp^2$ network of carbon-based nanomaterials. For example, in our works, we use Raman spectroscopy to characterize the structure of carbon bonds in hydrogenated DLC films deposited onto $Ti_6Al_4V$ alloy by using a CVD system. The measurement was performed at excitation wavelengths of 532 nm using an $Ar^+$ laser. The visible Raman spectra of DLC films synthesized using different hydrocarbon gasses are shown in Figure 2.8 in the range 1000–2000 $cm^{-1}$. It is observed that two broad bands centered at approximately 1350 and 1550 $cm^{-1}$, corresponding to the disorder (D) and graphitic (G) bands of disordered $sp^2$ carbon [70]. The Raman spectra were then analyzed by Gaussian fitting; the peak can be

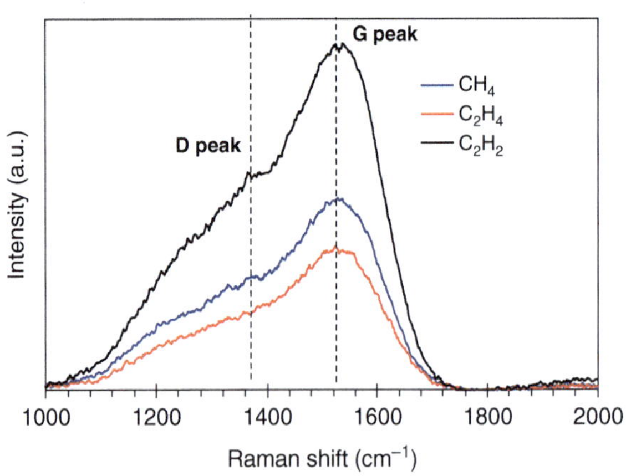


**Figure 2.8** Raman spectra of deposited DLC films using different hydrocarbon gas sources.

**Table 2.4** Variation of Raman D and G peak position and $I_D/I_G$ of DLC films.

| Hydrocarbon source | D peak ($cm^{-1}$) | G peak ($cm^{-1}$) | $I_D/I_G$ |
|---|---|---|---|
| $CH_4$ | 1361 | 1543 | 1.50 |
| $C_2H_4$ | 1357 | 1541 | 1.34 |
| $C_2H_2$ | 1338 | 1544 | 1.02 |

well fitted by two Gaussian distributions, situated approximately at 1540–1545 $cm^{-1}$, corresponding to the G band for crystalline graphite, while another peak appears at approximately 1338–1361 $cm^{-1}$ corresponding to the band D for disorder structure in the graphite. The fitted parameter, including G and D positions and intensity ratio $I_D/I_G$, is presented in Table 2.4. $I_D/I_G$ ratio was obtained by computing the area under the individually fit curves representing the graphite crystal's size in the DLC films. The low $I_D/I_G$ indicates a high graphitization degree, also known as high phase transformation from amorphous carbon into graphitic carbon.

Besides DLC films, a Raman spectrometer was also performed to quantify disorder graphite of other nanostructure materials, such as graphene, carbon fiber, CNTs, and other CNS, containing $sp^2$ dominant hybridization [29, 30, 71]. Besides disorder (D) bands that indicate the defects of the CNS quality, Raman spectra have also been used to estimate the number of graphitic layers. Figure 2.9 features Raman microscopy study of graphene mono and multilayer in determining the effect of substrate on atomic/electronic structures of graphene [72]. Graphene is dominantly recognized by band 2D (~2680 $cm^{-1}$), aside from D and G. $I_{2D}/I_G$ intensity ratio was obtained by computing the area under the individually fit curves representing the graphite crystal's size in the graphene films. The low ratio of $I_{2D}/I_G$ indicates a high number of the graphene layer, and vice versa, whereas $I_{2D}/I_G > 2$ for monolayer. Also, Figure 2.9c,d reveals that the symmetric and sharp 2D peak (~30 $cm^{-1}$) is the best indicator to determine monolayer graphene. A similar study has also been reported for the CNTs case, where they can distinguish single-, double-, and multi-walled CNTs in a rapid experiment using a similar technique [73].

### 2.3.2 Electron Microscopy

Electron microscopy is the most convenient method to identify the structural properties of the CNS by taking advantage of scattered electron signals from electron irradiation on the samples. Figure 2.10 shows the signals that can be utilized for various analyses. This section will focus on two common and convenient methods to analyze CNS, namely, scanning electron microscopy (SEM) and TEM. Both are valuable analytical tools to determine CNS information, including attached catalyst, composite, and impurities. Normally, they are equipped with commonly attached accessories known as energy dispersive X-ray (EDX) spectrometers. This is useful to get information regarding the elemental composition of the CNS, which can support the information obtained from the selected area diffraction (SAED) pattern by

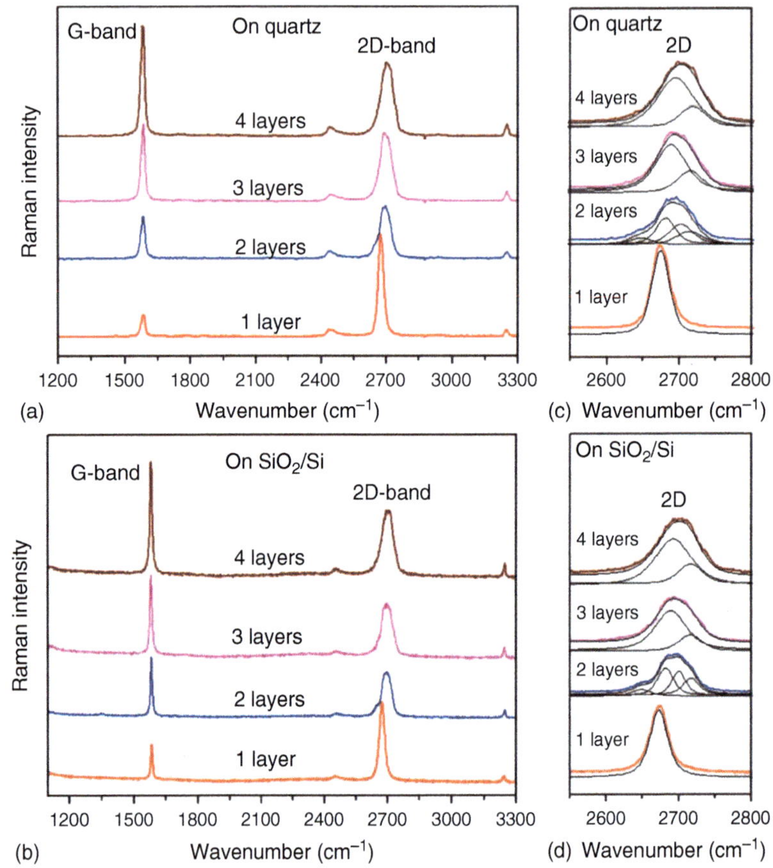


**Figure 2.9** The Raman spectra of the monolayer, bilayer, three layers, and four layers of graphene on quartz (a) and SiO2 (300 nm)/Si substrate (b). The enlarged 2D-band regions with curve fit are also shown in panels (c) and (d). Source: Reproduced from Wang et al. [72] with the permission of American Chemical Society.

TEM. For further details of the composition information, electron energy-loss spectroscopy (EELS) offers information about the materials' valence state and electronic structure under TEM observation. Generally, employable samples for EM characterization are in solid form and electron conducted. However, some other types of samples can still be characterized by complex sample preparation and special accessories of the EM system.

#### 2.3.2.1 Scanning Electron Microscopy

Generally, SEM utilizes the secondary electron generated from the incident electron to cover the information of morphological structure and topography for the sample for two-dimensional characterization. The depth (~50 nm: depends on materials) from which the relevant secondary electrons typically escape will provide surface and bulk information. SEM also can be equipped with a sample stage with a tilting

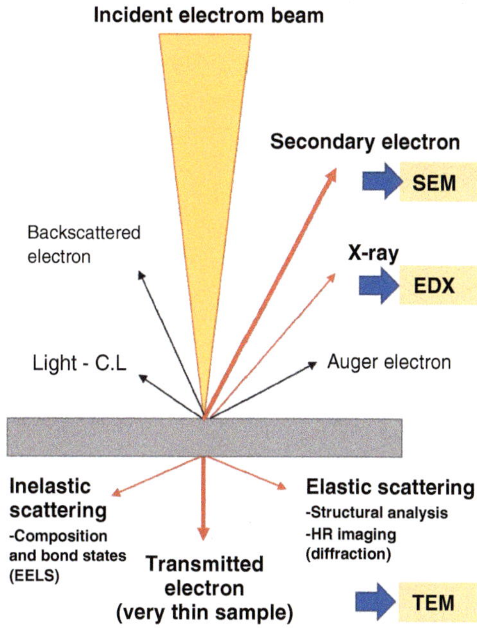


**Figure 2.10** Signals generated from electron-beam-irradiated sample.

function that can clarify more information for three-dimensional surface characterization, such as the length of CNFs and depth of pores, by *in situ* changing the angle of observation. For example, Tanemura et al. fabricated Ar ion-induced CNFs on glassy carbon surfaces [65]. From the SEM observation, the sample's surface morphology revealed the growth of cones with either aligned or nonaligned CNFs on their tips. The CNFs length and diameter were about 0.2–10 mm and 50 nm, respectively. The latter work showed that these fabrication techniques are a reliable method to produce densely distributed CNFs even with metal composites, such as silver (Ag), that reached up to $\sim 10^6\ mm^{-2}$, through careful observation and analysis of the surface morphology of SEM image [66, 74]. It was further confirmed as incorporating Ag particles in amorphous CNF matrix by TEM images (Figure 2.15a) and supported by SAED and EDX for the structural form and elemental information, as shown in Figure 2.11.

#### 2.3.2.2 Transmission Electron Microscopy

TEM and scanning transmission electron microscopy (STEM) utilized the transmitted electron through a thin sample to study the details of the presence of surface–interface. They can provide a wide and deep range of data about the specimen of interest up to atomic structure level, commonly known as high-resolution TEM (HRTEM). Both real space (image) and reciprocal space (diffraction) data, including chemical and electronic analytical information (derived from EDX or EELS), can be obtained from the same nanoscale area. For example, Hazan et al. worked on CNS synthesis from waste latex, which features various CNSs (amorphous, nanosphere, nanotubes, and graphitic film) confirmed through TEM observation [31]. Further analysis by EDX mapping reveals impurities elements

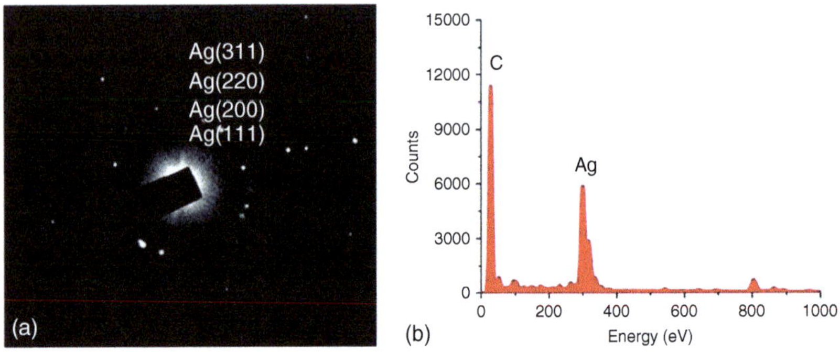


**Figure 2.11** (a) SAED and (b) EDX spectra corresponding to Ag-CNF. Source: (a) Yaakob et al. [74], Reproduced with permission from Royal Society of Chemistry, (b) Reproduced from Yaakob et al. [74] with the permission of The Royal Society of Chemistry.

contained in the sample other than carbon and iron (catalyst), which is sulfur, that hindered the CNS growth. Other than that, TEM is also equipped with a function in series of tilt angles, which can determine the overall shape of the nanostructure in three-dimension. Therefore, by exploring this function, the CNS can be studied not only by two-dimensional imaging which sometimes misinterprets the CNS's actual structures, but also in more details analysis at three-dimension. This idea has been brought out in work on studying graphene transformation from graphite through the passage of current. Depending on the tilt sequence, they revealed that the graphite did not transform to flat graphene but also included the joined formation on nanotubes and ribbon-like structures [75].

#### 2.3.2.3 *In Situ* Transmission Electron Microscopy

*In situ* TEM is a very convenient tool for live investigation of the carbon-based nanostructure's transformation and properties in detail. It is important to note that measuring the properties of bulk would not be the same with nanostructure because of different properties due to greater surface area per unit mass and quantum effects. Various studies have been done to explore these techniques to investigate individual CNS properties, such as mechanical, electrical, and field emission [76–78]. Along this journey, the researcher has noticed one of the crucial factors that affect those mentioned properties, which will also significantly affect their performance at the application level in the long run, that is growth, transformation, and function of the structural properties at the nanoscale during the measurement. For example, Heersche et al. reported that polycrystalline gold wires narrowed down during increment of applied bias voltage due to induced electromigration before breakdown [79]. The difference in electrode shape after the breakdown also reveals the transformation pattern dissimilarity with and without applied voltage bias. A similar case has also been reported by Oh et al. [80]. They observed an electromigration phenomenon in a current applied Cu line, which is a piece of valuable information, such as favored diffusion path of Cu that leads to Cu interconnect performance reliability in integrated circuit design.

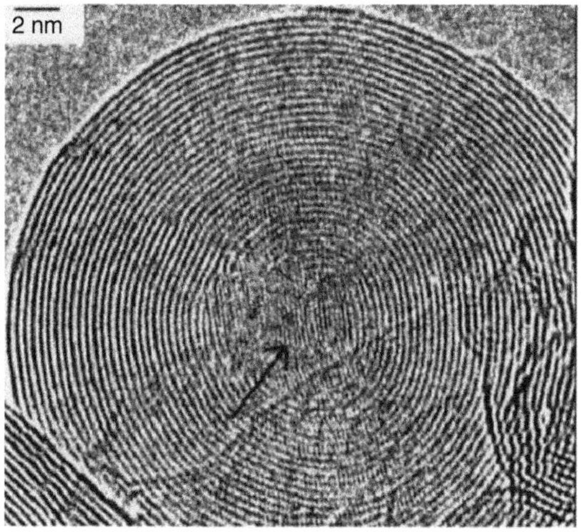


**Figure 2.12** Spherical concentric-shell carbon onion, generated under electron irradiation (1.25 MeV, 100 A $cm^{-2}$) of a polyhedral graphitic particle at 1000 K. In the core, a diamond crystal of 2 nm in size has formed. The image was taken shortly after the nucleation of the diamond crystal (arrowed). Source: Banhart [63], Reproduced with permission from AIP Publishing LLC.

Aside from the phenomenon study helping in preserving reliability and failure prevention of the properties of the materials, there is also an opportunity for exploration of transformation mechanisms in materials growth at the nanoscale during measurement. Figure 2.12 shows *in situ* structure transformation displayed a spherical concentric-shell graphitic structure transformed into the diamond crystal of 2 nm in size after two hours of electron beam irradiation inside TEM [63]. The solid-phase reaction carbon growth of CNTs has also been directly observed by using a special TEM sample holder for heating [81] and electrical measurement [82]. For example, Yusop et al. revealed that during *in situ* field emission (FE) measurement of Fe-incorporated CNF, the applied current rose significantly along with Joule heating on the CNFs, resulting in the agglomeration and electromigration of the Fe particles in CNF matrix [82]. They confirmed the structural transformation of amorphous-CNFs into graphitic bamboo-like nanotubes.

In a similar work, as illustrated in Figure 2.13, Yaakob et al. performed field emission measurements on individual Ag-CNF fabricated on graphite foil [74]. In this case, Ag-CNF showed different transformation characteristics during field emission measurement. Whereas the applied emission current rose significantly at 1 μA along with Joule heating on the CNFs, the Ag particles tend to evaporate (Figure 2.14), leaving a defective graphitic structure around Ag particles' initial place (Figure 2.15b). A similar case has been reported for the gold (Au) metal by Takahashi et al. [83]. They were confirmed through EELS spectra at the same spot before and after the FE measurement. No significant $\pi + \sigma$ plasmon peak was detected before FE, indicating a poor crystallinity of the carbon. However, after FE, high intensity of $\pi + \sigma$ plasmon peak was detected at 25.2 eV, suggesting the formation of a hollow graphitic structure. The mechanism can be explained as the rise of Joule heating during applied current increment escalates the graphitic transformation of the metal catalyst-incorporated amorphous carbon matrix. However, depending on catalyst characteristics, i.e. melting point and carbon solubility of metal, the way they omit from the CNF, and the graphitization quality will be

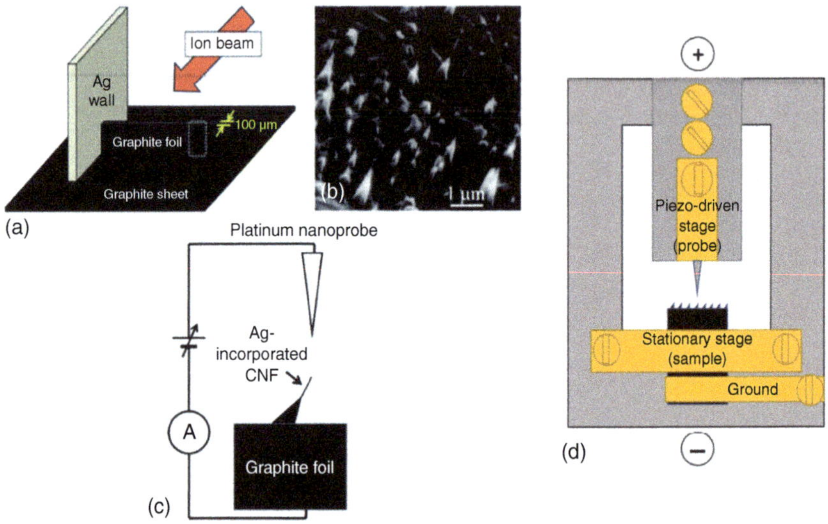


**Figure 2.13** (a) Schematic diagram of the experimental setup for Ag-CNF fabrication via ion irradiation. (b) SEM image of Ag-incorporated CNFs at the graphite edge. (c) Schematic diagram of the *in situ* FE measurement experimental setup. (d) Illustration of special TEM holder with nano-contacting capabilities. Source: (a, c, d) Reproduced from Yaakob et al. [74] with the permission of The Royal Society of Chemistry, (b) Yaakob et al. [74]. Reproduced with permission from Royal Society of Chemistry.

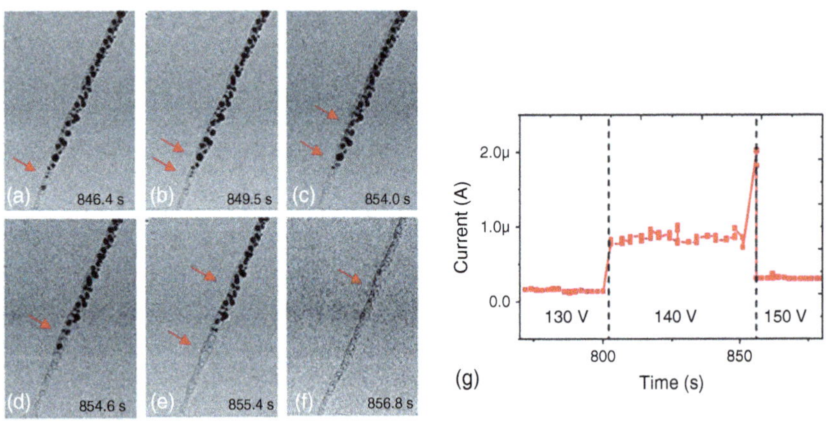


**Figure 2.14** (a–f) Time-lapsed HRTEM images of Ag-incorporated CNF during the field emission process. (g) Current properties of the fiber corresponding to the TEM images. Source: (a–f) Yaakob et al. [74]. Reproduced with permission from Royal Society of Chemistry, (g) Reproduced from Yaakob et al. [74] with the permission of The Royal Society of Chemistry.

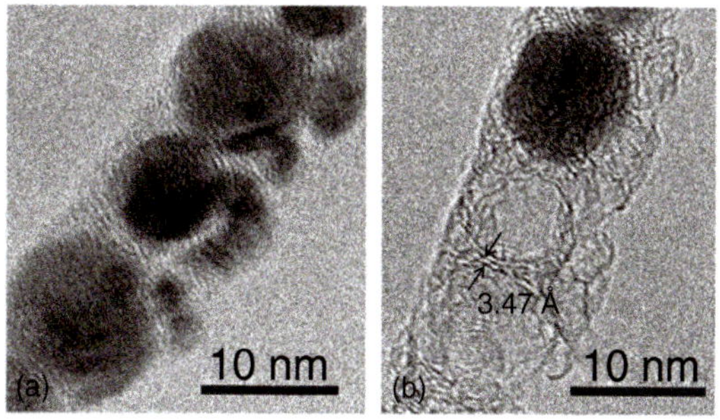


**Figure 2.15** High magnification images (a) before and (b) after the FE process. The arrows in (b) indicate the graphitization of the carbon around a silver particle. Source: Yaakob et al. [74]. Reproduced with permission from Royal Society of Chemistry.

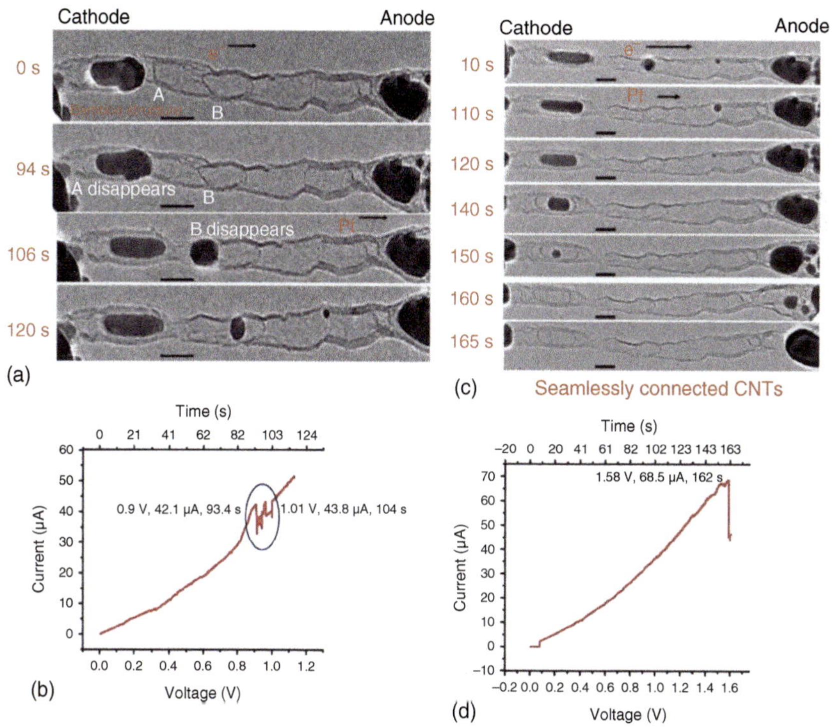


**Figure 2.16** (a) Video clip showing the nano-soldering process of CNTs with change in applied potential from 0.0 to 1.2 V. (all the scale bars are 10 nm). Bamboo-like structures A and B disappear after electromigration of Pt. (b) *I*–*V* curve of the nano-soldering process of CNTs by applying high bias voltage in the two-probe system (0.0–1.2 V). An irregular *I*–*V* curve (inside the circle) was observed during the formation of a seamless structure. (c) Screenshot from video showing flow of remaining Pt to the anode with complete soldering of CNTs (all scale bars are 10 nm). (d) *I*–*V* curve of the connected CNT showing Ohmic behavior. Source: (a, c) Sharma et al. [87]. Reproduced with permission from Elsevier, (b, d) Reproduced from Sharma et al. [87] with the permission of Elsevier.

different. For example, electromigration occurs for Fe, Pt, and Cu cases, while Au, Ag, Mo, Sn, and Ga end up with evaporation of the metal. Those materials employment have already been established in industrial applications especially electric and electronic sectors. These findings may further enhance the potential for a wide range of possibilities. For instance, Sn exhibits low-temperature graphitization at 180 °C, while commonly employed Ga and Mo catalytic behavior may benefit for high-temperature electronics and interconnect applications [84–86].

These features are very interesting and can be manipulated creatively in several applications. For example, Sharma et al. introduced nanosoldering on CNTs by manipulating the Pt catalyst electromigration of two non-connected nanotubes [87]. They introduced direct current to the CNTs up to 40 μA before the Pt particles agglomerate and migrate toward the counterpart following electron direction to the anode, resulting in seamless connection of CNTs with significant current increases. Then, the direct current was reapplied, which showed a smooth increment of the *I–V* characteristic before breakdown at 70 μA (see Figure 2.16). Another potential application is graphene interconnect fabrication from carbon–copper nanoneedle to produce graphene nanoribbon [77]. Rosmi et al. reported significant structural transformation of Cu–CNF (amorphous-C) into graphitic carbon during I-V measurement between 4.9 and 49.0 μA. The fabricated graphene nanoneedle via Joule heat demonstrated a high-current density of $10^6$ A cm$^{-2}$, which is on par with Cu in typical interconnect applications.

## 2.4 Summary

An overview of CNSs and their synthesis techniques has been described. One important challenge for CNS fabrication is implementing a controllable synthesis parameter to achieve appropriate CNS specific to the desired application. Characterization of the structure and composition of CNS is essential to understand the synthesis process, their properties, and alteration possibility for diverse applications. This chapter reviews CNS synthesis via CVD and ion irradiation techniques, showing their excellent suitability for a sustainable wide range of CNS production. The potential carbon precursor, related growth parameter, and graphitization mechanism have also been elaborated. Typical characterization techniques for CNS via Raman spectroscopy and electron microscopy also were discussed. Case in point from the recent literature concerning CNS research manifests that each technique offers unique advantages and emphasizes the importance of cost-effective techniques. They provide complementary information on the structure and properties in fast, accurate, and easy to interpret. Additionally, *in situ* transmission electron microscopy method offers an opportunity to obtain detailed information from structural transformation to properties enhancement. This detailed information obtained has significantly broadened their potential, especially in nanoelectronics applications.

## References


1 Jamaludin, S., Azmir, N.A., Ayob, A.F.M., and Zainal, N. (2020). COVID-19 exit strategy: transitioning towards a new normal. *Annals of Medicine and Surgery* 59: 165–170.

2 Moore, G.E. (1965). Cramming more components onto integrated circuits. *Electronics* 38: 8.

3 Reibold, M., Paufler, P., Levin, A.A. et al. (2006). Carbon nanotubes in an ancient Damascus sabre. *Nature* 444: 286.

4 Rossi, M. (2007). How can graphite and diamond be so different if they are both composed of pure carbon. *Scientific American* 287 (3): 103.

5 Savage, N. (2012). Super carbon. *Nature* 483: S30.

6 Georgakilas, V., Perman, J.A., Tucek, J., and Zboril, R. (2015). Broad family of carbon nanoallotropes: classification, chemistry, and applications of fullerenes, carbon dots, nanotubes, graphene, nanodiamonds, and combined superstructures. *Chemical Reviews* 115: 4744–4822.

7 Iijima, S. (1991). Helical microtubules of graphitic carbon. *Nature* 354: 56–58.

8 Novoselov, K.S., Geim, A.K., Morozov, S.V. et al. (2004). Electric field effect in atomically thin carbon films. *Science* 306: 666–669.

9 Lin, Y., Sun, X., Su, D.S. et al. (2018, 2018). Catalysis by hybrid $sp^2/sp^3$ nanodiamonds and their role in the design of advanced nanocarbon materials. *Chemical Society Reviews* 47: 8438–8473.

10 Banerjee, A., Bernoulli, D., Zhang, H. et al. (2018). Ultralarge elastic deformation of nanoscale diamond. *Science* 360: 300–302.

11 Vlasov, A., Ralchenko, V., Gordeev, S. et al. (2000). Thermal properties of diamond/carbon composites. *Diamond and Related Materials* 9: 1104–1109.

12 Piña-Salazar, E.Z., Sagisaka, K., Hattori, Y. et al. (2019). Electrical conductivity changes of water-adsorbed nanodiamonds with thermal treatment. *Chemical Physics Letters: X* 2: 100018.

13 Raccichini, R., Varzi, A., Passerini, S., and Scrosati, B. (2014). The role of graphene for electrochemical energy storage. *Nature Materials* 14: 271–279.

14 Németh, P., McColl, K., Smith, R.L. et al. (2020). Diamond-graphene composite nanostructures. *Nano Letters* 20: 3611–3619.

15 Habiba, K., Makarov, V.I., Weiner, B.R., and Morell, G. (2014). Fabrication of nanomaterials by pulsed laser synthesis. In: *Manufacturing Nanostructures* (ed. A. Waqar and N. Ali), 263–292. Manchester, UK: One Central Press (OCN) ISBN: 9781910086070.

16 Wang, Y., Lu, D., Wang, F. et al. (2020). A new strategy to prepare carbon nanotube thin film by the combination of top-down and bottom-up approaches. *Carbon* 161: 563–569.

17 Wyss, K.M., Beckham, J.L., Chen, W. et al. (2021). Converting plastic waste pyrolysis ash into flash graphene. *Carbon* 174: 430–438.

18 Karu, A.E. and Beer, M. (1966). Pyrolytic formation of highly crystalline graphite films. *Journal of Applied Physics* 37: 2179–2181.

19 Oberlin, A., Endo, M., and Koyama, T. (1976). Filamentous growth of carbon through benzene decomposition. *Journal of Crystal Growth* 32 (3): 335–349.

20 Kumar, M. and Ando, Y. (2010). Chemical vapor deposition of carbon nanotubes: a review on growth mechanism and mass production. *Journal of Nanoscience and Nanotechnology* 10 (6): 3739–3758.

21 Huang, Z.P., Carnahan, D.L., Rybczynski, J. et al. (2003). Growth of large periodic arrays of carbon nanotubes. *Applied Physics Letters* 82 (3): 460–462.

22 In, J.B., Grigoropoulos, C.P., Chernov, A.A., and Noy, A. (2011). Hidden role of trace gas impurities in chemical vapor deposition growth of vertically-aligned carbon nanotube arrays. *Applied Physics Letters* 98 (15): 153102.

23 Lee, C.-S. and Hyun, Y. (2016). Preparation and characterization of carbon nanofibers and its composites by chemical vapor deposition. In: *Chemical Vapor Deposition - Recent Advances and Applications in Optical, Solar Cells and Solid State Devices* (ed. S. Neralla), 3–21. IntechOpen.

24 Taleshi, F., Hosseini, A.A., Mohammadi, M., and Pashaee, M. (2013). Effect of hydrocarbon gas on synthesis and diameter of carbon nanotubes. *Indian Journal of Physics* 87 (9): 873–877.

25 Miseikis, V., Convertino, D., Mishra, N. et al. (2015). Rapid CVD growth of millimetre-sized single crystal graphene using a cold-wall reactor. *2D Materials* 2 (1): 014006.

26 Mamun, M.A.A., Furuta, H., and Hatta, A. (2018). Pulsed DC plasma CVD system for the deposition of DLC films. *Materials Today Communications* 14: 40–46.

27 Porro, S., Musso, S., Giorcelli, M. et al. (2007). Optimization of a thermal-CVD system for carbon nanotube growth. *Physica E* 37: 16–20.

28 Karthikeyan, S. and Mahalingam, P. (2010). Synthesis and characterization of multi-walled carbon nanotubes from biodiesel oil: green nanotechnology route. *International Journal of Green Nanotechnology: Physics and Chemistry* 2 (2): 39–46.

29 Rahman, S.F.A., Rusop, M., and Hashim, A.M. (2014). Growth of graphene on nickel using a natural carbon source by thermal chemical vapor deposition. *Sains Malaysiana* 43 (8): 1205–1211.

30 Hazan, M.A., Mohamad, S., Hamid, M.A. et al. (2020). Effect of reaction temperature on the growth of carbon nanotubes from waste natural rubber glove. *Pertanika Journal of Science and Technology* 28 (S2): 217–230.

31 Hazan, M.A., Mamat, M.S., Haroun, R.Z. et al. (2019). Carbon nanostructures grown from waste latex via chemical vapor deposition. *Malaysian Journal of Microscopy* 15 (1): 1–9.

32 Robaiah, M., Mahmud, M.A., Salifairus, M.J. et al. (2019). Synthesis and characterization of graphene from waste cooking palm oil at different deposition temperatures. *AIP Conference Proceedings* 2151: 020026.

**33** Tahir, N.A.M., Abdollah, M.F.B., Tamaldin, N. et al. (2020). Optimisation of graphene grown from solid waste using CVD method. *International Journal of Advanced Manufacturing Technology* 106: 211–218.

**34** Sharma, S., Kalita, G., Hirano, R. et al. (2014). Synthesis of graphene crystals from solid waste plastic by chemical vapor deposition. *Carbon* 72: 66–73.

**35** Rosmi, M.S., Shinde, S.M., Rahman, N.D.A. et al. (2016). Synthesis of uniform monolayer graphene on re-solidified copper from waste chicken fat by low pressure chemical vapor deposition. *Materials Research Bulletin* 83: 573–580.

**36** Shah, K.A. and Tali, B.A. (2016). Synthesis of carbon nanotubes by catalytic chemical vapour deposition: a review on carbon sources, catalysts and substrates. *Materials Science in Semiconductor Processing* 41: 67–82.

**37** Seah, C.-M., Chai, S.-P., and Mohamed, A.R. (2014). Mechanisms of graphene growth by chemical vapour deposition on transition metals. *Carbon* 70: 1–21.

**38** Son, M. and Ham, M.-H. (2017). Low-temperature synthesis of graphene by chemical vapor deposition and its applications. *FlatChem* 5: 40–49.

**39** Shang, S.M. and Zeng, W. (2013). Multidisciplinary know-how for smart-textiles developers. In: *Woodhead Publishing Series in Textiles* (ed. T. Kirstein), 92–128. Elsevier.

**40** Zou, Q., Yin, G., Gu, Y., and Xie, D. (2015). Effects of Substrate Temperature on Graphene Grown on Ni Foil via Atmospheric Pressure Chemical Vapor Deposition. In: *Proceedings of the 2015 2nd International Workshop on Materials Engineering and Computer Sciences*, 317–320. Atlantis Press.

**41** Ni, L., Kuroda, K., Zhou, L.-P. et al. (2006). Kinetic study of carbon nanotube synthesis over Mo/Co/MgO catalysts. *Carbon* 44 (11): 2265–2272.

**42** Khorrami, S.A. and Lotfi, R. (2016). Influence of carrier gas flow rate on carbon nanotubes growth by TCVD with Cu catalyst. *Journal of Saudi Chemical Society* 20 (4): 432–436.

**43** Ikuno, T., Katayama, M., Yamauchi, N. et al. (2004). Selective growth of straight carbon nanotubes by low-pressure thermal chemical vapor deposition. *Japanese Journal of Applied Physics* 43 (2): 860–863.

**44** Shudin, N.H., Mukri, M.A.R., Aziz, M. et al. (2021). The role of solid, liquid and gaseous hydrocarbon precursors on chemical vapor deposition grown carbon nanomaterials' growth temperature. *Synthetic Metals* 274: 116735.

**45** Pessoa, R.S., Fraga, M.A., Santos, L.V. et al. (2015). Plasma-assisted techniques for growing hard nanostructured coatings: an overview. In: *Anti-Abrasive Nanocoatings* (ed. M. Aliofkhazraei), 455–479. Woodhead Publishing, Elsevier.

**46** Kim, C.J., Jun, B.H., Kim, H.J. et al. (2005). Comparative study of YBCO film growth in the cold wall type and hot wall type MOCVD processes. *Physica C: Superconductivity and Its Applications* 426–431 (2): 915–919.

**47** Finnie, P., Li-Pook-Than, A., Lefebvre, J., and Austing, D.G. (2006). Optimization of methane cold wall chemical vapor deposition for the production of single walled carbon nanotubes and devices. *Carbon* 44 (15): 3199–3206.

**48** Ortega-Cervantez, G., Rueda-Morales, G., and Ortiz-López, J. (2009). Cold-wall CVD carbon nanotube synthesis on porous alumina substrates. *Journal of Materials Science: Materials in Electronics* 20: 403–407.

**49** Seidel, R., Duesberg, G.S., Unger, E. et al. (2004). Chemical vapor deposition growth of single-walled carbon nanotubes at 600 °C and a simple growth model. *The Journal of Physical Chemistry. B* 108 (6): 1888–1893.

**50** Shah, K.A., Najar, F.A., Sharda, T., and Sreenivas, K. (2018). Synthesis of multi-walled carbon nanotubes by thermal CVD technique on Pt–W–MgO catalyst. *Journal of Taibah University for Science* 12 (2): 230–234.

**51** Chakravarty, M., Sharma, R., Amit, K. et al. (2016). Synthesis and bio-compatibility study of thermal-CVD grown graphene. *International Journal of Nanoscience* 15 (5, 6): 1660016.

**52** Arjmandi-Tash, H., Lebedev, N., van Deursen, P.M.G. et al. (2017). Hybrid cold and hot-wall reaction chamber for the rapid synthesis of uniform graphene. *Carbon* 118: 438–442.

**53** Husairi, F.S., Zobir, S.A.M., Rusop, M., and Abdullah, S. (2013). Electrical properties of carbon nanotubes synthesis by double furnace thermal-CVD technique at different temperatures on porous silicon template. *IOP Conference Series: Materials Science and Engineering* 46: 012023.

**54** Hamedani, Y., Macha, P., Bunning, T.J. et al. (2016). Plasma-enhanced chemical vapor deposition: where we are and the outlook for the future. In: *Chemical Vapor Deposition - Recent Advances and Applications in Optical, Solar Cells and Solid State Devices* (ed. S. Neralla), 247–280. IntechOpen.

**55** Wang, H. and Moore, J.J. (2010). Different growth mechanisms of vertical carbon nanotubes by RF- or Dc-plasma enhanced chemical vapor deposition at low temperature. *Journal of Vacuum Science & Technology, B: Nanotechnology & Microelectronics: Materials, Processing, Measurement, & Phenomena* 28 (6): 1081–1085.

**56** Khalid, A., Mohamed, M.A., and Umar, A.A. (2017). Graphene growth at low temperatures using RF-plasma enhanced chemical vapour deposition. *Sains Malaysiana* 46 (7): 1111–1117.

**57** Qi, J.L., Zhang, L.X., Cao, J. et al. (2012). Synthesis of graphene on a Ni film by radio-frequency plasma-enhanced chemical vapor deposition. *Chinese Science Bulletin* 57 (23): 3040–3044.

**58** Fernandes, A.J.S., Pinto, M., Neto, M.A. et al. (2009). Nano carbon hybrids from the simultaneous synthesis of CNT/NCD by MPCVD. *Diamond and Related Materials* 18 (2-3): 160–163.

**59** Chockalingam, S., Bisht, A., Panwar, O.S. et al. (2015). Synthesis, structural and field emission properties of multiwall carbon nanotube-graphene-like nanocarbon hybrid films grown by microwave plasma enhanced chemical vapor deposition. *Materials Chemistry and Physics* 156: 38–46.

**60** Cho, C.W., Hong, B., and Lee, Y.Z. (2005). Wear life evaluation of diamond-like carbon films deposited by microwave plasma-enhanced CVD and RF plasma-enhanced CVD method. *Wear* 259 (1–6): 789–794.

**61** Yasuda, H. (2004). *Luminous Chemical Vapor Deposition and Interface Engineering*. New York: Marcel Dekker.

**62** Krasheninnikov, A.V. and Nordlund, K. (2010). Ion and electron irradiation-induced effects in nanostructured materials. *Journal of Applied Physics* 107: 071301.

**63** Banhart, F. (1997). The transformation of graphitic onions to diamond under electron irradiation. *Journal of Applied Physics* 81 (8): 3440.

**64** Feldman, L.C. and Mayer, J.W. (1986). *Fundamentals of Surface and Thin Film Analysis*. North-Holland: Elsevier Science.

**65** Tanemura, M., Okita, T., Yamauchi, H. et al. (2004). Room-temperature growth of a carbon nanofiber on the tip of conical carbon protrusions. *Applied Physics Letters* 84: 3831–3833.

**66** Yaakob, Y., Yusop, M.Z., Takahashi, C. et al. (2013). Room-temperature fabrication of Au- and Ag-incorporated carbon nanofibers by ion irradiation and their field emission properties. *Japanese Journal of Applied Physics* 52: 11NL01.

**67** Yaakob, Y., Kuwataka, Y., Yusop, M.Z.M. et al. (2015). Room- temperature growth of ion-induced Si- and Ge-incorporated carbon nanofibers. *Physica Status Solidi B: Basic Solid State Physics* 252 (6): 1345–1349.

**68** Sharma, S., Osugi, T., Elnobi, S. et al. (2020). Synthesis and characterization of Li–C nanocomposite for easy and safe handling. *Nanomaterials* 10: 1483.

**69** Araby, M.I., Rosmi, M.S., Vishwakarma, R. et al. (2017, 2017). Graphene formation at 150 °C using indium as catalyst. *RSC Advances* 7: 47353.

**70** Ferrari, A.C. and Robertson, J. (2000). Interpretation of Raman spectra of disordered and amorphous carbon. *Physical Review B* 61: 14095.

**71** Fauteux, C. and Pegna, J. (2004). Radial characterization of 3D-LCVD carbon fibers by Raman spectroscopy. *Applied Physics A: Materials Science & Processing* 78 (6): 883–888.

**72** Wang, Y.Y., Ni, Z.H., Yu, T. et al. (2008). Raman studies of monolayer graphene: the substrate effect. *Journal of Physical Chemistry C* 112: 10637–10640.

**73** Costa, S., Borowiak-palen, E., Kruszynska, M. et al. (2008). Characterization of carbon nanotubes by Raman spectroscopy. *Materials Science-Poland* 26 (2): 433–441.

**74** Yaakob, Y., Yusop, M.Z.M., Takahashi, C. et al. (2016). *In situ* transmission electron microscopy of Ag-incorporated carbon nanofibers: the effect of Ag nanoparticle size on graphene formation. *RSC Advances* 5: 5647–5651.

**75** Harris, P.J.F. (2020). Structures formed by passage of current through graphite: flat or three-dimensional? *Carbon* 156: 416–421.

**76** Wang, Z.L., Poncharal, P., and Heer, W.A. (2000). Measuring physical and mechanical properties of individual carbon nanotubes by *in situ* TEM. *Journal of Physics and Chemistry of Solids* 61: 1025–1103.

**77** Rosmi, M.S., Yaakob, Y., Yusop, M.Z.M. et al. (2016). *In situ* fabrication of graphene from a copper– carbon nanoneedle and its electrical properties. *RSC Advances* 6: 82459–82466.

**78** Yusop, M.Z., Kalita, G., Yaakob, Y. et al. (2014). Field emission properties of chemical vapor deposited individual graphene. *Applied Physics Letters* 104: 093501.

**79** Heersche, H.B., Lientschnig, G., O'Neill, K. et al. (2007). *In situ* imaging of electromigration-induced nanogap formation by transmission electron microscopy. *Applied Physics Letters* 91: 072107.

**80** Oh, Y.H., Kim, S.I., Kim, M. et al. (2017). Preferred diffusion paths for copper electromigration by *in situ* transmission electron microscopy. *Ultramicroscopy* 181: 160–164.

**81** Huang, X., Farra, R., Schlögl, R., and Willinger, M.G. (2019). Growth and termination dynamics of multiwalled carbon nanotubes at near ambient pressure: an *in situ* transmission electron microscopy study. *Nano Letters* 19: 5380–5387.

**82** Yusop, M.Z.M., Ghosh, P., Yaakob, Y. et al. (2012). *In situ* TEM observation of Fe-included carbon nanofiber: evolution of structural and electrical properties in field emission process. *ACS Nano* 6 (11): 9567–9573.

**83** Takahashi, C., Yaakob, Y., Yusop, M.Z.M. et al. (2014). Direct observation of structural change in Au-incorporated carbon nanofibers during field emission process. *Carbon* 75: 277–280.

**84** Araby, M.I., Sharma, S., Elnobi, S. et al. (2019). Temperature dependence of catalytic activity in graphene synthesis for Sn nanoparticles. *Journal of Materials Science: Materials in Electronics* 30: 12796–12803.

**85** Jaisi, B.P., Sharma, S., Elnobi, S. et al. (2020). Graphitization of gallium-incorporated carbon nanofibers and cones: *in situ* and *ex situ* transmission electron microscopy studies. *Physica Status Solidi B: Basic Solid State Physics* 257: 2000309.

**86** Sharma, S., Jaisi, B.P., Araby, M.I. et al. (2019, 2019). The Mo catalyzed graphitization of amorphous carbon: an *in situ* TEM study. *RSC Advances* 9: 34377–34381.

**87** Sharma, S., Rosmi, M.S., Yaakob, Y. et al. (2017). *In situ* TEM synthesis of carbon nanotube Y-junctions by electromigration induced soldering. *Carbon* 132: 165–171.

# 3

# Functional Carbon-Based Nanomaterials and Sensor Applications

*Muhammad A. S. M. Haniff and Chang Fu Dee*

*Institute of Microengineering and Nanoelectronics (IMEN), Universiti Kebangsaan Malaysia, 43600 Bangi, Selangor, Malaysia*

## 3.1 Introduction to Low-Dimensional Carbon-Based Nanomaterials

Carbon is one of the robust chemical elements in group IV of the second period of the periodic table with the ability to hybridize at the atomic level. Due to the four valence electrons in the outermost shell, carbon atom(s) can be linked via single, double, or triple covalent bonds with other atoms to form $sp^3$-, $sp^2$-, or sp-hybridized carbons. For example, when carbon is bonded to four other atoms (with no lone electron pairs), the hybridization is $sp^3$ in a tetrahedral arrangement. Meanwhile, a carbon atom bound to three other atoms (two single bonds, one double bond) is $sp^2$ hybridized and forms a flat trigonal or triangular arrangement with an angle of 120° between bonds. Whereas, the sp hybridization refers to a carbon atom bound to two other atoms (two double bonds or one single and one triple bond) in a linear arrangement with an angle of 180° between bonds. These hybridizations of carbon can determine their discrete properties as a unique feature known as "allotropy."

From bulk perspectives, three classical carbon allotropes, i.e. amorphous, diamond, and graphite, are commonly known to exist in nature and are being used in a wide range of applications [1]. Until now, the list of known carbon allotropes with at least one dimensional in nanometer size has been continuously expanded since the discovery of fullerenes ($C_{60}$, $C_{70}$, and $C_{84}$) by Sir H. Kroto et al. in 1985 as the first member of the low-dimensional carbon-based nanomaterials (LDCBNMs) family [2], followed by the discovery of the carbon nanotube (CNT) in 1991 [3], the graphene in 2004 [4], and many discovered allotropes (see Figure 3.1) [5, 6]. LDCBNMs normally refer to zero-dimensional (0-D), one-dimensional (1-D), and two-dimensional (2D) carbon allotropes with at least one dimensional between 1 and 100 nm in size.

LDCBNMs, which mainly include fullerene, CNTs, graphene, and their hybrid derivatives, have been a topic of great interest, both in the scientific and technological communities. The first three kinds of LDCBNMs mainly consist of $sp^2$ carbon atoms as a seamless network of conjugated $\pi$ electrons, while their derivatives

*Enhanced Carbon-Based Materials and Their Applications*, First Edition.
Edited by Poh Choon Ooi, Mengying Xie, and Chang Fu Dee.

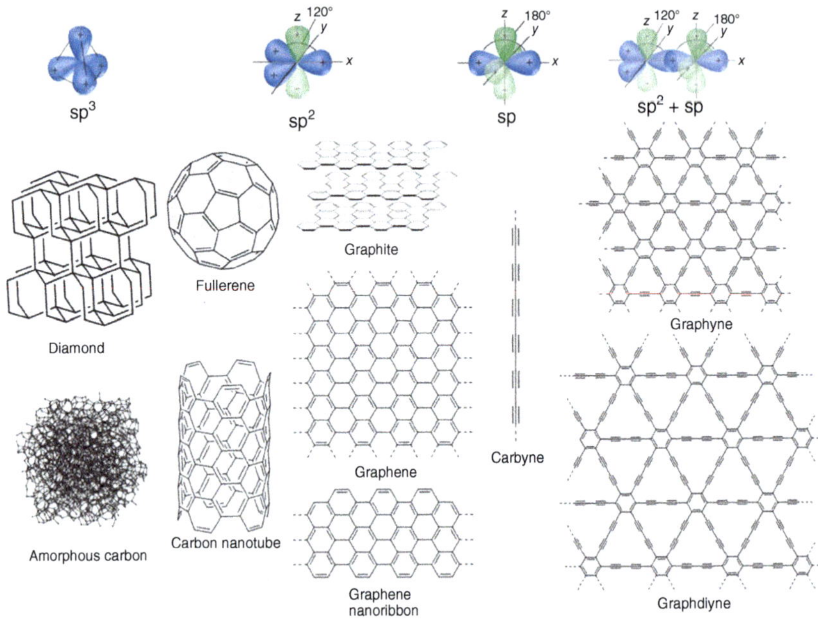


**Figure 3.1** Variety of synthetic carbon allotropes with differently hybridized carbon. Source: Chen et al. [5]. Reproduced with permission of John Wiley & Sons.

possess partial $sp^2$ and $sp^3$ carbon atoms mixed with defects and heteroatoms. In the past decade, LDCBNMs have been extensively explored and boosted the research in the field of carbon-related sensor applications, such as gas, strain–pressure, electrochemical, and optoelectronic sensors, owing to their outstanding optical, electrical, optoelectronic, mechanical, and chemical properties with respect to the bulk counterparts.

Fullerenes, as 0-D LDCBNMs, are polyhedral carbon cages in which $sp^2$-hybridized carbon is directly bonded to three neighbors in an arrangement of five- and six-membered rings. $C_{60}$ (also known as buckminsterfullerene or buckyball) with 60 carbon atoms was initially discovered in 1985 [2], and the scientists were awarded Nobel Prizes a few years later in 1991. Years later, higher (e.g. $C_{70}$, $C_{76}$, and $C_{78}$) and lower order fullerenes (e.g. $C_{24}$ and $C_{50}$) were successfully isolated [6]. Fullerenes constitute a large group of all-carbon molecules and present a theoretically unlimited number of structures depending on the number of hexagonal and pentagonal rings. Among all fullerenes, $C_{60}$ is the most abundant, stable, soluble, and least expensive fullerene; therefore, it has been the most extensively studied in the past few decades [7]. The $C_{60}$ structure has a diameter of 7.09 Å. It consists of two types of bonds, namely, those at the junction of two hexagonal rings (6–6 bonds with a mean C–C distance of 1.391 Å) and those at the junction of a pentagonal and a hexagonal ring (5–6 bonds with a mean C–C distance of 1.449 Å). Figure 3.2 illustrates the crystal structure of the $C_{60}$ molecule [8]. It possesses 20 hexagonal and 12 pentagonal rings as the basis of an icosahedral symmetry closed cage structure. The crystal structure of $C_{60}$ exhibits poor electron delocalization over the

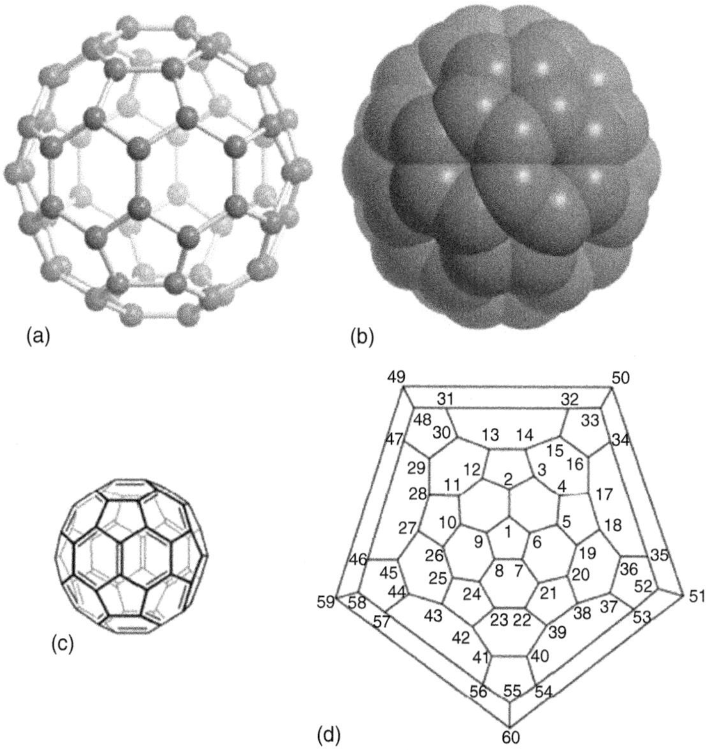


**Figure 3.2** Schematic representation of $C_{60}$. (a) Ball and stick model, (b) space-filling model, (c) VB formula, and (d) Schlegel diagram with the numbering of the C atoms. Source: Hirsch and Brettreich [8]. Reproduced with permission of John Wiley & Sons.

spherical surface due to the partial rehybridization of the $sp^2$ carbon atoms leading to a higher contribution from the s-orbitals. As a result, $C_{60}$ behaves like those in electron-deficient arenes and alkenes rather than electron-rich aromatic systems and reacts readily with electron-rich species [8]. In general, since $C_{60}$ is a good electron acceptor and versatile molecule, it can form charge-transfer complexes with electron-rich solvents and undergo a large number of reactions, including cycloadditions and nucleophilic additions.

CNTs, as 1-D LDCBNMs, are allotropes of carbon with a cylindrical nanostructure that constitute a hexagonal network of carbon atoms and can be thought of as a layer of graphite rolled up into a cylinder. CNTs have been intensively studied in many applications due to their excellent mechanical, thermal, and electronic properties since their discovery in 1991 by Sumio Iijima [3]. The extraordinary properties of CNTs bring many inventions to develop advanced materials and nanoscale devices with enhanced functional characteristics. For example, CNTs created from the $sp^2$ bonds among the individual carbon atoms can exhibit higher tensile strength than steel and are even stronger than the $sp^3$ bond found in the diamond. The strong $sp^2$ bonds between the carbon atoms also allow CNTs to withstand higher electric currents and temperatures than copper, thereby making CNTs excellent electrical and thermal conductors. There are three common production methods of CNTs: arc discharge, laser ablation, and chemical vapor deposition (CVD). The CNTs can be produced of either single-walled CNTs consisting of a single graphite sheet seamlessly wrapped into a cylindrical tube with a diameter of 0.4–3 nm, or multi-walled carbon nanotubes which comprise an array of concentric cylinders with a diameter of 1.4–100 nm, respectively. In addition, depending on the chirality of their atomic structure (e.g. armchair, zigzag, and chiral), CNTs can behave like metallic or semiconducting with an inversely proportional bandgap to the nanotube diameter. The chiral vector of a single-walled nanotube's structure is represented by a pair of indices, $n$, and $m$. These two integers correspond to the number of unit vectors along with the two directions in the honeycomb crystal lattice of the carbon network [9]. The nanotube configuration is known as armchair when $n = m$, zigzag when $m = 0$, and chiral for other configurations (see Figure 3.3a–d) [10]. The armchair configuration is always metallic (highly conducting), while other configurations are semiconductors.

Most prior fundamental studies on the properties of CNTs were conducted on resultant nanotubes produced by high-temperature arc discharge and laser ablation due to their high crystallinity and straightness [11, 12]. The present drawbacks of these methods are that the nanotubes have to be produced separately, and post-processing steps are further required to functionalize them. Alternatively, CVD has been considered the most promising method to enable precise control of aligned CNTs at much lower temperatures [13], although the CVD-grown CNTs

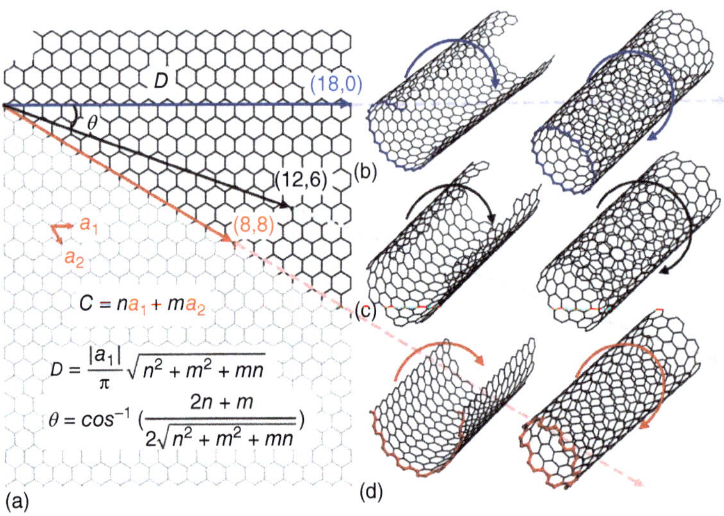


**Figure 3.3** Atomic structure of single-walled carbon nanotube. (a) Schematic representation of a two-dimensional graphene sheet illustrating lattice vectors $a_1$ and $a_2$, and the roll-up chiral vector $C = na_1 + ma_2$. (b–d) Different single-walled carbon nanotubes depending on the rolling direction and the length of the vector can be obtained. The chiral angle $\theta$ and the diameter $D$ of the single-walled carbon nanotube as functions of the chiral indexes ($n$, $m$) are also shown. Source: Qiu and Ding [10]. Reproduced with permission of American Chemical Society.

might possess a certain degree of defects or disorders. The CVD process involves the chemical breakdown of a hydrocarbon precursor on a substrate. The CNTs grown via CVD are strongly dependent on controllable variables, such as catalysts, supportive layers, the ratio of feed gases, and temperatures, thus enabling the easy control of the growth process [14–17]. The advantages of this CVD are that the yield is relatively high, the chirality of the nanotube is consistent (e.g. metallic or semiconducting nanotube), and the growth area size is theoretically arbitrary [10, 18–20].

Graphene is a 2D allotrope of carbon consisting of a single flat layer of $sp^2$-hybridized carbon atoms just one-atom-thick arranged in a honeycomb lattice structure. It is known that carbon atoms have a total of six electrons (two in the inner shell and four in the outer valence shell). Each carbon atom is bonded to three other carbon atoms with strong covalent bonds, which are difficult to break, leaving one of the four electrons in each carbon atom's outer valence shell available for conduction and free to interact with other atoms or molecules for further functionalization. These highly mobile electrons are called $\pi$ electrons or $\pi$ orbitals ($\pi$ and $\pi^*$) are located above and below the graphene sheet. The $\pi$ orbitals overlap and support to enhance carbon-to-carbon bonds in graphene. Because of this, graphene has been known as an electrical and thermal conductor, which is a great alternative to copper. Also, graphene is the strongest material ever discovered due to its compact structure and strong $sp^2$ bonds between carbon atoms. It exhibits a high tensile strength of 130 GPa and is 200 times stronger than steel yet incredibly lightweight than paper. The theory of graphene was first predicted in 1947 by P. R. Wallace [21]. A few decades later, it was discovered in 2004 by K. Novoselov and A. Geim using a simple scotch peel method (mechanical exfoliation of graphite). Both scientists were awarded the Nobel Prize a few years later in 2010 for this breakthrough [4]. Since then, graphene has gained great attention among researchers due to its unique properties in terms of optical, electronic, mechanical, chemical, and biochemical [22].

## 3.2 Modification of Low-Dimensional Carbon-Based Nanomaterials

Pristine LDCBNMs are hydrophobic in nature and have very poor sensing performance when interacting with analytes due to their seamless network of conjugated $\pi$ electrons. Because of this, LDCBNMs modification is needed to alter sensitivity and selectivity toward target analytes. Meanwhile, it is also critical for improving solubility and processability. Modification of LDCBNMs (e.g. fullerene, CNTs, graphene, and their hybrid derivatives) can be achieved via covalent (chemical) or noncovalent (physical) interaction. Covalent functionalization uses chemical reactions to attach functional groups covalently to the conjugated surfaces of LDCBNMs and changes the intrinsic optical and electronic properties. In contrast, the noncovalent functionalization incorporates the adsorption of small-molecule surfactants to the surface of LDCBNMs without changing the intrinsic properties of LDCBNMs at all.

During covalent functionalization, defects are introduced on the surface of LDCBNMs. The $sp^2$- and/or $sp^3$-hybridized carbon atoms of LDCBNMs and some of the oxygenated functional groups associated with the defects may act as primary target sites for surface functionalization, which is important for the surface energy changes to improve the solubility and processability of LDCBNMs with other substances. Many efforts to modify carbon-based nanomaterials (CBNMs) have been made to generate distinctive functional group profiles and study their surface chemistry upon the heterogeneity of the chemical reactivity toward sensor applications.

Many routes have been explored to form robust covalent bonds between the carbon atoms in LDCBNMs and organic molecules, giving rise to a variety of functionalized LDCBNMs with distinguished properties that prevent agglomeration, enhance host compatibility and processability. Among the LDCBNMs, the fundamentals of fullerene chemistry have been thoroughly investigated and studied during the last decades. A wide range of the covalent functionalization routes was subsequently applied to CNTs and graphene using the same synthetic protocols used for the functionalization of fullerenes. It is known that the hydrophobic nature of fullerenes makes its solubility in organic (nonpolar) solvents, such as toluene, chloroform, and benzene quite high. At the same time, it is very challenging to avoid aggregation of fullerenes in polar solvents, such as aqueous dispersions. Poor solubility in water, low processability, and other drawbacks of fullerenes have been addressed by a convenient route of covalent functionalization, leading to water-soluble fullerene derivatives, which have great potential for applications in biochemistry, biology, and medicine [23]. The bonding of fullerene has delocalized $\pi$ molecular orbitals extending throughout the structure, and the carbon atoms are of a mixture of $sp^2$- and $sp^3$-hybridized systems. Altering the partial $sp^2$- to $sp^3$-hybridized carbon is the driving factor for the functionalization of fullerene. The basic chemical modification of $C_{60}$ via covalent functionalization, including free-radical reactions, cyclopropanation, nucleophilic–electrophilic addition, or cycloaddition reactions, such as [1+2], [2+2], [3+2], and [4+2] reactions, can covalently conjugate various molecules to $C_{60}$ [23, 24]. Due to the electron-deficient nature, $C_{60}$ behaves as a $2\pi$ electron-deficient dienophiles and 1,3-dipolarophiles, which experience a variety of cycloaddition reactions, such as [1+2], [4+2] (Diels–Alder reactions), [3+2], and [2+2]. For instance, room-temperature and *in situ* [2+1] nitrene cycloaddition reactions between sodium azide, 2,4,6-trichloro-1,3,5-triazine, and fullerenes can efficiently produce triazine-functionalized $C_{60}$ with a preserved $\pi$-conjugated system [25]. The attachment of these molecules to the fullerenes overcomes the hydrophobicity and insolubility of $C_{60}$ in the water while retaining their unique properties.

On the other hand, chemical modification of water-soluble fullerene derivatives with attached hydroxyl (–OH), carboxyl (–COOH), and amine (–$NH_2$) groups via chemical modification route has also been extensively explored in the past few years to increase hydrophilicity, and water solubility for chemical and biological applications, and are also beneficial in enabling the interaction between fullerenes with biomolecules for sensor applications [24, 26]. The derived fullerenes can offer stable

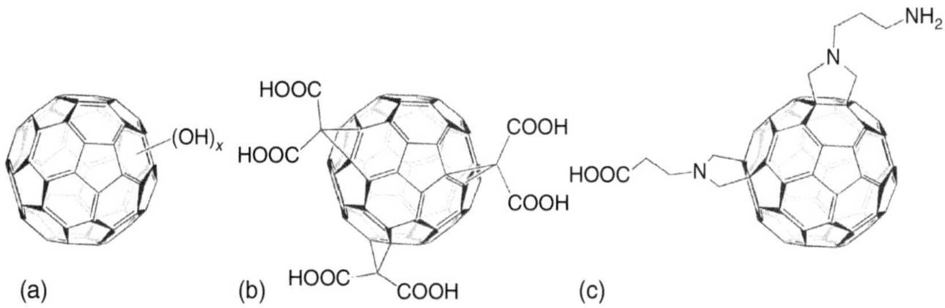


**Figure 3.4** Schematic representations of (a) hydroxylfullerene, (b) carboxyfullerene, and (c) amino acidic fullerene. Source: Goodarzi et al. [24]. Reproduced with permission of Elsevier.

molecular dispersed suspensions or encapsulation of $C_{60}$ in order to prevent aggregation between the fullerene cores. Hydroxyl fullerene or fullerenol, $C_{60}(OH)_x$, where multiple hydroxyl groups are chemically attached to the surface of $C_{60}$ ($2 \leq x \leq 44$), has a simple structure, convenient for practical use, low toxicity, and possible for further functionalization with other molecules (see Figure 3.4a). The solubility of fullerenol is dependent on the number of introduced hydroxyl groups by various degrees of hydroxylation with sodium hydroxide and hydrogen peroxide [27, 28]. Generally, a fullerenol with fewer than 12 hydroxyl groups shows very poor water solubility, whereas fullerenols with more numerous hydroxyl groups (16, 20–24, or 36–40) are soluble in water. The specific behavior of fullerenol molecules is an outcome of their structural flexibility, the rotation of the hydroxyl groups around the axes going through the C—O bonds, and the distribution of the hydroxyl groups across numerous carbon sites of the fullerenes.

Fullerene epoxide, $C_{60}O_x$ ($x \geq 2$), having epoxide groups chemically attached to the surface of $C_{60}$ is also anticipated to play an important role in functionalizing fullerenes with various nucleophilic substitution reactions, in which carbon–oxygen bond cleavage of the epoxy ring occurs with a frontside attack of a nucleophile [29]. Carboxyfullerene ($C_{60}$–COOH), where multiple carboxyl groups are chemically attached to the carbon cage, is one of the most promising classes of water-soluble fullerenes due to its practical use and low toxicity (see Figure 3.4b). Hirsch and his coworkers introduced highly water-soluble fullerenes by attaching carboxyl groups in dendrimeric mode to the surface of $C_{60}$ via hydrogenolysis and nucleophilic cyclopropanation reactions [30, 31]. Meanwhile, another chemical modification of fullerenes is the addition reaction between the fullerene and amine groups under ambient conditions to obtain fullerene derivatives [32]. A key step in the addition reaction is the exothermic formation of charge-transfer complexes between fullerenes and amine groups (see Figure 3.4c), which can be simply adopted for further modification with biomolecules to form stable immobilized $C_{60}$–biomolecule conjugates for biosensor devices [33–35].

The covalent functionalization of CNTs has an almost similar route to that of fullerenes which can be organic additions, including cycloadditions, free-radical reactions, and nucleophilic–electrophilic additions, and oxidation reactions with inorganic oxidants. Although the methods for covalent functionalization of CNTs have been well-established for the past few decades, the challenges to preserve their intrinsic properties for robust applications still remain a concern. Defect and sidewall functionalization of CNTs are the most promising strategy for the development of the covalently functionalized CNTs due to the capability of CNTs to undergo chemical reactions on the defects and sidewalls of nanotubes. These enable the more combination of material properties of CNTs with other substances and therefore help to solubilize them and facilitate their studies. A wide variety of defects, such as pentagon–heptagon defects, Stone–Wales (or 7-5-5-7) defects, $sp^3$ defects, and vacancies with dangling bonds and open tube ends terminated by carboxyl, hydroxyl, or carbonyl functionalities located at the nanotube termini and also in the nanotube sidewalls, are the useful target sites for further functionalization reactions [36].

Among various covalent reactions, oxidation is one of the most common to functionalize CNTs with the aim of building sensing platforms. The highly introduced carboxyl functionalities via strong oxidation reaction with a mixture of concentrated nitric and sulfuric acid in a liquid phase or refluxing in concentrated nitric acid represent primary target sites for further functionalization, as they facilitate the covalent coupling of other molecules through the creation of amide and ester bonds [37, 38]. The carboxyl functionalities predominantly occur at the more open tube ends or defects of CNTs rather than their sidewalls. However, the oxidation reactions with strong inorganic oxidants may lead to severe structural damage and cause more perturbation of the $\pi$-surface of CNTs.

Alternatively, a green method of oxidation reaction using $K_2FeO_4$ was introduced to provide minimal perturbation to the intrinsic properties of CNTs, in which the oxidation reaction of CNTs occurs selectively at the $sp^3$ defects while retaining most of the $sp^2$ carbon network in the carboxylated CNTs [39]. Using this method, it is suggested that strong and stable anchoring of functional groups such as amine functionalities can be precisely attached to the CNTs via their termini or sidewalls of oxidized nanotubes to ensure long-term stability and reproducibility in an aqueous solution. For instance, carboxylated CNTs can react with biomolecules (e.g. peptides, antibodies, and insulin) via EDC or EDC–NHS coupling reaction to form stable primary amine-terminated CNTs [40], as shown in Figure 3.5. The bioconjugated CNTs with biomolecules are beneficial for biomedical applications, such as sensing, imaging, drug delivery, and therapy [41–44].

Covalent functionalization with organic species is the most investigated form of graphene functionalization. This is due to the fact that covalent functionalization causes the most significant changes to the electronic properties of graphene (e.g. the opening of a bandgap and shifting of the Fermi level) and allows for the attachment of a wide variety of functional groups on the surface. Therefore, the covalent functionalization can change pristine graphene into chemical-sensitive and water-soluble graphene derivatives and becomes attractive for sensor applications. Covalently functionalized graphene is associated with the rehybridization of $sp^2$ carbon into $sp^3$ carbon in several ways, including free-radical reactions, cycloadditions, halogenations, hydrogenations, and oxidations. Through the free-radical reactions, the graphene surface with the most common addition of

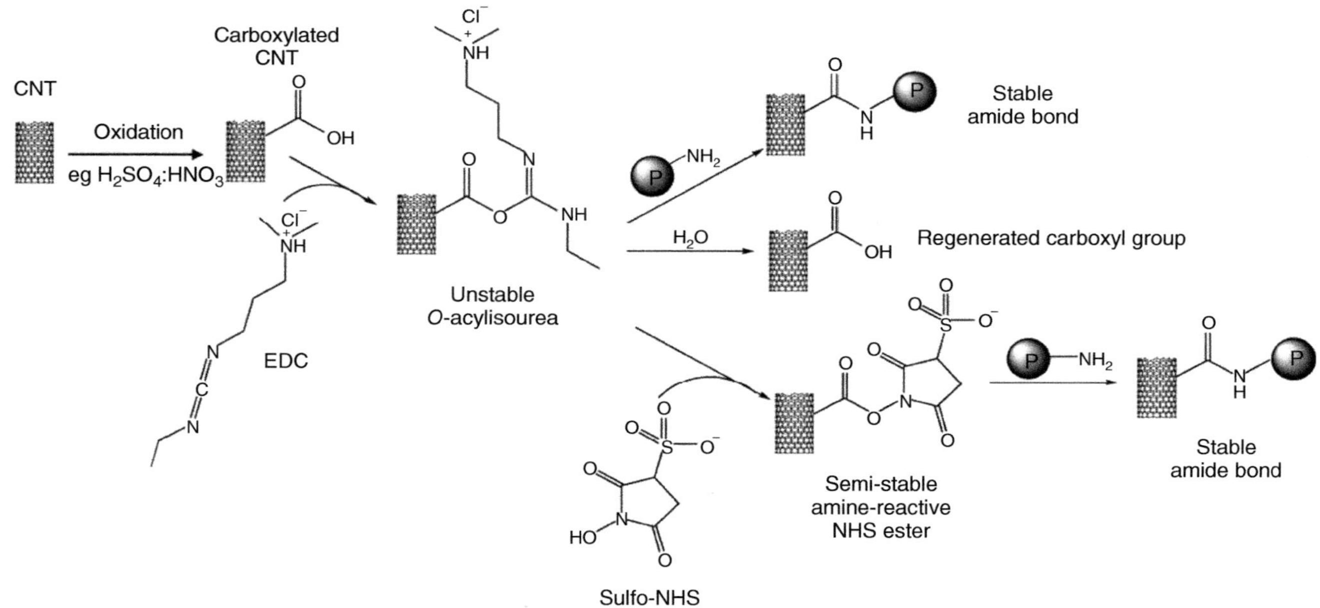


**Figure 3.5** Schematic representations of theoretical conjugation of proteins to carboxylated carbon nanotubes using 1-ethyl-3-(3-dimethylaminopropyl)carbodiimide (EDC) in the presence or absence of sulfo-*N*-hydroxysuccinimide (sulfo-NHS). Source: Gao and Kyratzis [40]. Reproduced with permission of American Chemical Society.

aryl radicals can be generated from the reduction of diazonium salts due to its high controllability and stability in the ambient environment [45–47]. The degree of reaction is strongly dependent on the number of graphene layers. It has been shown that single-layer graphene has a much higher reaction than bilayer and multilayer graphene [48]. The aryl radical reacts with the aromatic structure of the graphene network to form a covalent bond with one of the carbon atoms, thus changing the hybridization from $sp^2$ to $sp^3$ carbon on both the basal plane and edges of graphene. This diazonium chemistry reaction, for instance, has been utilized to anchor alkynyl functionalities on the graphene surface (see Figure 3.6), thus enabling further covalent immobilization for a wide range of biorecognition molecules [49–51].

The oxidized graphene derivatives with hydroxyl, carboxyl, carbonyl, epoxide, ketone, quinine, and phenol functionalities have great potential for further covalent functionalization due to their strong hydrophilicity and good dispersibility in numerous solvents [52]. Different methods have been described for oxidizing the graphene derivatives using different oxidants [53]. Oxidized graphene can be easily synthesized by acid treatments starting from raw graphite, for example, the Hummers' method ($KMnO_4$, $NaNO_3$, and $H_2SO_4$) [54]. This synthesis method has been modified to improve the efficiency of the oxidation process with a substantial amount of hydrophilic oxidized graphene [55]. Further, controlled oxidized graphene can also be prepared via direct thermal oxidation in the air [56] and electrochemical oxidation approach by applying the corresponding voltage or current to the graphene or graphite in $HNO_3$ [57] or $Na_3C_6H_5O_7$ medium [58]. Through the oxidation reactions, the π-conjugated system in graphene lattice is disrupted, and the newly introduced $sp^3$-hybridized carbon domains with oxygen functional groups on the basal plane and edges of graphene. With the help of well-established EDC or EDC–NHS coupling reactions, the negatively charged carboxylic groups of the oxidized graphene at the edges are possible to be activated and therefore allowing further functionalization with a variety of bioreceptors (e.g. antibody, ssDNA, and enzyme) for biosensor applications [59–62]. For instance, covalently functionalized carboxylic groups of graphene oxide (GO) with the amine of glucose

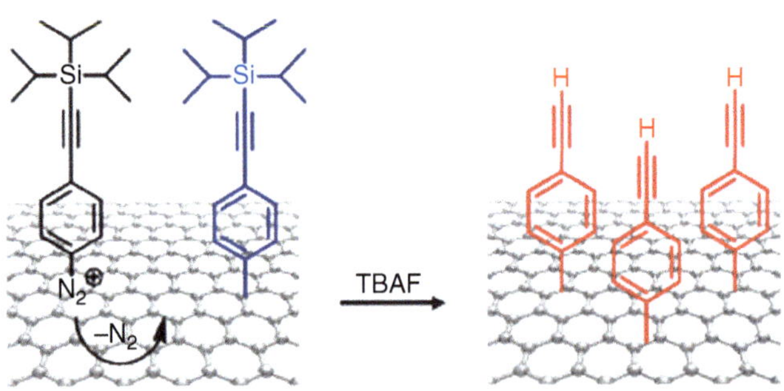


**Figure 3.6** Schematic representation of electrochemically driven anchoring of 4-[(triisopropylsilyl)ethylenyl]benzene diazonium tetrafluoroborate diazonium salt, followed by chemical deprotection forming alkynyl-terminated graphene. Source: Mishyn et al. [49]. Reproduced with permission of Royal Society of Chemistry.

oxidase ($GO_X$) have been used in a biocompatible GO–$GO_X$ biosensor platform for human glucose detection, and these biosensor chips offer great potential for use in *in vivo* clinical diagnostics, such as diabetic retinopathy [63, 64]. In addition, covalently functionalized GO with a biorecognition of anti-heat shock protein (antiHSPs) has also been developed on the label-free immunosensor platform to detect tumor marker HSPs in real human blood serum [65]. The functionalized GO-based HSPs biosensors exhibited a broad linear range, good repeatability, and reproducibility, suggesting strong and stable immobilized GO-antiHSPs conjugates for use in the high-sensitive determination of HSPs as low as sub-femtogram scales.

## 3.3 Plasma-Based Synthesis of Heteroatom-Doped Graphene

Many methods have been applied for the synthesis of heteroatom-doped graphenes, such as nitrogen (N), boron (B), phosphorus (P), sulfur (S), and the first three halogen groups (F, Cl, and Br). In this section, we first emphasize on plasma-based method, including *in situ* plasma-assisted growth and doping approaches and post-growth plasma treatment approaches for large-scale synthesis and processing of graphene on arbitrary substrates as it offers several advantages, such as low reaction temperature, high purity of atmosphere, and high control of the doped graphene processing with enhanced physical and chemical properties. Plasma used in most surface modification is low-temperature nonequilibrium plasma generated at atmospheric or low pressure by igniting a gas discharge [66]. In the case of heteroatom-doped graphene, the plasma provides the ability to tune both physical and chemical properties of graphene simultaneously by surface modification (i.e. chemical doping with heteroatoms) from a rich chemical environment, including a mixture of highly reactive species (radicals, ions) in order to achieve new or improved properties, such as bandgap engineering or doping. Furthermore, the dopants can be effectively substituted into the graphene lattice to form stable covalent bonding between the carbon and dopant atoms. However, the substitutional doping of graphene with such heteroatoms disrupts ideal $sp^2$ hybridization of carbon atoms, hence locally inducing defects to alter the physical and chemical properties significantly. Depending on the distinct nature of each dopant, new or improved properties may arise that are advantageous for particular graphene-based applications, including sensing and electronics [67].

### 3.3.1 *In Situ* Plasma-Assisted Growth and Doping

Many methods have been established for doped-graphene synthesis. Among these methods, plasma-enhanced chemical vapor deposition (PECVD) is the most practical route to synthesizing large, continuous, and uniform heteroatom-doped graphene on arbitrary substrates at low temperatures [68]. In the PECVD method, as shown in Figure 3.7, the synergistic effect of plasma-assisted and catalytic-assisted growth mechanisms makes it a convenient route to synthesize graphene films and

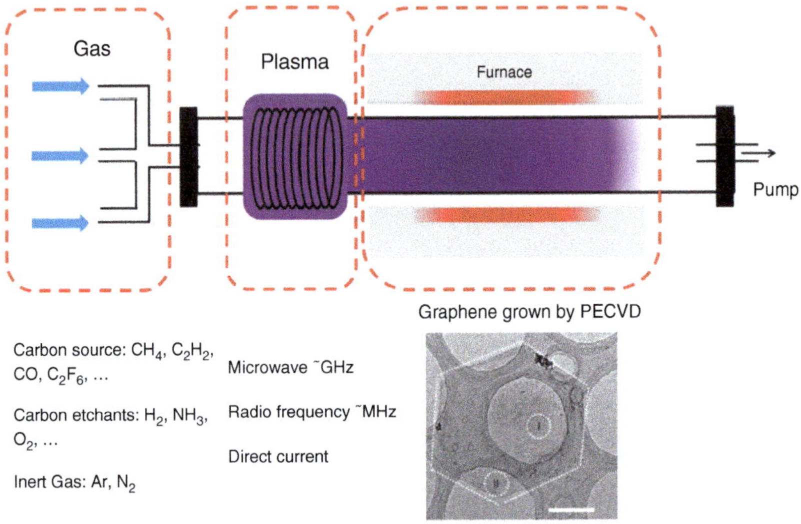


**Figure 3.7** Schematic of the PECVD system. Source: Li et al. [68], John Wiley & Sons, CC BY 4.0.

to dope heteroatoms during the formation of graphene films on metals or even directly on dielectric surfaces, particularly to incorporate heteroatoms directly into the graphene lattices. The heteroatom-doping can occur by introducing solid, liquid, or gaseous precursors containing heteroatoms along with the carbon sources into the growth reaction furnace or chamber. The bonding configuration of heteroatom-doped graphene varies with different types of catalysts and precursors.

In the case of plasma-assisted N-doping, recent studies have demonstrated the use of gaseous precursors, such as methane ($CH_4$) and nitrogen ($N_2$) or ammonia ($NH_3$), as the C and N sources to produce N-doped graphene in the presence of a metal catalyst (Cu) and dielectric surfaces ($SiO_2$, $Al_2O_3$, h-BN, and glass) [69–72]. The growth temperature and the gas ratio between $CH_4$ and $N_2$ or $NH_3$ considerably influence the N content and the N-bonding configuration. The N-doped graphene grown at high temperatures (e.g. 950 °C) primarily exhibits less N content in graphitic-N form (N atom has three carbon neighbors in a hexagonal ring) [69]. Meanwhile, a few studies on the growth of N-doped graphene on dielectric surfaces at low temperatures (e.g. 435 and 760 °C) have demonstrated the dominance of pyridinic-N form in graphene films (N atom has two carbon neighbors in a hexagonal ring) [70, 71]. Similarly, the pyridinic-N is also preferred in graphene films and leads to the dominant doping configuration at a high $N_2$ flow rate (or high $N_2/CH_4$ ratio) [72]. The increased content of pyridinic-N over pyrrolic-N and graphitic-N could be attributed to the dehydrogenation of N—H bonds of pyrrolic-N for forming $sp^2$ N–C and the increased formation of aromatic carbon rings under high-dilution ratio conditions [73].

A recent study of plasma-assisted graphene growth using liquid benzene ($C_6H_6$) and $NH_3$ has reported that the pyridinic-N and pyrrolic-N are easier to form than graphitic-N at low temperatures [74]. As illustrated in Figure 3.8, few-layered

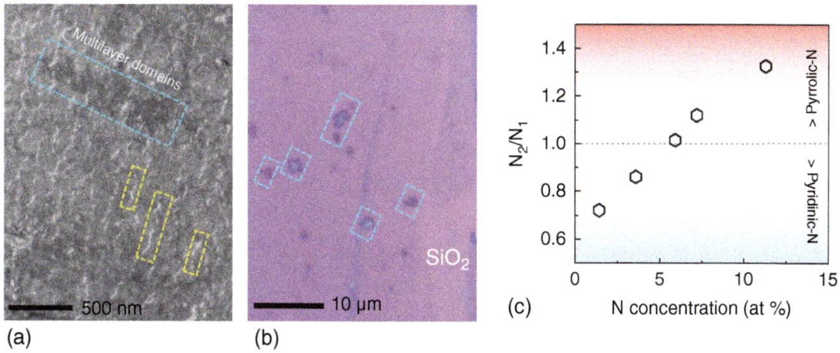


**Figure 3.8** Growth of N-doped nanocrystalline graphene by plasma-assisted chemical vapor deposition. (a) Scanning electron microscopy and (b) optical image of as-transferred N-doped nanocrystalline graphene films on $SiO_2$/Si substrate. (c) Ratio of pyrrolic-N to pyridinic-N content ($N_2/N_1$) as a function of atomic N concentration. Source: (a) Zainal Ariffin et al. [74]. Reproduced with permission of American Chemical Society, (b, c) Zainal Ariffin et al. [74]. Reproduced with permission of American Chemical Society.

N-doped nanocrystalline graphene synthesized on nickel (Ni) films at 100 °C exhibited a high N content, from 1.42 to 11.28 at%, in which the dominant doping configuration of N-doped graphene changes from pyridinic-N to pyrrolic-N at higher plasma power or $NH_3$ flow rate conditions. Upon introducing the precursor into the reaction chamber at 100 °C, the uniform plasma generated on the Ni surface continuously dissociates $C_6H_6$–$NH_3$ molecules into highly reactive $C_xH_y$ species (radicals, ions) ($x < 6$, $y < 6$), $NH_y$ species ($y < 3$) and generates reactive H species in a controlled manner. During the nanocrystalline graphene synthesis, possible chemical reaction can be expressed by the following equilibrium: $C_6H_6(g) + 2NH_3(g) + Ar(g) \rightleftharpoons$ N-doped graphene(s) $+ 6H_2(g) + Ar(g) - Q$, in which $Q$ is the heat. The H atoms promote the activation of C- and N-related radicals and lead to the formation of C—N bonds into the carbon lattices.

At a low plasma power or low $NH_3$ flow rate, fewer H species are likely to initiate the nucleation of small graphitic clusters from the assembly of $C_6$ rings on the Ni surface and result in high coverage of the nanocrystalline graphene film on the Ni surface. In parallel, more $C_xH_y$ species in the form of $C_6$ rings are generated to form covalent bonds with $NH_y$ species at the edges and graphitic basal plane during the adsorption of $C_xH_y$–$NH_y$ species on the Ni surface, hence resulting in more pyridinic-N ($sp^2$-coordinated, two N—C bonds in a hexagon) than pyrrolic-N ($sp^3$-coordinated, two N—C bonds in a pentagon). In contrast, at a high plasma power or high $NH_3$ flow rate, the densities of $C_6H_6$ and $NH_3$ molecules tend to increase more intensely, and more H species are generated during the $C_6H_6$–$NH_3$ dissociation reactions, probably leading to significant edge etching. $NH_y$ species tend to attach to the edges and form NH-terminated edges to increase the energy barrier for C species attachment. On the other hand, some of the $C_6H_6$ molecules tend to partially dissociate into C radical fragments or broken graphitic $sp^3$ carbon instead of forming $C_6$ rings to assemble the nanocrystalline graphene layer. Therefore, strong interaction between the $NH_y$ species and the graphene edges is expected to take place, especially at the edges or defects, resulting in more pyrrolic-N than pyridinic-N due to the strong influence of reactive $NH_y$ species [74].

Since N and B atoms are symmetric owing to their comparable atomic sizes and valence electron numbers with the C atom, they are expected to share the same behavior in chemically doping graphene films. However, compared to N-doping, much less progress has been made in B-doped graphene films so far, especially using plasma-assisted processes. In the case of the CVD method, B-doped graphene was grown on Cu foil via graphitization of adsorbed carbonaceous and boron species by the sole solid feedstock of phenylboronic acid ($C_6H_7BO_2$) in a low-pressure system [75]. Meanwhile, the growth of B-doped graphene on Ni substrate via segregation of B from bulk Ni was also reported with the use of triethylborane ($C_6H_{15}B$) as the B-containing precursor [76]. However, unlike N-doping, the major challenge of B-doping seems to be the difficulty in controlling high-quality B-doped graphene growth due to B-containing precursors being highly reactive to water from the ambient. Furthermore, the highly reactive B species often form metal boride compounds on the metal catalyst, thus hindering the capabilities to grow B-doped graphene films in a one-step process [77, 78]. To overcome this drawback, a recent study of metal-free growth of B-doped graphene on glass has been conducted using diborane ($B_2H_6$) as the B dopant through a hot-filament CVD method [79]. The as-grown B-doped graphene on glass was found to be uniform on a large scale, in which the B content can be simply changed from 0 to 5.3 at% by modulating the $B_2H_6$ flow rate. Furthermore, the B-doped graphene grown with high B content of 5.3 at% also showed the optimal transparent conductive performance superior to that graphene prepared by other state-of-the-art methods.

As for the plasma-assisted process, unlike N-doped graphene, the growth of B-doped, P-doped, S-doped, and halogen-doped graphene in a one-step process is still rare, and some of the main challenges are still urgent to be solved shortly. This is mainly ascribed to the ambiguous mechanism of the plasma-assisted process, in which the etching effect often occurs together with the *in situ* graphene growth and doping. Furthermore, the plasma-assisted process as rapid and efficient processing of graphene on insulating substrates is still at a promising state, and lots of opportunities for further progress exist in the near future.

### 3.3.2 Post-Growth Plasma Treatment

Conventional approaches for achieving heteroatom-doped graphene often depend on the post-growth thermal treatment in different atmospheres. Here, post-growth plasma treatment offers an alternate convenience strategy to simply introduce heteroatom dopants into graphene lattice with a short reaction time and low-power consumption at low temperatures or even room temperature. With plasma, heteroatom radicals are capable of substituting the carbon atom, hence altering the overall electronic properties of graphene. In the case of N-doped graphene, ammonia ($NH_3$) and nitrogen ($N_2$) are effective for doping graphene with N atoms [80–83]. Compared to thermal treatment, the plasma can introduce a higher number of graphitic-N species. For instance, upon $N_2$ plasma treatment, N atoms are likely to substitute C atoms and dope at the in-plane graphitic sites rather than the edge sites [84]. In addition, Lin et al. achieved N-doping onto the graphene surface via N plasma exposure with controlled N-bonding configuration using N ions or thermalized N atoms source [82]. Low-energy N ions exposure is inclined to induce graphitic-N species on the in-plane and strong n-type doping. In contrast, thermalized N atoms exposure may favor the formation of pyridinic-N in the presence of defects with minor n-type doping. The honeycomb structure of graphene and the $sp^2$ configuration, including graphitic-N, are mostly preserved in the N-doped graphene. This is confirmed by the LEED patterns of N-doped graphene and pristine graphene, in which the graphene spots are almost comparable (see Figure 3.9a,b). Kato et al. demonstrated a room-temperature N-doping of graphene by mild $NH_3$ plasma [86]. Interestingly, the N atoms are selectively doped near the edge of graphene without drastically perturbing the in-plane properties of graphene. Later, Lin et al. introduced a low-temperature N-doping of graphene using the mixture plasma of ozone ($O_3$) and $N_2$ to control the N-doping configuration [85]. Upon the plasma treatment, $O_3$, along with the N-containing active species, such as $N^+$ and $N_2{}^+$ ions, reacts with graphene at a substrate temperature of 100 °C, producing selective graphitic-N and pyridinic-N defects in graphene lattice network (see Figure 3.9c–h).

On the other hand, Haniff et al. demonstrated one-step plasma doping of CVD-grown pristine graphene with N using a combined ammonia ($NH_3$)/argon (Ar) plasma at a substrate temperature of 200 °C [87]. During the downstream

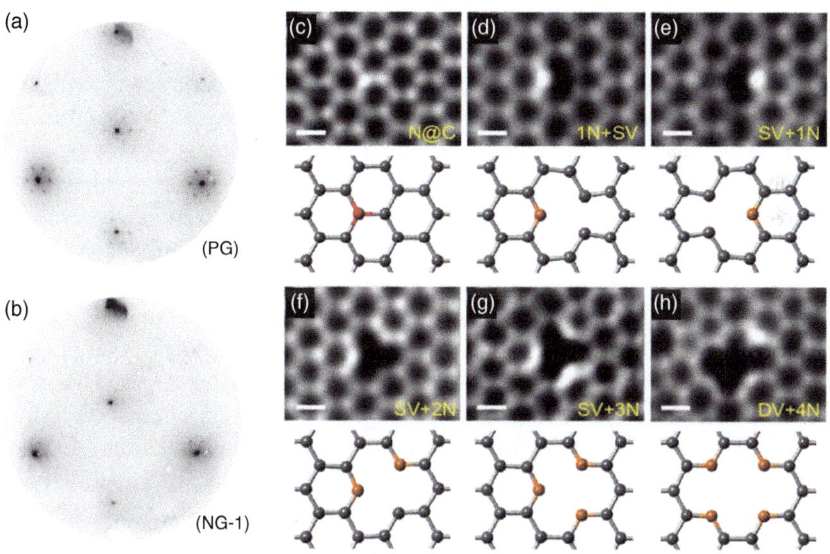


**Figure 3.9** Low-energy electron diffraction (LEED) patterns of (a) pristine graphene (PG) and (b) N-doped graphene (NG-1) with a primary electron energy of 100 eV. Source: Lin et al. [82]. Reproduced with permission of Elsevier. Annular dark-field images and atomic models of (c) graphitic-N (N@C), (d, e) single pyridinic-N, 1N + SV, and SV + 1N showing at two possible equivalent N sites, (f) SV + 2N, (g) SV + 3N, and (h) DV + 4N. The graphitic-N is presented by a red ball while pyridinic-N by an orange ball. Source: Lin et al. [85]. Reproduced with permission of American Chemical Society.

plasma treatment, the N-containing active species (e.g. excited $NH_3$, $NH_2$, NH, H radicals, and $NH_3^+$, $H^+$ ions) and $Ar^+$ ions react with graphene not only to produce covalently functionalized graphene with pyridinic-N, pyrrolic-N, and graphitic-N but also vacancy defects in the in-plane graphene. To introduce a controllable number of vacancy defects, McManus et al. reported a two-step method to induce plasma doping of graphene with N in a controllable manner [88]. The two-step method incorporates Ar plasma exposure first to create vacancy defects in the graphene lattice and then $NH_3/H_2$ plasma exposure to simultaneously restore the defects and dope the graphene with N atoms. In this case, the $sp^2$ configuration is partly restored due to N atoms being incorporated into the lattice as graphitic-N, rather than mainly at the edges as pyridinic-N and pyrrolic-N.

As for the B-doped graphene, Tang et al. introduced controllable doping in exfoliated graphene by reaction with the ion atmosphere of trimethylboron ($C_3H_9B$) decomposed by microwave plasma, in which the B atoms are substituted into graphene homogeneously without any segregation domains [89]. The distributions of the C and B elements were determined by the EELS mapping X, showing that the C and B elements coexist in the same regions. Through this process, the B content can be tuned over a wide range of 0–13.85 at% by controlling the ion reaction from 0 to 20 minutes. As a result, tunable bandgaps of B-doped graphene ranging from 0 to 0.54 eV can be achieved with the increase of B content. The B-doping strategy for tunable bandgaps described here is similar to the case of N-doping, in which sufficient N-doping leads to a bandgap opening around the Dirac point due to local A–B sublattice asymmetry induced by the substitution of N atoms in the graphene lattice [90].

Similar to N-doping and B-doping, graphene with n-type or p-type semiconducting behavior can also be achieved by P-doping and S-doping, but with a much wider bandgap at the same doping concentration [91]. In terms of post-growth P-doping, Shin et al. reported a two-step P-doping of CVD-grown graphene by inductively coupled plasma (ICP) and triphenylphosphine ($C_{18}H_{15}P$) treatment [92]. The P atoms are mostly substituted into the plasma-induced defects, resulting in a rise of $sp^3$-hybridized carbon associated with the P—C and P—O bonds. The scanning transmission electron microscopy (STEM) showed that the honeycomb structure of P-doped graphene is almost preserved and comparable to that of raw graphene. The spatially resolved EELS mapping by STEM indicated the C and P elements coexist in the same region, suggesting that P atoms are homogeneously distributed in the graphene without any segregation domains. According to the electrical transfer characteristics, the P-doped graphene field-effect transistor exhibited a negative Dirac voltage of −54 V, leading to n-type semiconducting behavior due to the electron-donating character of the P atom. On the other hand, in the case of post-growth S-doping, Liang et al. introduced a facile route to dope CVD-grown graphene with S by annealing in hydrogen sulfide ($H_2S$) atmosphere [93]. By increasing the annealing time, the S content can be increased from 0.55 to 1.46 at%. Electrical measurements of S-doped graphene field-effect transistor exhibited a sign of p-type semiconducting behavior due to the electron-withdrawing character of the S atom. Meanwhile, Wong et al. reported the efficient synthesis of

S-doped microwave-exfoliated graphene by subjecting graphite oxides to microwave plasma in the presence of $H_2S$ [94]. By varying the degree of exfoliation of the graphite oxides, the S content can be tuned from 0.62 to 1.11 at%. The high degree of exfoliation here enables the penetration of $H_2S$ into the exfoliated layers, thus making S-doping more efficient.

Post-growth plasma treatment for p-type halogen-doping (e.g. F, Cl, and Br) in graphene is particularly effective because halogen atoms are highly reactive. Baraket et al. introduced F-doped graphene using sulfur hexafluoride ($SF_6$) plasma as a safer and more convenient alternative to fluorine ($F_2$) and xenon difluoride ($XeF_2$) sources [95]. This method used electron-beam-generated plasma in the $SF_6$ atmosphere to control the F-doping in the CVD-grown monolayer graphene without any substantial etching or altering of the graphene lattice. The F-containing reactive species, such as $SF_5^+$ and $SF_3^+$, react with the graphene surface to form C–CF, C–F, and C–$F_2$ functional groups upon the plasma treatment. The Raman results indicated a rising number of the $sp^3$-hybridized carbon in the plasma-treated graphene (PG) structure caused by incorporated functionalities. Later, Yang et al. reported the layer-dependent fluorination of mechanically exfoliated graphene (one to three layers) by $SF_6$ plasma treatment [96]. The Raman results indicated that F-doping is preferential to monolayer than bilayer and trilayer graphene for rapid plasma exposure. Meanwhile, in the case of prolonged $CF_4$ plasma treatment, Shen et al. reported that the thinning effect of F-doped few-layer graphene could only occur on the topmost layer. They predicted p-type doping by theoretical simulation [97]. Further studies on the F-doping of multilayer graphene were also demonstrated. Sherpa et al. reported the F-doped epitaxial graphene (~20 layers) by $SF_6$ plasma treatment and investigated the effect of polarity of carbon—fluorine bonds on the work function [98]. The findings indicated that the work function of F-doped epitaxial graphene could be tuned by modulating the proportion of ionic and covalent C—F bonds as well as the F content.

As for the Cl-doping, Wu et al. reported low-power plasma-assisted Cl-doping on CVD-grown graphene without the generation of significant defects in chlorine ($Cl_2$) atmosphere [99]. According to the electrical measurements, the Cl-doped graphene at 8.5 at% exhibited p-type doping characteristic accompanied by a conductance improvement, indicating nondestructive Cl-doping. Pham et al. demonstrated a novel two-step method for enhancing Cl-doping on CVD-grown graphene in Cl plasma, in which pre-doping and post-doping of CVD-graphene were conducted before and after the transfer, respectively. Using a combined Cl doping, a high Cl-doping of 47.2 at% and low sheet resistance of 240 Ω $sq^{-1}$, without sacrificing the optical transmittance, were achieved [100]. Later, Pham et al. extended the Cl-doping of CVD-graphene by introducing the Cl-trapped layer-by-layer graphene and achieved a low sheet resistance of 72 Ω $sq^{-1}$ and high hole mobility of 3970 $cm^2\ V^{-1}\ s^{-1}$. As for the Br-doping, unlike F and Cl, a large-sized Br atom is likely to interact with graphene via physisorption and charge-transfer complex without disrupting the $sp^2$ carbon formation [101]. Zheng et al. reported the chemically exfoliated Br-doped graphene by using a microwave-sparks-assisted method in a liquid state of bromine ($Br_2$). The structure and electronic properties of Br-doped graphene can only be improved after thermal annealing [102]. Meanwhile, Mansour et al. reported the Br-doping on CVD-grown few-layer graphene using vapor or liquid $Br_2$ in an inert atmosphere [103]. The electrical measurements indicated that the B-doped few-layer graphene has a low sheet resistance of ~180 Ω $sq^{-1}$ with p-type doping due to the physisorption of Br atoms on the graphene surface.

### 3.3.3 Properties of Heteroatom-Doped Graphene

Functional nanomaterials based on doped graphene have gained much interest. Numerous heteroatoms (such as N, B, P, S, F, Cl, and Br) can be effectively modified, thus enabling graphene with distinctive properties. The doping of graphene with heteroatoms can significantly alter the bandgap in graphene from a metal to a semiconductor, depending on the doping level. When heteroatoms are substituted into graphene lattice, a transition to p-type or n-type semiconducting graphene can be induced because of the electron transfer between the carbon (C) atoms and the heteroatoms. For the past few years, considerable studies have focused on doping graphene with the N and B heteroatoms due to their similar atomic size as that of C atom. The doping of N or B heteroatoms into graphene lattice resulted in an n-type or a p-type semiconductor, respectively. In general, doping graphene with N heteroatom generates three common bonding configurations known as pyridinic-N, pyrrolic-N, and graphitic-N (or quaternary-N) [104]. Pyridinic-N refers to the N atom that contributes one p-electron to the π system, has a lone-pair electron in the plane of the ring, and is bonded to the two C atoms. As for the pyrrolic-N, the N atom contributes two p-electrons to the π system, has two lone-pair electrons, and contains N—H bonds. Meanwhile, graphitic-N refers to the N atom that replaces the C atom in the hexagonal ring. Because of the similar bond lengths of C–N (1.41 Å) and C–C (1.42 Å), $sp^2$-bonded pyridinic-N and graphitic-N have a marginal influence on the graphene structure, whereas the $sp^3$-bonded pyrrolic-N disrupts the planar structure of graphene. It was reported that N-doping could induce a bandgap opening of 0.16 eV in N-doped graphene, and its n-type behavior can be systematically tuned by adjusting the level of N-doping [74, 105, 106]. Similar to N-doped graphene, doping graphene with B atom can form two in-plane bonding moieties that are equivalent to pyridinic-N and graphitic-N. In contrast, another bonding is an out-of-plane configuration in which the B atom sticks out from a single C atom at the vacant sites, and this causes structural distortion [107]. As the B atom forms $sp^2$ hybridization in the carbon lattices, the planar structure of graphene is retained even though the B—C bond (~1.50 Å) is a bit longer than a C—C bond (1.42 Å). With one less valence electron to the adjacent C, the B-doping induces a charge polarization in the graphene lattice and p-type behavior by downshifting the Fermi level from the Dirac point. Different bandgaps of B-doped graphene from 0 to 0.54 eV can be induced by tuning the level of B-doping [89].

Doping graphene with P atoms also exhibits the same three bonding configurations as N-doping. However, as the P atom is larger than the N atom, P-doping causes more structural distortion. Changing the hybridization from $sp^2$ to $sp^3$, allows the P atom to form a pyramidal overhang with three C atoms at 1.33 Å from the graphene plane [108]. The P-doping can induce a small bandgap opening of 0.15 eV with n-type behavior by upshifting the Fermi level toward the conduction band edge of graphene. Theoretically, doping graphene with S can occur in two ways: the formation of defect sites and S=S bond rupture to form several bonding configurations, where the C—S bond (1.78 Å) is ~25% longer than a C—C bond. The commonly observed configurations are the thiophenic (generating a pentagonal vacancy) and bonding similar to pyridinic and graphitic arrangements as N-doping [109]. According to the formation energy using the DFT method, adsorption of S onto graphene lattice results in the most stable structure; S prefers to substitute C atoms at the zigzag edge, and the Stone–Wales defects can facilitate S-doping because of the changed local charge distribution in the C lattice [16]. Doping graphene with one S atom can induce a bandgap opening of 0.3 eV, but that with two S atoms increases the metallic properties of graphene [110, 111].

In the case of heteroatom-doped graphene with halogens, such as F, Cl, and Br, it results in severe distortions in the geometric and electronic structures of graphene (bonds with one C in graphene and stick out from the basal plane) as halogen-doping changes the hybridization of carbon from planar $sp^2$ to trigonal $sp^3$ [112]. Larger halogen atoms induce (such as Br) more significant structural defects than that of the smaller atoms (such as F) as F is one of the most reactive species, the F-doping results in the most stable and strong F—C bond, where it sticks out the basal plane and stretches the C—C bond length to ~1.58 Å. Depending on the F-doping coverage, F-doped graphene can be employed as semiconductors or insulators with a tunable bandgap up to ~3.0 eV [113–115]. In the case of Cl-doped graphene, Cl—C bond is less stable than F—C bond as Cl atom on graphene has much lower binding energy and longer bond length than that of F. Similar to F-doped graphene, Cl-doping at high coverage can alter graphene from a semimetal into an insulator due to the large disruption of graphene network [116]. Previous studies showed that graphene doped with F and Cl are direct bandgap semiconductors [117]. Meanwhile, in the case of Br-doped graphene, Br-doping is likely to interact with graphene via surface physisorption and intercalation without disrupting its basal plane, which results in p-type behavior with an indirect bandgap [103, 118].

## 3.4 Doping Modulation in Graphene for Optoelectronic Applications

Graphite is one of the most inactive and inert materials in terms of the chemical reaction. Therefore, its applications are limited only to conventional fields, such as pencils and lubricants. This situation changed a few decades ago when the nanotechnology revolution took place. Plenty of carbon allotropes have been discovered, including ball, tube, ribbon, buds, and sheet shapes commonly named CNT, fullerene, graphene, and conjoined CNT/fullerene/graphene structures. Carbon atoms that exhibit different morphs reveal salient properties in robustness, optical transparency, mechanical flexibility, environmental stability, and extremely high mobility.

Plenty of carbon-based allotrope's devices for optical and optoelectronic applications have been developed, such as graphene photodetectors [119], graphene optical biosensors [44], light-emitting devices, photodetectors, solar cells, ultrafast lasers, touch screens, based on CNTs [120] and fullerenes [121, 122]. Introduction of doping can lead the material toward wider device applications, including catalytic systems [123–125], organic thin-film transistors (OTFTs) [126], MWCNT field-effect transistor [127], photovoltaics and photodetector [128, 129], sensors [130], organic light-emitting diodes [131, 132], and supercapacitor [133, 134].

N-doped $C_{60}$ was used as Ohmic contact [128] to the hole transport layer ($C_{60}$ mixing with a fluorinated zinc phthalocyanine derivative) to achieve efficient organic tandem solar cells based on small molecules [135]. N-doped graphene [136] and graphene quantum dots (GQDs) [137] have been achieved by Usachov et al. and Tang et al. By doping with N, a new level has been introduced to the electronic structure, which will be useful for tuning of optoelectronic property of the material [137]. Single-atom doping of the more electronegative B and N compared to the host carbon matrix has been studied as well as their excellent response to the EUV region [138].

Currently, the most studied form of carbon-based material is graphene; it is a zero-overlap semimetal with extremely high conductivity. However, it has a high dark current due to its high mobility. Hence, due to high dark current, graphene-based devices cannot achieve a high on/off current ratio. The semimetallic property of graphene does not provide high conductance in its intrinsic condition. Therefore, property modification is needed to turn this material into a useful device. Its pristine state with zero bandgap limits its direct application. Doping is one method used for inducing bandgap to it by incorporation of foreign atoms. Absorption of molecules/atoms on/in a material can result in an exchange of charges between the host and foreign molecules/atoms. Those absorptions can be chemisorption or physisorption. The concentration of charge carriers can be then altered by tuning the Fermi level. Therefore, most of the electronic and optoelectronic properties can be enhanced by doping. Usually, it results in a redox reaction that alters the charge concentration. Injection and extraction of electrons from the material result in Fermi level engineering, called n-type and p-type doping [139].

Azobenzene chromophore functionalized single-walled CNTs have been demonstrated to be used as a nanoscale color detector [140]. The chromophore is able to detect low intensity at a narrow wavelength region and convert to electrical read-out to SWCNT. Methyl-orange-functionalized graphene shows switchable isomerization behavior to light, which can respond in femtoseconds regime, enabling conductance control under UV and white light illuminations [141]. Kim et al. showed that azobenzene chromophore-tethered graphene demonstrated the modulation of hole concentration up to ~18% under the illumination of UV and white light (see Figure 3.10a,b) [142]. Graphene–azobenzene–Au hybrid compound has been developed by Margapoti et al. [143], which is photoswitchable upon illumination with UV light. It provides the possibility of performing optically programmable carrier dynamics and transport in graphene.

Hydrogenation of graphene can result in the rehybridization of the carbon bonding state from $sp^2$ to $sp^3$. A bandgap of ~1 eV can be achieved by 8% of

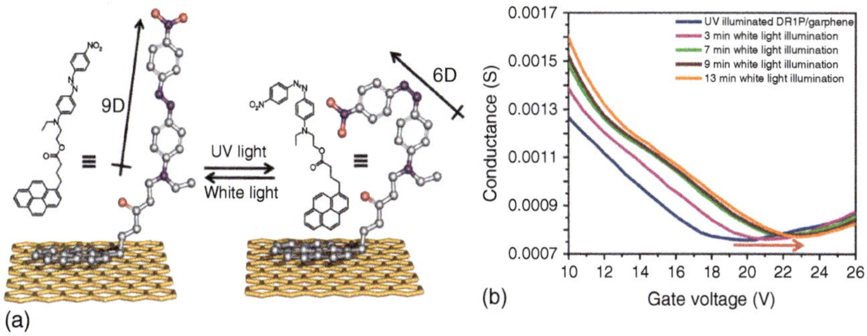


**Figure 3.10** (a) Schematic representation of photoisomerization of DR1P on graphene. (b) FET device characterization shows conductance as a function of gate voltage when the UV illuminated DR1P/graphene devices are exposed to increasing duration of white light. Source: Kim et al. [142]. Reproduced with permission of American Chemical Society.

hydrogen incorporation in graphene [144]. Zhang et al. showed phosphorus-doped carbon nitride solid that improves photocurrent generation up to a factor of 5 [145]. Sulfur-doped GQDs prepared from fructose and sulfuric acid and potassium-doped GQDs have been studied by Tang et al. [146] with tuned electronic and optical properties, which can be used for future optoelectronic applications. Besides physical doping, a new concept of "carrier doping" modulation has been demonstrated by Javier García de Abajo, namely, electronic "doping." The Fermi level shift can be induced by electrostatic doping, enabling the control over optical absorption and inelastic light scattering [147, 148].

## 3.5 Imperfections in Graphene for Strain–Pressure-Sensing Applications

Strain and pressure sensors are one of the most attractive research directions for human health monitoring, human motion detection, and human–machine interaction in recent years. Considerable progress in the invention of the wearable strain–pressure sensors has been made by researchers, which provides new opportunities for enhancing sensing capabilities. As an emerging 2D material, graphene has been considered a novel sensing material for strain–pressure sensors owing to its high sensitivity and gauge factor associated with remarkable mechanical and electronic properties. A pioneering study by Pereira and Neto investigated the piezoresistive effect of graphene and proposed that the piezoresistive effect in graphene is strongly related to the graphene lattice distortion, which leads to a modified electronic band structure [149]. The findings suggested that the tunable gauge fields induced by local strain can generate confined states, quantum wires, and collimation in graphene. Later, Huang et al. demonstrated the electronic–mechanical coupling in suspended graphene devices under large strain deformation. The electrical measurements indicated that the small gauge factor of 1.9 is due to the semimetal behavior of graphene [150]. Zhu et al. first reported a graphene-based piezoresistive pressure sensor device using monolithic integration in silicon-based on standard semiconductor processing [151]. The meander patterns made of CVD-grown multilayer graphene were placed on the maximum tensile strain region of silicon nitride ($SiN_x$) suspended membrane to enable the maximum resistance changes upon applying pressures (see Figure 3.11a,b). The electromechanical results indicated that the pressure sensitivity of $3.4 \times 10^{-5}$ $kPa^{-1}$ with a gauge factor of 1.6 was achieved. On the other hand, Smith et al. introduced the electronic–mechanical coupling sensing in suspended graphene, particularly CVD-grown monolayer graphene, as both effective sensing and membrane for novel nanoelectromechanical devices (see Figure 3.11c,d) [152]. The electromechanical measurements suggested that the gauge factors vary depending on the pressure range, with a maximum value of 4.33 and an average of 2.92. Similarly, comparable gauge factors were also reported by others, ranging from 1.6 to 6.1, suggesting that graphene in its pristine form has minor resistance changes under strain or pressure [153–155].

Due to the little piezoresistive effect existing in pristine graphene, further advancements in altering the physical-chemical properties of graphene have been made by researchers for enhancing the strain–pressure sensor sensitivity. Wang et al. reported a bucking approach to form nanoscale periodical buckling of graphene-like ripples on stretchable polydimethylsiloxane (PDMS) substrate (see Figure 3.12a–c) [156]. Unlike pristine graphene, the graphene-like ripples can afford large strain deformation up to 30%, thus making it suitable for high-level strain sensors. The graphene-like ripples can also act as an additional source of scattering for charge carriers, thus increasing resistance at its prestrain state. Meanwhile, Zhao et al. introduced ultrasensitive strain sensors based on piezoresistive effect in PECVD-grown nanocrystalline graphene [157]. A considerable gauge factor exceeding 300 was achieved from the electromechanical measurements, indicating the giant piezoresistive effect associated with the tunneling mechanism between the adjacent nanographene islands. Interestingly, unlike pristine graphene, the gauge factor of nanocrystalline graphene can be further modulated by altering the original tunneling gap. The higher gauge factor of graphene, resulting from a larger tunneling gap, may reduce the probability of conduction path drastically, thus leading to increased resistance (or too small current). Later, Hempel et al. reported a novel mechanism of strain gauges based on percolative graphene flake networks that offer a promising alternative to conventional strain gauges combining tunable gauge factors with scalable processability [158]. The findings indicated that the morphology-dependent resistance correlates well with the gauge factors, which can be tuned to exceed 150. For instance, the highest gauge factor can be expected for devices composed of low-density graphene flakes, even without a piezoresistive effect.

Considering the morphological properties of graphene, it is also important to realize the piezoresistive effect in graphene using other alternative ways for strain–pressure sensors. Recently, Haniff et al. have reported an enhanced piezoresistive effect in controllable defective CVD-grown graphene and used it for flexible pressure-sensing applications [159]. From the electromechanical measurements,

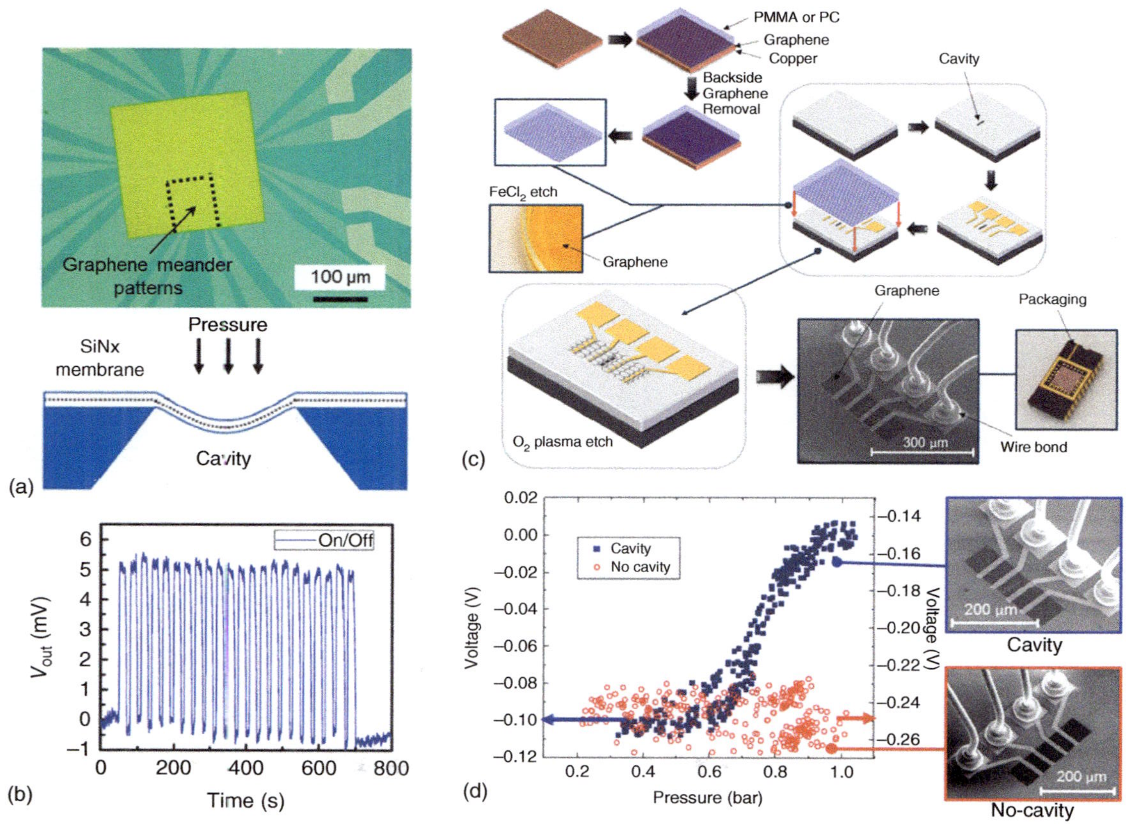


**Figure 3.11** Graphene-based piezoresistive pressure sensors. (a, b) Optical image of graphene piezoresistors on silicon nitride square membrane with a real-time response of cycling test under applied differential pressure. Source: Zhu et al. [151]. Reproduced with permission of the American Institute of Physics. (c, d) Schematic representations of a fabrication sequence of the suspended graphene pressure sensor device. The voltage changes in relation to the pressure difference of a device with a cavity indicate the piezoresistive effect in the graphene membrane. Source: Smith et al. [152]. Reproduced with permission of American Chemical Society.

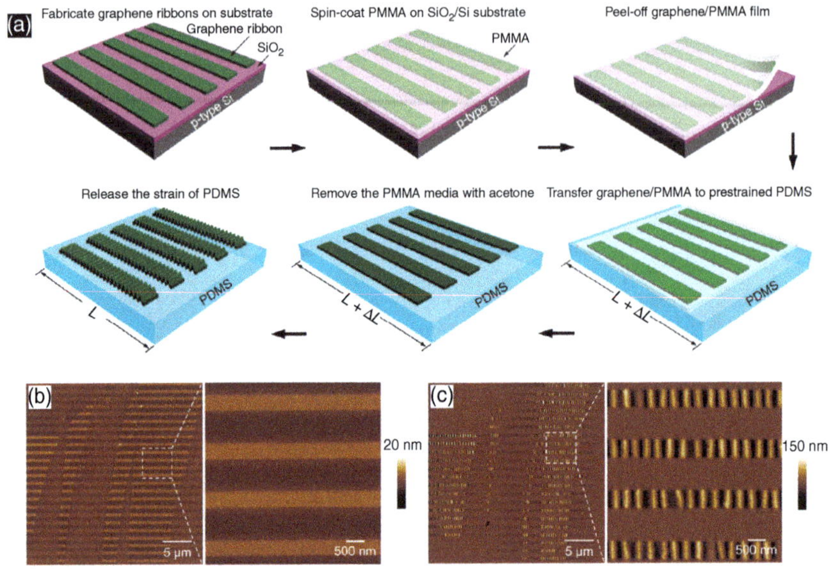


**Figure 3.12** Buckled graphene ribbon strain sensors. (a) Schematic representation of the fabrication process of buckled graphene ribbon arrays on PDMS substrate. AFM images of (b) flat graphene ribbons before buckling and (c) buckled graphene ribbons after releasing a prestrain of 30% on the PDMS substrate. Source: Wang et al. [156]. Reproduced with permission of American Chemical Society.

a linear piezoresistive effect and a high-pressure sensitivity of $10^{-3}$ $kPa^{-1}$ were observed for the applied pressure within the range of 0–50 kPa. It was also revealed that the graphene grown with a high density of defects has a fourfold higher sensitivity than a low density of defects. In the case of high-density defects, the charge-carrier transportation of defective graphene does not work similarly to that of pristine graphene. The piezoresistive effect mechanism is limited to the lattice deformation of the $sp^2$ carbon network and the defects. Here, the deformation of high-density localized defects in the graphene lattice has shown a more prominent effect on the overall resistance changes, resulting in a high-pressure sensitivity due to the significant quantum confinement in the presence of vacancies and grain boundaries.

In addition, Haniff et al. reported a piezoresistive effect in plasma-doping of CVD-grown N-doped graphene by $NH_3$–Ar plasma treatment for enhancing strain–pressure sensing [87]. From the electrical measurements, both untreated graphene (UG) and PG exhibited linear current–voltage and Ohmic characteristics, indicating the good quality of the contacts between the graphene and metal electrodes (see Figure 3.13a,b). However, the initial resistance of the PG device is relatively higher than that of the UG device because of the larger charge scatterings originating from the tunneling barriers or the structural disorders caused by the plasma treatment. The higher resistance of the PG devices can be possibly contributed by the four components (see Figure 3.13c): (i) the intrinsic resistance of the metal electrode $R_{electrode}$; (ii) the intrinsic resistance of the graphene itself

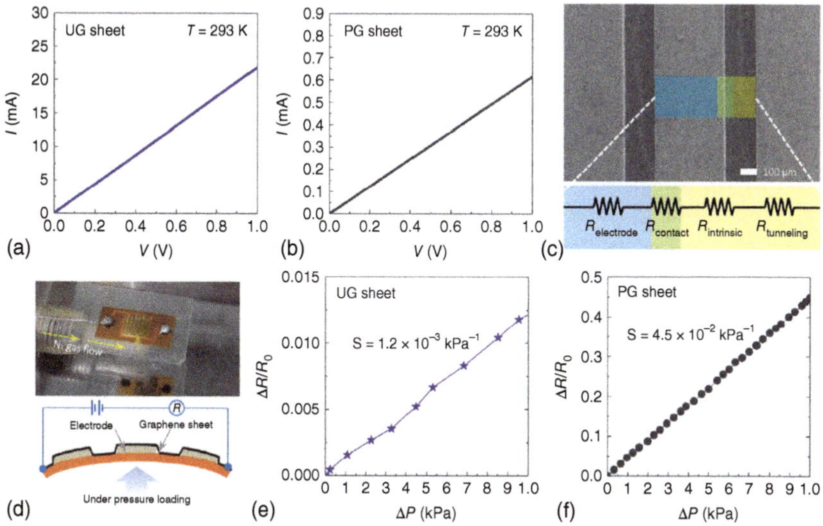


**Figure 3.13** Current–voltage characteristics of the fabricated pressure sensors with (a) the UG sheet and (b) the PG sheet. (c) Possible resistance components of the device. The linear $I$–$V$ curve indicates the ohmic contact characteristics of the graphene sheet incorporated into the device. (d) Schematic representation of the fabricated device showing the deformation of both the substrate and graphene under pressure loading. (e and f) Relative change in resistance in response to applied differential pressure utilizing the UG and PG sheets as functional piezoresistive materials. Source: Haniff et al. [87]. Reproduced with permission of American Chemical Society.

by the modulation of the graphene bandgap created from the graphitic-N $R_{intrinsic}$; (iii) the tunneling resistance caused by the defects and disorders created by the pyridinic-N, pyrrolic-N, or vacancies $R_{tunneling}$; and (iv) the resistance due to the interface between the metal electrode–graphene contacts $R_{contact}$. From the electromechanical measurements (see Figure 3.13d–f), the PG device exhibited a pressure sensitivity of $4.5 \times 10^{-2}$ kPa$^{-1}$, which is about 1 order of magnitude higher than that of the UG device ($1.2 \times 10^{-3}$ kPa$^{-1}$). In the case of graphene on a flexible substrate, the applied pressure causes the strain or bending displacement of the combined graphene substrate. Under the applied tensile strain of 0.2%, a gauge factor of 56 associated with the tunneling-dominated conduction through the localized defects was observed, indicating the enhanced piezoresistive effect in N-doped graphene.

## 3.6 Structural Defect in Graphene for Gas-Sensing Applications

A good sensor comprises three factors, i.e. sensitivity, stability, and selectivity, which can be abbreviated as "3S." Basically, if viewed from a microscopic angle, it should have significantly less binding energy where the analytes can be easily bound to the surface for sensing. It should be able to easily detach and prevent sensing poisoning (stability and reusability). Unique receptor sites at the sensor surface are typically designed based on the type of analyte for the purpose of selectivity [160, 161].

To date, low-dimensional structures with high surface-area-to-volume materials are preferred for gas-sensing applications. Low-dimensional structures, 0D, 1D, and 2D structures, can provide maximum surface-area-to-volume ratio and high chemical sensitivity, which will be a great advantage. Graphene is mainly composed of honeycomb-like building blocks. The conventional crystal lattice is hexagonal shape while those octagonal, pentagonal, heptagonal carbon shapes and Stone–Wales (SW) morphs [162, 163] are considered localized defects [164]. Defects incorporated in a part of the honeycomb structure of low-dimensional carbon-based materials always pose higher chemical reactivity than their perfect crystal counterpart [165]. Defects in graphene can cause the current confinement in the defect area [164]. The simulation results showed that the defect-induced gas adsorption on graphene leads to the enhancement of $NO_2$ detection. Experimental proof has been carried out by Liu et al. on a graphene-based transistor for methanol and ethanol gas [166]. The defects in graphene will act as scattering centers. The defects-induced charges will enhance the absorption of counter-charge, enhancing molecule absorption.

Pristine C–C network in graphene sheet and rolled up sheet of CNT are less active for interaction with surrounding molecules. The formation of defects can introduce dangling bonds that will act as active sites to react with molecules closer to them. Defects can be induced by radiation and other chemical and physical means [167, 168]. High-energy ion beam irradiation can induce single or multiple until clustered material defects. Control irradiation is used to manipulate the defects in a controlled manner, which could be used for bandgap engineering [169]. Nevertheless, excessive

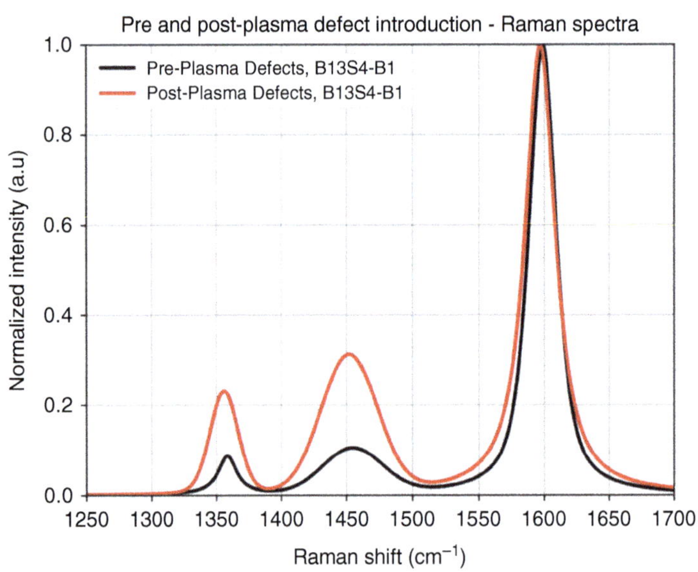


**Figure 3.14** Raman spectra of pre-plasma treated device (black curve) and the same device post-plasma treatment (red curve). Source: Hankins et al. [182]. Reproduced with permission of the American Institute of Physics.

radiation can totally destroy the crystal lattice and result in an entirely de-crystalline amorphous structure [170].

Lahiri et al. [171] have achieved dedicated octagonal and pentagonal $sp^2$-hybridized carbon rings across graphene sheets. It can act as a topological one-dimensional well-ordered defect that can provide peculiar behavior for future all-carbon electronics. The first-principle density functional theory predicted those types of controlled defects [172, 173]. The capability of defects to absorb gas species has been widely demonstrated by simulation [174–176], especially for $H_2$ [174] and $NO_2$ [164, 177, 178].

Nebogatikova et al. have found that HF solutions of 3–7% have the highest capability of breaking the C–C bonding to form defects that can functionalize and fluorinate the pure carbon region within a carbon-based material [179]. The introduced defects were confirmed by Raman spectra [180]. Ar, C, H, and N ions and Ar plasma irradiation have been used to induce defects in the carbon nanostructure [167, 169, 180]. Based on Kalbac's work [181], the plasma method induced a uniform defect density into the exposed nanotube. Hankins et al. showed a two-time increase in Raman D peak before and after the plasma treatment (see Figure 3.14) [182].

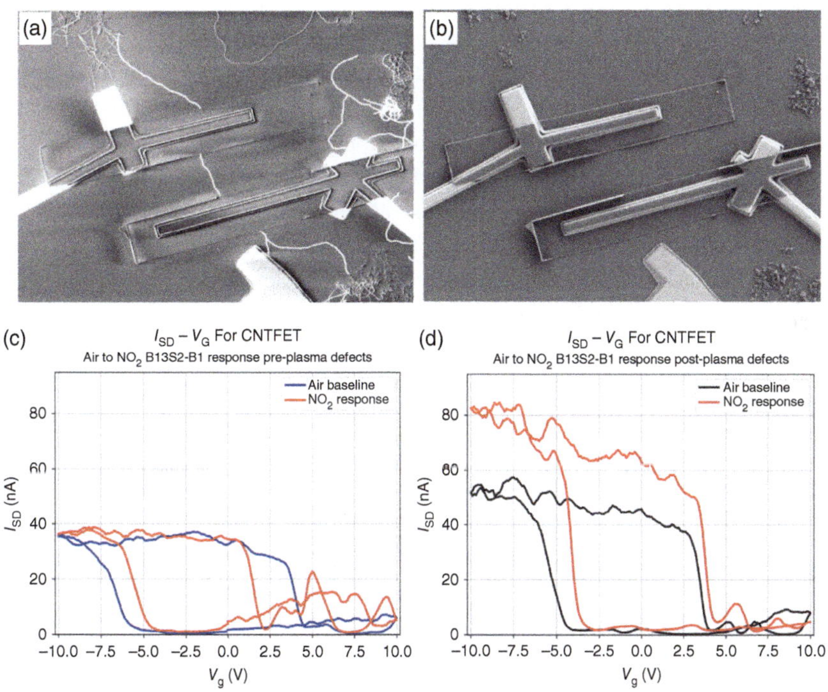


**Figure 3.15** Fully passivated CNTFET-based gas sensors. (a, b) SEM images of the fabricated CNTFET device, showing electrodes covered by $Al_2O_3$ with CNT connected beneath the electrodes using the in-lens detector and the secondary electron detector to view surface topography. (c, d) $I_{SD}$–$V_G$ response for CNTFET device in $NO_2$ for pre-plasma and post-plasma treatment. An obvious increase in $I_{SD}$ is presented in the post-plasma measurement. Source: Hankins et al. [182]. Reproduced with permission of the American Institute of Physics.

Carbon nanotube field-effect transistors (CNTFETs)-based gas sensors for $CO_2$, $NH_3$, and $H_2O$ [183] rely on gas–carbon interaction. Calculations have shown that the existence of vacancies can attract molecule species to form physical and chemical binding to the CNT [182]. Hankins et al. have studied the effect of a defect in CNTFET gas sensors, as shown in Figure 3.15a,b. A plasma-treated CNTFET has gained a much higher source-drain response compared to the untreated sample, as shown in Figure 3.15c,d. In addition, there is a clear shift of on-state current response $I_{SD}$ when exposed to $NO_2$ gas compared to ambient air. Introducing the defect into the CNT structure has proven to enhance the sensing capability of the CNTFET. This has been shown in the Raman spectroscopy of the sample before and after plasma treatment (see Figure 3.14). However, the actual mechanism involved in this sensing process is still unclear to date. A more detailed study is necessary for any future device on precise control and fabrication of a defect-activated device.

## References


1 Pierson, H.O. (1993). The element carbon. In: *Handbook of Carbon, Graphite* (ed. H.O. Pierson), 11–42. Oxford: Diamonds and Fullerenes, William Andrew Publishing.

2 Kroto, H.W., Heath, J.R., O'Brien, S.C. et al. (1985). $C_{60}$: Buckminsterfullerene. *Nature* 318: 162–163.

3 Iijima, S. (1991). Helical microtubules of graphitic carbon. *Nature* 354: 56–58.

4 Novoselov, K.S., Geim, A.K., Morozov, S.V. et al. (2004). Electric field effect in atomically thin carbon films. *Science* 306: 666–669.

5 Chen, Z., Molina-Jirón, C., Klyatskaya, S. et al. (2017). 1D and 2D graphdiynes: recent advances on the synthesis at interfaces and potential nanotechnological applications. *Annalen der Physik* 529: 1700056.

6 Hirsch, A. (2010). The era of carbon allotropes. *Nature Materials* 9: 868–871.

7 Zeinalipour-Yazdi, C.D., Pullman, D.P., and Catlow, C.R.A. (2016). The sphere-in-contact model of carbon materials. *Journal of Molecular Modeling* 22: 40.

8 Hirsch, A. and Brettreich, M. (2004). Parent fullerenes. In: *Fullerenes: Chemistry and Reactions*, 1–48. Wiley.

9 Odom, T.W., Huang, J.-L., Kim, P., and Lieber, C.M. (1998). Atomic structure and electronic properties of single-walled carbon nanotubes. *Nature* 391: 62–64.

10 Qiu, L. and Ding, F. (2021). Understanding single-walled carbon nanotube growth for chirality controllable synthesis. *Accounts of Materials Research* 2: 828–841.

11 Journet, C., Maser, W.K., Bernier, P. et al. (1997). Large-scale production of single-walled carbon nanotubes by the electric-arc technique. *Nature* 388: 756–758.

12 Thess, A., Lee, R., Nikolaev, P. et al. (1996). Crystalline ropes of metallic carbon nanotubes. *Science* 273: 483–487.

**13** Wei, B.Q., Vajtai, R., Jung, Y. et al. (2002). Organized assembly of carbon nanotubes. *Nature* 416: 495–496.

**14** Cheung, C.L., Kurtz, A., Park, H., and Lieber, C.M. (2002). Diameter-controlled synthesis of carbon nanotubes. *The Journal of Physical Chemistry B* 106: 2429–2433.

**15** Liu, C. and Cheng, H.-M. (2013). Carbon nanotubes: controlled growth and application. *Materials Today* 16: 19–28.

**16** Chen, Y. and Zhang, J. (2014). Chemical vapor deposition growth of single-walled carbon nanotubes with controlled structures for nanodevice applications. *Accounts of Chemical Research* 47: 2273–2281.

**17** Ando, Y., Zhao, X., Sugai, T., and Kumar, M. (2004). Growing carbon nanotubes. *Materials Today* 7: 22–29.

**18** Mubarak, N.M., Abdullah, E.C., Jayakumar, N.S., and Sahu, J.N. (2014). An overview on methods for the production of carbon nanotubes. *Journal of Industrial and Engineering Chemistry* 20: 1186–1197.

**19** Sanchez-Valencia, J.R., Dienel, T., Gröning, O. et al. (2014). Controlled synthesis of single-chirality carbon nanotubes. *Nature* 512: 61–64.

**20** He, M., Wang, X., Zhang, S. et al. (2019). Growth kinetics of single-walled carbon nanotubes with a (2*n*, *n*) chirality selection. *Science* 5: eaav9668.

**21** Wallace, P.R. (1947). The band theory of graphite. *Physical Review* 71: 622–634.

**22** Geim, A.K. and Novoselov, K.S. (2007). The rise of graphene. *Nature Materials* 6: 183–191.

**23** Diederich, F. and Thilgen, C. (1996). Covalent fullerene chemistry. *Science* 271: 317–324.

**24** Goodarzi, S., Da Ros, T., Conde, J. et al. (2017). Fullerene: biomedical engineers get to revisit an old friend. *Materials Today* 20: 460–480.

**25** Beiranvand, Z., Kakanejadifard, A., Donskyi, I.S. et al. (2016). Functionalization of fullerene at room temperature: toward new carbon vectors with improved physicochemical properties. *RSC Advances* 6: 112771–112775.

**26** Pochkaeva, E.I., Podolsky, N.E., Zakusilo, D.N. et al. (2020). Fullerene derivatives with amino acids, peptides and proteins: from synthesis to biomedical application. *Progress in Solid State Chemistry* 57: 100255.

**27** Kokubo, K., Matsubayashi, K., Tategaki, H. et al. (2008). Facile synthesis of highly water-soluble fullerenes more than half-covered by hydroxyl groups. *ACS Nano* 2: 327–333.

**28** Afreen, S., Kokubo, K., Muthoosamy, K., and Manickam, S. (2017). Hydration or hydroxylation: direct synthesis of fullerenol from pristine fullerene [$C_{60}$] via acoustic cavitation in the presence of hydrogen peroxide. *RSC Advances* 7: 31930–31939.

**29** Fusco, C., Seraglia, R., Curci, R., and Lucchini, V. (1999). Oxyfunctionalization of non-natural targets by dioxiranes. 3.[1] Efficient oxidation of buckminsterfullerene $C_{60}$ with methyl(trifluoromethyl)dioxirane. *The Journal of Organic Chemistry* 64: 8363–8368.

**30** Lamparth, I. and Hirsch, A. (1994). Water-soluble malonic acid derivatives of $C_{60}$ with a defined three-dimensional structure. *Journal of the Chemical Society, Chemical Communications* 1727–1728.

**31** Brettreich, M. and Hirsch, A. (1998). A highly water-soluble dendro[60] fullerene. *Tetrahedron Letters* 39: 2731–2734.

**32** Hirsch, A., Li, Q., and Wudl, F. (1991). Globe-trotting hydrogens on the surface of the fullerene compound $C_{60}H_6(N(CH_2CH_2)_2O)_6$. *Angewandte Chemie International Edition* 30: 1309–1310.

**33** Han, J., Zhuo, Y., Chai, Y. et al. (2013). Multi-labeled functionalized $C_{60}$ nanohybrid as tracing tag for ultrasensitive electrochemical aptasensing. *Biosensors and Bioelectronics* 46: 74–79.

**34** Afreen, S., Muthoosamy, K., Manickam, S., and Hashim, U. (2015). Functionalized fullerene ($C_{60}$) as a potential nanomediator in the fabrication of highly sensitive biosensors. *Biosensors and Bioelectronics* 63: 354–364.

**35** Liébana, S. and Guido, A. (2016). Drago, bioconjugation and stabilisation of biomolecules in biosensors. *Essays in Biochemistry* 60: 59–68.

**36** Georgakilas, V., Otyepka, M., Bourlinos, A.B. et al. (2012). Functionalization of graphene: covalent and non-covalent approaches, derivatives and applications. *Chemical Reviews* 112: 6156–6214.

**37** Wong, S.S., Joselevich, E., Woolley, A.T. et al. (1998). Covalently functionalized nanotubes as nanometre-sized probes in chemistry and biology. *Nature* 394: 52–55.

**38** Hamon, M.A., Hu, H., Bhowmik, P. et al. (2001). End-group and defect analysis of soluble single-walled carbon nanotubes. *Chemical Physics Letters* 347: 8–12.

**39** Zhang, Z.-Y. and Xu, X.-C. (2015). Nondestructive covalent functionalization of carbon nanotubes by selective oxidation of the original defects with $K_2FeO_4$. *Applied Surface Science* 346: 520–527.

**40** Gao, Y. and Kyratzis, I. (2008). Covalent immobilization of proteins on carbon nanotubes using the cross-linker 1-ethyl-3-(3-dimethylaminopropyl) carbodiimide–A critical assessment. *Bioconjugate Chemistry* 19: 1945–1950.

**41** Liu, Z., Tabakman, S.M., Chen, Z., and Dai, H. (2009). Preparation of carbon nanotube bioconjugates for biomedical applications. *Nature Protocols* 4: 1372–1381.

**42** Liu, Z., Tabakman, S., Welsher, K., and Dai, H. (2009). Carbon nanotubes in biology and medicine: *in vitro* and *in vivo* detection, imaging and drug delivery. *Nano Research* 2: 85–120.

**43** Tîlmaciu, C.-M. and Morris, M.C. (2015). Carbon nanotube biosensors. In: *Frontiers in Chemistry*, vol. 3, 1–21.

**44** Maiti, D., Tong, X., Mou, X., and Yang, K. (2019). Carbon-based nanomaterials for biomedical applications: a recent study. *Frontiers in Pharmacology* 9: 1–16.

**45** Lomeda, J.R., Doyle, C.D., Kosynkin, D.V. et al. (2008). Diazonium functionalization of surfactant-wrapped chemically converted graphene sheets. *Journal of the American Chemical Society* 130: 16201–16206.

**46** Zhu, H., Huang, P., Jing, L. et al. (2012). Microstructure evolution of diazonium functionalized graphene: a potential approach to change graphene electronic structure. *Journal of Materials Chemistry* 22: 2063–2068.

**47** Huang, P., Jing, L., Zhu, H., and Gao, X. (2013). Diazonium functionalized graphene: microstructure, electric, and magnetic properties. *Accounts of Chemical Research* 46: 43–52.

**48** Sharma, R., Baik, J.H., Perera, C.J., and Strano, M.S. (2010). Anomalously large reactivity of single graphene layers and edges toward electron transfer chemistries. *Nano Letters* 10: 398–405.

**49** Mishyn, V., Rodrigues, T., Leroux, Y.R. et al. (2021). Controlled covalent functionalization of a graphene-channel of a field effect transistor as an ideal platform for (Bio)sensing applications. *Nanoscale Horizons* 6: 819–829.

**50** Zhu, C., Du, D., and Lin, Y. (2017). Graphene-like 2D nanomaterial-based biointerfaces for biosensing applications. *Biosensors and Bioelectronics* 89: 43–55.

**51** Eissa, S., Tlili, C., L'Hocine, L., and Zourob, M. (2012). Electrochemical immunosensor for the milk allergen β-lactoglobulin based on electrografting of organic film on graphene modified screen-printed carbon electrodes. *Biosensors and Bioelectronics* 38: 308–313.

**52** Criado, A., Melchionna, M., Marchesan, S., and Prato, M. (2015). The covalent functionalization of graphene on substrates. *Angewandte Chemie International Edition* 54: 10734–10750.

**53** Araújo, M.P., Soares, O.S.G.P., Fernandes, A.J.S. et al. (2017). Tuning the surface chemistry of graphene flakes: new strategies for selective oxidation. *RSC Advances* 7: 14290–14301.

**54** Hummers, W.S. and Offeman, R.E. (1958). Preparation of graphitic oxide. *Journal of the American Chemical Society* 80: 1339–1339.

**55** Marcano, D.C., Kosynkin, D.V., Berlin, J.M. et al. (2010). Improved synthesis of graphene oxide. *ACS Nano* 4: 4806–4814.

**56** Surwade, S.P., Li, Z., and Liu, H. (2012). Thermal oxidation and unwrinkling of chemical vapor deposition-grown graphene. *The Journal of Physical Chemistry C* 116: 20600–20606.

**57** Itkis, M.E., Wang, F., Ramesh, P. et al. (2011). Enhanced photosensitivity of electro-oxidized epitaxial graphene. *Applied Physics Letters* 98: 093115.

**58** Abdelkader, A.M., Kinloch, I.A., and Dryfe, R.A.W. (2014). High-yield electro-oxidative preparation of graphene oxide. *Chemical Communications* 50: 8402–8404.

**59** Peña-Bahamonde, J., Nguyen, H.N., Fanourakis, S.K., and Rodrigues, D.F. (2018). Recent advances in graphene-based biosensor technology with applications in life sciences. *Journal of Nanobiotechnology* 16: 75.

**60** Eissa, S., N'diaye, J., Brisebois, P. et al. (2020). Probing the influence of graphene oxide sheets size on the performance of label-free electrochemical biosensors. *Scientific Reports* 10: 13612.

**61** Jiang, Z., Feng, B., Xu, J. et al. (2020). Graphene biosensors for bacterial and viral pathogens. *Biosensors and Bioelectronics* 166: 112471.

**62** Song, Y., Luo, Y., Zhu, C. et al. (2016). Recent advances in electrochemical biosensors based on graphene two-dimensional nanomaterials. *Biosensors and Bioelectronics* 76: 195–212.

**63** Liu, Y., Yu, D., Zeng, C. et al. (2010). Biocompatible graphene oxide-based glucose biosensors. *Langmuir* 26: 6158–6160.

**64** Song, Y., Qu, K., Zhao, C. et al. (2010). Graphene oxide: intrinsic peroxidase catalytic activity and its application to glucose detection. *Advanced Materials* 22: 2206–2210.

**65** Özcan, B. and Sezgintürk, M.K. (2016). Graphene oxide based electrochemical label free immunosensor for rapid and highly sensitive determination of tumor marker HSP70. *Talanta* 160: 367–374.

**66** Alfred, G. (1994). *Fundamentals of Plasma, Cold Plasma Materials Fabrication: From Fundamentals to Applications*, 1–23. IEEE.

**67** Ullah, S., Shi, Q., Zhou, J. et al. (2020). Advances and trends in chemically doped graphene. *Advanced Materials Interfaces* 7: 2000999.

**68** Li, M., Liu, D., Wei, D. et al. (2016). Controllable synthesis of graphene by plasma-enhanced chemical vapor deposition and its related applications. *Advanced Science* 3: 1600003.

**69** Terasawa, T.-O. and Saiki, K. (2012). Synthesis of nitrogen-doped graphene by plasma-enhanced chemical vapor deposition. *Japanese Journal of Applied Physics* 51: 055101.

**70** Boas, C.R.S.V., Focassio, B., Marinho, E. et al. (2019). Characterization of nitrogen doped graphene bilayers synthesized by fast, low temperature microwave plasma-enhanced chemical vapour deposition. *Scientific Reports* 9: 13715.

**71** Wei, D., Peng, L., Li, M. et al. (2015). Low temperature critical growth of high quality nitrogen doped graphene on dielectrics by plasma-enhanced chemical vapor deposition. *ACS Nano* 9: 164–171.

**72** Zhai, Z., Shen, H., Chen, J. et al. (2019). Direct growth of nitrogen-doped graphene films on glass by plasma-assisted hot filament CVD for enhanced electricity generation. *Journal of Materials Chemistry A* 7: 12038–12049.

**73** Zhai, Z., Shen, H., Chen, J. et al. (2018). Evolution of structural and electrical properties of carbon films from amorphous carbon to nanocrystalline graphene on quartz glass by HFCVD. *ACS Applied Materials & Interfaces* 10: 17427–17436.

**74** Zainal Ariffin, N.H., Mohammad Haniff, M.A.S., Syono, M.I. et al. (2021). Low-temperature nitrogen doping of nanocrystalline graphene films with tunable pyridinic-N and pyrrolic-N by cold-wall plasma-assisted chemical vapor deposition. *ACS Omega* 6: 23710–23722.

**75** Wang, H., Zhou, Y., Wu, D. et al. (2013). Synthesis of boron-doped graphene monolayers using the sole solid feedstock by chemical vapor deposition. *Small* 9: 1316–1320.

**76** Gebhardt, J., Koch, R.J., Zhao, W. et al. (2013). Growth and electronic structure of boron-doped graphene. *Physical Review B* 87: 155437.

**77** Cattelan, M., Agnoli, S., Favaro, M. et al. (2013). Microscopic view on a chemical vapor deposition route to boron-doped graphene nanostructures. *Chemistry of Materials* 25: 1490–1495.

**78** Usachov, D.Y., Fedorov, A.V., Petukhov, A.E. et al. (2015). Epitaxial B-graphene: large-scale growth and atomic structure. *ACS Nano* 9: 7314–7322.

**79** Zhai, Z., Shen, H., Chen, J. et al. (2020). Metal-free synthesis of boron-doped graphene glass by hot-filament chemical vapor deposition for wave energy harvesting. *ACS Applied Materials & Interfaces* 12: 2805–2815.

**80** Lin, Y.-C., Lin, C.-Y., and Chiu, P.-W. (2010). Controllable graphene N-doping with ammonia plasma. *Applied Physics Letters* 96: 133110.

**81** Shao, Y., Zhang, S., Engelhard, M.H. et al. (2010). Nitrogen-doped graphene and its electrochemical applications. *Journal of Materials Chemistry* 20: 7491–7496.

**82** Lin, Y.-P., Ksari, Y., Prakash, J. et al. (2014). Nitrogen-doping processes of graphene by a versatile plasma-based method. *Carbon* 73: 216–224.

**83** Rybin, M., Pereyaslavtsev, A., Vasilieva, T. et al. (2016). Efficient nitrogen doping of graphene by plasma treatment. *Carbon* 96: 196–202.

**84** Zeng, J.-J. and Lin, Y.-J. (2014). Tuning the work function of graphene by nitrogen plasma treatment with different radio-frequency powers. *Applied Physics Letters* 104: 233103.

**85** Lin, Y.-C., Teng, P.-Y., Yeh, C.-H. et al. (2015). Structural and chemical dynamics of pyridinic-nitrogen defects in graphene. *Nano Letters* 15: 7408–7413.

**86** Kato, T., Jiao, L., Wang, X. et al. (2011). Room-temperature edge functionalization and doping of graphene by mild plasma. *Small* 7: 574–577.

**87** Haniff, M.A.S.M., Hafiz, S.M., Huang, N.M. et al. (2017). Piezoresistive effect in plasma-doping of graphene sheet for high-performance flexible pressure sensing application. *ACS Applied Materials & Interfaces* 9: 15192–15201.

**88** McManus, J.B., Hennessy, A., Cullen, C.P. et al. (2017). Controlling defect and dopant concentrations in graphene by remote plasma treatments. *Physica Status Solidi B* 254: 1700214.

**89** Tang, Y.-B., Yin, L.-C., Yang, Y. et al. (2012). Tunable band gaps and p-type transport properties of boron-doped graphenes by controllable ion doping using reactive microwave plasma. *ACS Nano* 6: 1970–1978.

**90** Rani, P. and Jindal, V.K. (2013). Designing band gap of graphene by B and N dopant atoms. *RSC Advances* 3: 802–812.

**91** Denis, P.A. (2010). Band gap opening of monolayer and bilayer graphene doped with aluminium, silicon, phosphorus, and sulfur. *Chemical Physics Letters* 492: 251–257.

**92** Shin, D.-W., Kim, T.S., and Yoo, J.-B. (2016). Phosphorus doped graphene by inductively coupled plasma and triphenylphosphine treatments. *Materials Research Bulletin* 82: 71–75.

**93** Liang, C., Wang, Y., and Li, T. (2015). Synthesis of sulfur-doped p-type graphene by annealing with hydrogen sulfide. *Carbon* 82: 506–512.

**94** Wong, C.H.A., Sofer, Z., Klímová, K., and Pumera, M. (2016). Microwave exfoliation of graphite oxides in $H_2S$ plasma for the synthesis of sulfur-doped graphenes as oxygen reduction catalysts. *ACS Applied Materials & Interfaces* 8: 31849–31855.

**95** Baraket, M., Walton, S.G., Lock, E.H. et al. (2010). The functionalization of graphene using electron-beam generated plasmas. *Applied Physics Letters* 96: 231501.

**96** Yang, H., Chen, M., Zhou, H. et al. (2011). Preferential and reversible fluorination of monolayer graphene. *The Journal of Physical Chemistry C* 115: 16844–16848.

**97** Shen, C., Huang, G., Cheng, Y. et al. (2012). Thinning and functionalization of few-layer graphene sheets by $CF_4$ plasma treatment. *Nanoscale Research Letters* 7: 268.

**98** Sherpa, S.D., Levitin, G., and Hess, D.W. (2012). Effect of the polarity of carbon-fluorine bonds on the work function of plasma-fluorinated epitaxial graphene. *Applied Physics Letters* 101: 111602.

**99** Wu, J., Xie, L., Li, Y. et al. (2011). Controlled chlorine plasma reaction for noninvasive graphene doping. *Journal of the American Chemical Society* 133: 19668–19671.

**100** Pham, V.P., Kim, K.H., Jeon, M.H. et al. (2015). Low damage pre-doping on CVD graphene/Cu using a chlorine inductively coupled plasma. *Carbon* 95: 664–671.

**101** Yaya, A., Ewels, C.P., Suarez-Martinez, I. et al. (2011). Bromination of graphene and graphite. *Physical Review B* 83: 045411.

**102** Zheng, J., Liu, H.-T., Wu, B. et al. (2012). Production of graphite chloride and bromide using microwave sparks. *Scientific Reports* 2: 662.

**103** Mansour, A.E., Dey, S., Amassian, A., and Tanielian, M.H. (2015). Bromination of graphene: a new route to making high performance transparent conducting electrodes with low optical losses. *ACS Applied Materials & Interfaces* 7: 17692–17699.

**104** Wang, H., Maiyalagan, T., and Wang, X. (2012). Review on recent progress in nitrogen-doped graphene: synthesis, characterization, and its potential applications. *ACS Catalysis* 2: 781–794.

**105** Panchakarla, L.S., Subrahmanyam, K.S., Saha, S.K. et al. (2009). Synthesis, structure, and properties of boron- and nitrogen-doped graphene. *Advanced Materials* 21: 4726–4730.

**106** Zhang, C., Fu, L., Liu, N. et al. (2011). Synthesis of nitrogen-doped graphene using embedded carbon and nitrogen sources. *Advanced Materials* 23: 1020–1024.

**107** Agnoli, S. and Favaro, M. (2016). Doping graphene with boron: a review of synthesis methods, physicochemical characterization, and emerging applications. *Journal of Materials Chemistry A* 4: 5002–5025.

**108** Wang, H.-M., Wang, H.-X., Chen, Y. et al. (2013). Phosphorus-doped graphene and (8, 0) carbon nanotube: structural, electronic, magnetic properties, and chemical reactivity. *Applied Surface Science* 273: 302–309.

**109** Duan, X., O'Donnell, K., Sun, H. et al. (2015). Sulfur and nitrogen co-doped graphene for metal-free catalytic oxidation reactions. *Small* 11: 3036–3044.

**110** Denis, P.A. (2013). Concentration dependence of the band gaps of phosphorus and sulfur doped graphene. *Computational Materials Science* 67: 203–206.

**111** Denis, P.A., Faccio, R., and Mombru, A.W. (2009). Is it possible to dope single-walled carbon nanotubes and graphene with sulfur? *ChemPhysChem* 10: 715–722.

**112** Wang, X., Sun, G., Routh, P. et al. (2014). Heteroatom-doped graphene materials: syntheses, properties and applications. *Chemical Society Reviews* 43: 7067–7098.

**113** Robinson, J.T., Burgess, J.S., Junkermeier, C.E. et al. (2010). Properties of fluorinated graphene films. *Nano Letters* 10: 3001–3005.

**114** Nair, R.R., Ren, W., Jalil, R. et al. (2010). Fluorographene: a two-dimensional counterpart of teflon. *Small* 6: 2877–2884.

**115** Cheng, S.H., Zou, K., Okino, F. et al. (2010). Reversible fluorination of graphene: evidence of a two-dimensional wide bandgap semiconductor. *Physical Review B* 81: 205435.

**116** Li, B., Zhou, L., Wu, D. et al. (2011). Photochemical chlorination of graphene. *ACS Nano* 5: 5957–5961.

**117** Karlický, F., Datta, K.K.R., Otyepka, M., and Zbořil, R. (2013). Halogenated graphenes: rapidly growing family of graphene derivatives. *ACS Nano* 7: 6434–6464.

**118** Fan, X., Liu, L., Kuo, J.-L., and Shen, Z. (2010). Functionalizing single- and multi-layer graphene with Br and $Br_2$. *The Journal of Physical Chemistry C* 114: 14939–14945.

**119** Mueller, T., Xia, F., and Avouris, P. (2010). Graphene photodetectors for high-speed optical communications. *Nature Photonics* 4: 297–301.

**120** Kim, H., Sheps, T., Collins, P.G., and Potma, E.O. (2009). Nonlinear optical imaging of individual carbon nanotubes with four-wave-mixing microscopy. *Nano Letters* 9: 2991–2995.

**121** Blau, W.J., Byrne, H.J., Cardin, D.J. et al. (1991). Large infrared nonlinear optical response of $C_{60}$. *Physical Review Letters* 67: 1423–1425.

**122** Yang, S.-C., Gong, Q., Xia, Z. et al. (1992). Large third-order nonlinear optical properties of $C_{70}$ fullerene in the infrared regime. *Applied Physics B* 55: 51–53.

**123** Choi, B., Lee, J., Lee, S. et al. (2013). Generation of ultra-high-molecular weight polyethylene from metallocenes immobilized onto n-doped graphene nanoplatelets. *Macromolecular Rapid Communications* 34: 533–538.

**124** Han, J., Sa, Y.J., Shim, Y. et al. (2015). Coordination chemistry of $[Co(acac)_2]$ with N-doped graphene: implications for oxygen reduction reaction reactivity of organometallic $Co–O_4–N$ species. *Angewandte Chemie International Edition* 54: 12622–12626.

**125** Jang, D., Lee, S., Kim, S. et al. (2018). Production of P,N Co-doped graphene-based materials by a solution process and their electrocatalytic performance for oxygen reduction reaction. *ChemNanoMat* 4: 118–123.

**126** Oh, J.H., Wei, P., and Bao, Z. (2010). Molecular n-type doping for air-stable electron transport in vacuum-processed *n*-channel organic transistors. *Applied Physics Letters* 97: 243305.

**127** Xiao, K., Liu, Y., Hu, P.A. et al. (2005). n-Type field-effect transistors made of an individual nitrogen-doped multiwalled carbon nanotube. *Journal of the American Chemical Society* 127: 8614–8617.

**128** Riede, M., Uhrich, C., Widmer, J. et al. (2011). Efficient organic tandem solar cells based on small molecules. *Advanced Functional Materials* 21: 3019–3028.

**129** Drechsel, J., Männig, B., Kozlowski, F. et al. (2005). Efficient organic solar cells based on a double p-i-n architecture using doped wide-gap transport layers. *Applied Physics Letters* 86: 244102.

**130** Schedin, F., Geim, A.K., Morozov, S.V. et al. (2007). Detection of individual gas molecules adsorbed on graphene. *Nature Materials* 6: 652–655.

**131** Pfeiffer, M., Leo, K., Zhou, X. et al. (2003). Doped organic semiconductors: physics and application in light emitting diodes. *Organic Electronics* 4: 89–103.

**132** Reineke, S., Lindner, F., Schwartz, G. et al. (2009). White organic light-emitting diodes with fluorescent tube efficiency. *Nature* 459: 234–238.

**133** Park, G., Park, S.K., Han, J. et al. (2014). Finely tuning oxygen functional groups of graphene materials and optimizing oxygen levels for capacitors. *RSC Advances* 4: 36377–36384.

**134** Yun, S., Kang, S.-O., Park, S., and Park, H.S. (2014). $CO_2$-activated, hierarchical trimodal porous graphene frameworks for ultrahigh and ultrafast capacitive behavior. *Nanoscale* 6: 5296–5302.

**135** Maennig, B., Drechsel, J., Gebeyehu, D. et al. (2004). Organic p-i-n solar cells. *Applied Physics A* 79: 1–14.

**136** Usachov, D., Vilkov, O., Grüneis, A. et al. (2011). Nitrogen-doped graphene: efficient growth, structure, and electronic properties. *Nano Letters* 11: 5401–5407.

**137** Tang, L., Ji, R., Li, X. et al. (2013). Energy-level structure of nitrogen-doped graphene quantum dots. *Journal of Materials Chemistry C* 1: 4908–4915.

**138** Hage, F.S., Kapetanakis, M.D., Idrobo, J.-C. et al. (2019). Atomic-scale spectroscopic imaging of the extreme-UV optical response of B- and N-doped graphene. *Advanced Functional Materials* 29: 1901819.

**139** Johannsen, J.C., Ulstrup, S., Crepaldi, A. et al. (2015). Tunable carrier multiplication and cooling in graphene. *Nano Letters* 15: 326–331.

**140** Zhou, X., Zifer, T., Wong, B.M. et al. (2009). Color detection using chromophore-nanotube hybrid devices. *Nano Letters* 9: 1028–1033.

**141** Peimyoo, N., Li, J., Shang, J. et al. (2012). Photocontrolled molecular structural transition and doping in graphene. *ACS Nano* 6: 8878–8886.

**142** Kim, M., Safron, N.S., Huang, C. et al. (2012). Light-driven reversible modulation of doping in graphene. *Nano Letters* 12: 182–187.

**143** Margapoti, E., Strobel, P., Asmar, M.M. et al. (2014). Emergence of photoswitchable states in a graphene–azobenzene–Au platform. *Nano Letters* 14: 6823–6827.

**144** Haberer, D., Vyalikh, D.V., Taioli, S. et al. (2010). Tunable band gap in hydrogenated quasi-free-standing graphene. *Nano Letters* 10: 3360–3366.

**145** Zhang, Y., Mori, T., Ye, J., and Antonietti, M. (2010). Phosphorus-doped carbon nitride solid: enhanced electrical conductivity and photocurrent generation. *Journal of the American Chemical Society* 132: 6294–6295.

**146** Tang, L., Ji, R., Tian, P. et al. (2017). Functionalization of graphene by size and doping control and its optoelectronic applications. *Proceedings of SPIE* 10177: 101770B.

**147** Chen, C.F., Park, C.H., Boudouris, B.W. et al. (2011). Controlling inelastic light scattering quantum pathways in graphene. *Nature* 471: 617–620.

**148** Wang, F., Zhang, Y., Tian, C. et al. (2008). Gate-variable optical transitions in graphene. *Science* 320: 206–209.

**149** Pereira, V.M., Castro Neto, A.H., and Peres, N.M.R. (2009). Tight-binding approach to uniaxial strain in graphene. *Physical Review B* 80: 045401.

**150** Huang, M., Pascal, T.A., Kim, H. et al. (2011). Electronic–mechanical coupling in graphene from *in situ* nanoindentation experiments and multiscale atomistic simulations. *Nano Letters* 11: 1241–1246.

**151** Zhu, S.-E., Ghatkesar, M.K., Zhang, C., and Janssen, G.C.A.M. (2013). Graphene based piezoresistive pressure sensor. *Applied Physics Letters* 102: 161904.

**152** Smith, A.D., Niklaus, F., Paussa, A. et al. (2013). Electromechanical piezoresistive sensing in suspended graphene membranes. *Nano Letters* 13: 3237–3242.

**153** Lee, Y., Bae, S., Jang, H. et al. (2010). Wafer-scale synthesis and transfer of graphene films. *Nano Letters* 10: 490–493.

**154** Lee, C., Wei, X., Kysar, J.W., and Hone, J. (2008). Measurement of the elastic properties and intrinsic strength of monolayer graphene. *Science* 321: 385–388.

**155** Bae, S.-H., Lee, Y., Sharma, B.K. et al. (2013). Graphene-based transparent strain sensor. *Carbon* 51: 236–242.

**156** Wang, Y., Yang, R., Shi, Z. et al. (2011). Super-elastic graphene ripples for flexible strain sensors. *ACS Nano* 5: 3645–3650.

**157** Zhao, J., He, C., Yang, R. et al. (2012). Ultra-sensitive strain sensors based on piezoresistive nanographene films. *Applied Physics Letters* 101: 063112.

**158** Hempel, M., Nezich, D., Kong, J., and Hofmann, M. (2012). A novel class of strain gauges based on layered percolative films of 2D materials. *Nano Letters* 12: 5714–5718.

**159** Mohammad Haniff, M.A.S., Muhammad Hafiz, S., Wahid, K.A.A. et al. (2015). Piezoresistive effects in controllable defective HFTCVD graphene-based flexible pressure sensor. *Scientific Reports* 5: 14751.

**160** Rocha, A.R., Rossi, M., Fazzio, A., and da Silva, A.J.R. (2008). Designing real nanotube-based gas sensors. *Physical Review Letters* 100: 176803.

**161** Amorim, R.G., Fazzio, A., da Silva, A.J.R., and Rocha, A.R. (2013). Confinement effects and why carbon nanotube bundles can work as gas sensors. *Nanoscale* 5: 2798–2803.

**162** Hashimoto, A., Suenaga, K., Gloter, A. et al. (2004). Direct evidence for atomic defects in graphene layers. *Nature* 430: 870–873.

**163** Meyer, J.C., Kisielowski, C., Erni, R. et al. (2008). Direct imaging of lattice atoms and topological defects in graphene membranes. *Nano Letters* 8: 3582–3586.

**164** de Souza, F.A.L., Amorim, R.G., Prasongkit, J. et al. (2018). Topological line defects in graphene for applications in gas sensing. *Carbon* 129: 803–808.

**165** Brito, W.H., Kagimura, R., and Miwa, R.H. (2011). Hydrogenated grain boundaries in graphene. *Applied Physics Letters* 98: 213107.

**166** Liu, Y., Liu, H., Chu, Y. et al. (2018). Defect-induced gas adsorption on graphene transistors. *Advanced Materials Interfaces* 5: 1701640.

**167** Ishaq, A., Iqbal, S., Ali, N. et al. (2013). $H^+$, $N^+$, and $Ar^+$ ion irradiation induced structure changes of carbon nanostructures. *New Carbon Materials* 28: 81–86.

**168** Ishaq, A., Usman, M., Dee, C.F. et al. (2013). Effect for hydrogen, nitrogen, phosphorous, and argon ions irradiation on ZnO NWs. *Journal of Nanoparticle Research* 15: 1467.

**169** Farooq, A., Aisida, S.O., Jalil, A. et al. (2021). C ions irradiation induced defects analysis and effects on optical properties of $TiO_2$ nanoparticles. *Journal of Alloys and Compounds* 863: 158635.

**170** Dee, C.F., Ahmad, I., Yan, L. et al. (2011). Amorphization of ZnO nanowires by proton beam irradiation. *Nano* 06: 259–263.

**171** Lahiri, J., Lin, Y., Bozkurt, P. et al. (2010). An extended defect in graphene as a metallic wire. *Nature Nanotechnology* 5: 326–329.

**172** Botello-Méndez, A.R., Cruz-Silva, E., López-Urías, F. et al. (2009). Spin polarized conductance in hybrid graphene nanoribbons using 5–7 defects. *ACS Nano* 3: 3606–3612.

**173** Okada, S., Nakada, K., Kuwabara, K. et al. (2006). Ferromagnetic spin ordering on carbon nanotubes with topological line defects. *Physical Review B* 74: 121412.

**174** Singla, M. and Jaggi, N. (2021). Synergistic effect of Cu decoration and N doping in divacancy defected graphene nanoribbons on hydrogen gas sensing properties: DFT study. *Materials Chemistry and Physics* 273: 125093.

**175** Tit, N., Said, K., Mahmoud, N.M. et al. (2017). Ab-initio investigation of adsorption of CO and $CO_2$ molecules on graphene: role of intrinsic defects on gas sensing. *Applied Surface Science* 394: 219–230.

**176** Yang, S., Lei, G., Xu, H. et al. (2019). A DFT study of CO adsorption on the pristine, defective, In-doped and Sb-doped graphene and the effect of applied electric field. *Applied Surface Science* 480: 205–211.

**177** Zhang, B., Liu, G., Cheng, M. et al. (2018). The preparation of reduced graphene oxide-encapsulated $\alpha$-$Fe_2O_3$ hybrid and its outstanding $NO_2$ gas sensing properties at room temperature. *Sensors and Actuators B* 261: 252–263.

**178** Kim, H.W., Na, H.G., Kwon, Y.J. et al. (2017). Microwave-assisted synthesis of graphene–$SnO_2$ nanocomposites and their applications in gas sensors. *ACS Applied Materials & Interfaces* 9: 31667–31682.

**179** Nebogatikova, N.A., Antonova, I.V., Volodin, V.A., and Prinz, V.Y. (2013). Functionalization of graphene and few-layer graphene with aqueous solution of hydrofluoric acid. *Physica E: Low-dimensional Systems and Nanostructures* 52: 106–111.

**180** Frank, O., Dresselhaus, M.S., and Kalbac, M. (2015). Raman spectroscopy and *in situ* Raman spectroelectrochemistry of isotopically engineered graphene systems. *Accounts of Chemical Research* 48: 111–118.

**181** Kalbac, M., Hsieh, Y.-P., Farhat, H. et al. (2010). Defects in individual semiconducting single wall carbon nanotubes: raman spectroscopic and *in situ* raman spectroelectrochemical study. *Nano Letters* 10: 4619–4626.

**182** Hankins, A., Willard, T.C., Liu, A.Y., and Paranjape, M. (2020). Role of defects in the sensing mechanism of CNTFET gas sensors. *Journal of Applied Physics* 128: 084501.

**183** Hosseingholipourasl, A., Ariffin, S.H.S., Koloor, S.S.R. et al. (2020). Analytical prediction of highly sensitive CNT-FET-based sensor performance for detection of gas molecules. *IEEE Access* 8: 12655–12661.

# 4

# Fabrication Techniques of Resistive Switching Carbon-Based Memories

*Poh Choon Ooi and M. F. Mohd Razip Wee*

*Institute of Microengineering and Nanoelectronics (IMEN), Universiti Kebangsaan Malaysia, 43600 Bangi, Selangor, Malaysia*

## 4.1 Introduction – Emerging Carbon-Based Memory Technologies

Non-volatile memories (NVMs) are an indispensable component and backbone of electronic integrated circuits due to their ability to retain the stored information even after power is removed. The introduction of solid-state NVMs based on silicon technologies has tremendously improved the speed and energy efficiency of modern computing systems, ascribed to their superior data storage properties compared to the magnetic hard disk drives in use since the 1950s [1]. To date, flash memories, for instance, NOR and NAND flash, are currently widely used as NVM devices in the modern electronics era, either in stand-alone or embedded chips. However, despite the success of flash memory, the requirement for low programming voltages (<10 V), short programming time (<10 μs), and higher endurance has driven the semiconductor industry to search for new storage-class memory technology [2, 3]. In addition, it should be fabricated with lower cost and the capability to scale down the size of memory cells in the nanometer dimension.

The increasing demand for a faster digital device with lower power consumption has led to the current trend to scale down the dimensions of the electronic device, as predicted by Moore's Law [4]. However, this trend has reached its limit in the recent decade primarily because of significant technological and fundamental issues of scaling techniques, particularly related to silicon-based material devices. To overcome this emerging problem and simultaneously move toward the "beyond Si" era, researchers need to search for alternative materials with better intrinsic properties. Indeed, the III–V compound semiconductor has appeared as a promising material with its merits of high electron mobility that is compatible with silicon complementary metal–oxide–semiconductor (CMOS) process technologies. III–V materials, particularly GaAs and GaN, have played an essential role in the development of optoelectronic devices widely used in RF telecommunication [5], high-density data storage [6], and solid-state lighting [7]. Nevertheless, this III–V material is usually composed of rare-earth costly elements with minimal resources on Earth.

*Enhanced Carbon-Based Materials and Their Applications*, First Edition.
Edited by Poh Choon Ooi, Mengying Xie, and Chang Fu Dee.

In view of the limited rare-earth elements, carbon, attributed to its unique properties physically, chemically in multiple forms, and more importantly, the most abundant material on the Earth, has been explored by the researchers for its potential NVM applications to fulfill the ever-growing requirement for lower cost, a smaller dimension with higher density information storage, greater endurance, and higher speed device [8]. The ability to store terabytes of data with remarkable energy efficiency using NVMs has been demonstrated possible through the use of various low-dimensional $sp^2$-hybridized carbon nanostructures [1]. In addition to unique chemical and physical properties, their atomic-scale thickness, transparency, and flexibility features are highly desirable for developing information-storage devices that can be integrated with wearable and smart electronic gadgets [8].

Global researchers have examined multiple vertical stacking of memory cell layers' concepts; innovative polymer-based device designs by incorporating various carbon nanomaterials, such as graphene quantum dots (GQDs) and carbon nanotubes, to achieve mechanically flexible and optically transparent devices. Meanwhile, the electrical memory performance could also be enhanced by incorporating other types of 2D nanomaterials to create the heterostructure with the carbon nanostructures family [9]. The multilayer stacking structure can be fabricated using cost-effective solution-processing techniques, for instance, spray coating, spin coating, and inkjet printing. In addition, the plasma polymerization technique is also demonstrated to be used for insulator deposition in the NVM stacking structure. Both deposition methods are predominantly suitable for developing memory devices on flexible plastic substrates and integrating them into wearable electronic gadgets. Here, we focus on the dielectric materials, carbon nanomaterials, and various fabrication techniques used for the NVM memory devices. Solution-processing and plasma polymerization methods are the attention in this chapter, with the electrical performance for a brief discussion.

## 4.2 Memristor-Based Memory

Leon Chua introduced the term memristor (the short form for memory resistor) in 1971 as the essential missing fourth fundamental circuit element besides the several passive components known as a resistor, capacitor, and inductor, allowing to define the linkage between the magnetic flux to the charge through a simple two-terminal device with the ability to memorize the current curve [10]. However, this discovery led to the experimental demonstration only in 2008 by Strukov and coworkers from HP Labs with a simple device made of titanium dioxide that well agreed with the theoretical idea without needing an internal power supply. In their work, as shown in Figure 4.1, they introduced a nanoscale semiconductor film of titanium dioxide in between two metal electrodes (platinum), leading to the memristor's behavior even though those are just standard electronic materials [11]. Since then, this simple metal/insulator/metal (MIM) stacking structure has been explored tremendously in the literature with a wide variety of resistive layers. A significant effort has been directed toward binary-transition-metal oxides (e.g. $TiO_2$) [12, 13], transition-metal

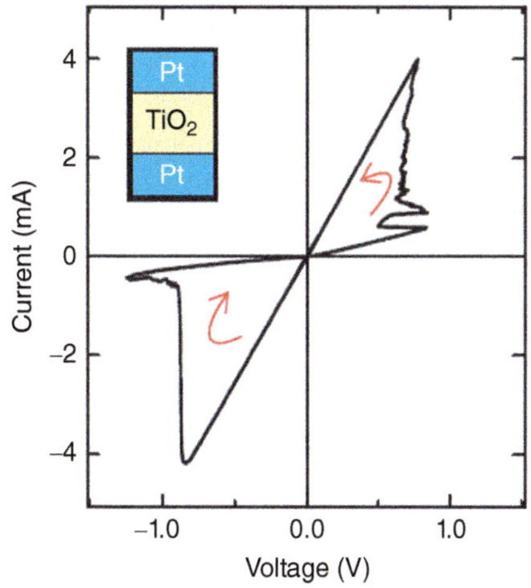


**Figure 4.1** An experimental current–voltage (*I*–*V*) plot of a Pt–$TiO_{2-x}$–Pt device, as shown in the inset. Source: Reproduced from Strukov et al. [11] with the permission of Springer Nature.

oxides with perovskite structure (e.g. $SrTiO_3$) [14], multiferroic (e.g. $BiFeO_3$) [15], large bandgap high-k dielectrics (e.g. $Al_2O_3$) [16], and non-oxide thin-layered materials specifically chalcogenides materials (e.g. $In_2Se_3$ and $MoS_2$) [17]. In addition, various types of organic-based materials have also been reported showing memristive behavior when integrated into the two-terminal MIM structure. For example, carbon nanomaterials, especially GQDs, can be embedded in a polymeric matrix to act as the resistive-switching medium in the memristor-based memories device. The use of carbon nanomaterials as an alternative approach to conventional metal oxide inorganic semiconductor-based material could favor the development of wearable, flexible, and transparent electronics systems due to their merit flexible and transparent properties [9, 18]. It is a non-volatile memory ability that allows the storage of the data even when the power is switched off, which is one of the essential features for the next generation of resistive random-access memories (RRAM) [19], programmable logic [20], and neuromorphic systems [21].

## 4.3 Substrate Options

The fabrication of carbon-based non-volatile memristors can be constructed on various substrates, including $SiO_2$, quartz, glass, and plastic, for instance, polyethylene terephthalate (PET) and polyethersulfone (PES). $SiO_2$ and quartz substrates can withstand very high temperatures, even higher than that of glass; hence, they are suitable for fabrication that involves a high-temperature process. On the other hand, glass and plastic substrates are low cost and transparent, while the latter also offer mechanical flexibility, which is desirable for flexible electronics applications. However, generally, plastics are made up of a wide range of synthetic or semisynthetic materials that utilize polymers as the primary substance. Consequently, polymer substrate is unable to withstand high temperature, especially PET, which starts to soften at 170 °C and is, therefore, not suitable to be used if a typical plasma-enhanced chemical vapor high-temperature deposition process is involved [22].

## 4.4 Effect of Electrode Materials

MIM, the two-terminal device, is of current interest owing to its simple structure and ease in fabrication and scalability [23]. The selection of top electrode (TE) and bottom electrode (BE) materials plays a crucial role in improving the performance of the resistance-switching memory devices. Jang studied the effect of electrode material on characteristics of non-volatile resistive memory consisting of metal/$Ag_2S$/metal structure with $Ag_2S$ nanoparticles as the active material. He fabricated Ag/$Ag_2S$/Ag, Au/$Ag_2S$/Ag, and Pt/$Ag_2S$/Ag devices. The bottom Ag and Au electrodes were deposited using the thermal evaporation method, and Pt electrodes were formed by dc/rf magnetron sputtering. The stable memory characteristics of memory endurance and retention are significantly improved to more than $10^3$ cycles and $10^4$ seconds, respectively, using Au and Pt inert BEs [23]. Subsequently, Liu et al. also investigated electrode work function effects on RRAM with gelatin biodielectric layer. They constructed metal/gelatin/tungsten (W) resistive switching (RS) devices with a silver (Ag), tungsten (W), copper (Cu), and gold (Au) as the TE, respectively. The TEs were deposited by vacuum thermal evaporation and magnetron sputtering to deposit the BE. It is found that the fabricated Ag/gelatin/W and Au/gelatin/W RRAM devices with larger work function differences of TE and BE metals have higher and better stable on/off resistance ratios of $>10^2$ and $>10^5$, respectively [24]. Meanwhile, for the development of two-terminal carbon-based NVM devices, Ooi et al. constructed the structure of silver nanowires (AgNWs)/nanocomposite/reduced graphene oxide (rGO) on a quartz substrate by using a simple solution process technique. The transparent rGO BE was derived from the synthesized graphene oxide (GO) flakes. GO was deposited on the substrate by a spin-coating method and annealed at a high temperature to form rGO with the aid of $C_2H_2/H_2$ plasma treatment. The TE AgNWs electrodes were deposited by spray coating, and the nanocomposite was a mixture of GQDs and GO in ethanol. They observed that the electron emission would be preferable from the lower work function of AgNWs TE rather than the higher work function of rGO BE. The on/off ratio of the current bistability for the devices was recorded to be $10^3$ with retention stability up to $1 \times 10^4$ seconds [25]. All the mentioned studies reported that observed switching behaviors could be associated with the metallic conductive filament formation in the devices.

## 4.5 Fabrication Methods of Metal/Insulator/Metal Structure

Solution-processing techniques and plasma polymerization deposition are promising, simple methods to produce low-cost carbon-based electronics. The discussion on solution-processing techniques will mainly focus on spin coating, spray coating, dip coating, and inkjet-printing methods.

### 4.5.1 Spin Coating

The most common method to fabricate a carbon-based memristor or NVM is the spin-coating method usually employed in semiconductor technology to deposit a photoresist onto the wafer. It is a simple and fast approach to depositing uniform film at a small scale from a few nanometers to a few micrometers thickness. To achieve homogeneous thin-film deposition, the user spun the solution of a specific material at high speeds. The centripetal force and the surface tension of the liquid simultaneously create an even coverage on the substrate. The layer thickness can be controlled by adjusting the spin-coating speed, and the excess solvent is evaporated during the heat treatment process [26]. The crystallinity structure of the deposited thin film will be improved during the thermal treatment stage. Note that in the case of a high surface tension substrate, oxygen plasma pre-treatment can be performed to improve the hydrophobicity and increase the substrate's surface wettability.

Nevertheless, the spin-coating technique is generally suitable for an even and rigid substrate but improper for a flat and nonrigid substrate, typically papers, textiles, and membranes, since the solution can dissolve, swell, or degrade the substrate. Because of its low cost and ease of preparation, the spin-coating method is the most popular among the solution process methods. It is commonly used to deposit carbon-based materials, such as GO, rGO, graphene nanoflakes, graphene, and GQDs, to form the resistive switching (RS) layer [27-29]. For example, Lin et al. fabricated mechanically flexible and optically transparent GQD-based NVM devices on the indium–tin–oxide (ITO)-coated PET substrate [30]. As shown in Figure 4.2a, GQDs as the charge-trapping layer was embedded between polymethylsilsesquioxane (PMSSQ) insulator layers. GQDs and PMSSQ layers were deposited by a spin-coating method, and the top silver nanowires (AgNWs) electrodes were formed by utilizing the spray-coating method. Figure 4.2b shows the scanning electron microscope (SEM) image of the NVM device. The electrical characterization of the devices observed reprogrammable bistable memory hysteresis windows with the $10^4$ ON/OFF ratio, and the NVM devices were stable up to $1 \times 10^4$ seconds. The possible reasons for bistable switching could be charge trapping in the GQDs

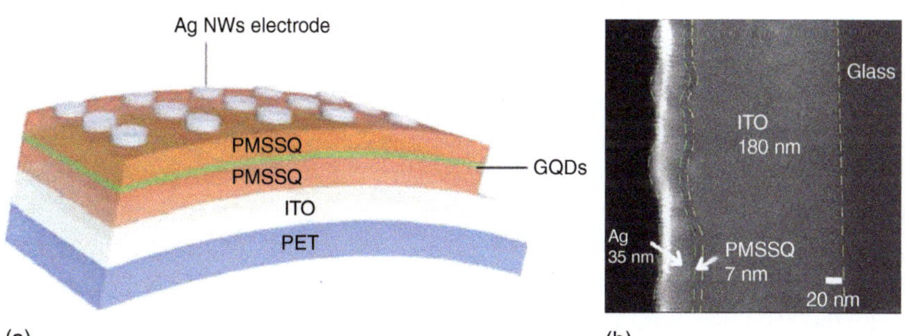


**Figure 4.2** Cross-sectional structures of the as-fabricated (a) NVM device and (b) SEM image of the NVM device. Source: Lin et al. [30]. Reproduced with permission from IEEE.

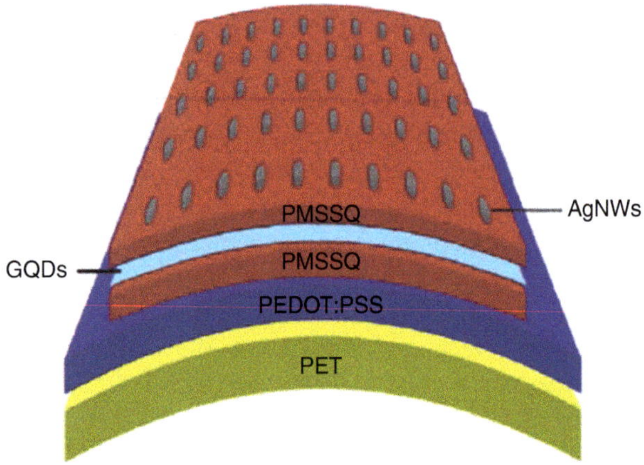


**Figure 4.3** Schematic diagram of the fabricated NVM device. Source: Reproduced from Ooi et al. [31] with the permission of Elsevier.

site or the formation and rupture of silver filaments. Additionally, the devices also showed stable electrical performance when they were under mechanical bending at different curvature radii [30]. The bistable switching behavior was speculated to be dominated by various carrier transport mechanisms, especially Schottky emission (SE), Poole–Frenkel (PF) emission, the trapped-charge limited-current (TCLC) model, and the space-charge limited-current (SCLC) mechanism.

Afterward, in the same study, Ooi et al. demonstrated the feasibility of ITO-free, all-solution-processed NVM devices [31]. To realize fully solution-processed devices, they replaced the bottom ITO electrode with a conductive polymer, namely, poly(3,4-ethylene dioxythiophene)-poly(styrene sulfonate) (PEDOT:PSS). They constructed a similar device structure consisting of AgNWs/PMSSQ/GQDs/PMSSQ/PEDOT:PSS on the PET substrate, as shown in Figure 4.3. The bottom conductive PEDOT:PSS, PMSSQ, and GQD layers were spin-coated, while the top AgNW electrode was spray coated to construct the all-solution-processed NVM devices. The device also recorded a $10^4$ ON/OFF ratio of the current bistability and stability up to $1 \times 10^4$ seconds [31].

Alternatively, it is also reported that carbon-based materials can be utilized as insulating and conductive layers. Consequently, Ooi et al. demonstrated the feasibility of solution process write-once-read-many-times (WORM) graphene-based memory devices [25]. Figure 4.4 shows the fabricated device structure on the quartz substrate. The GO flakes were spin-coated on the quartz substrate and plasma treated with $C_2H_2/H_2$ gas at 700 °C for the reduction of GO to obtain a conductive rGO thin film. The nanocomposite consists of GDQs and GO mixture. GQDs can be used as a charge-trapping layer and GO as the insulator matrix. The nanocomposite mixture and AgNWs TEs were deposited by spin-coating and spray-coating methods, respectively. The fabricated simple WORM structure shows a distinguishable $10^3$ ON/OFF ratio and retention stability up to $1 \times 10^4$ seconds.

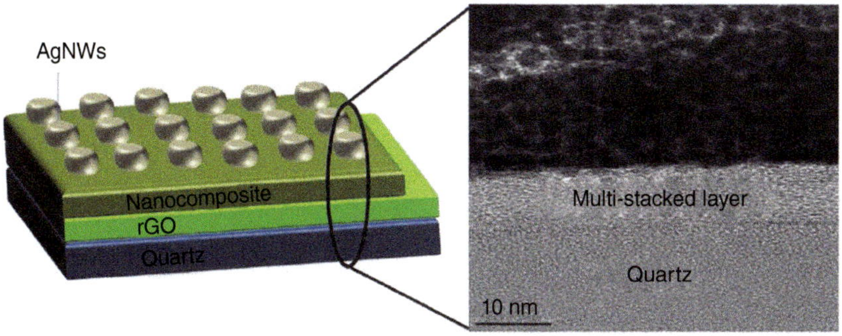


**Figure 4.4** Schematic diagram of the fabricated tristable NVM device. The inset shows a TEM image of the NVM device. Source: Ooi et al. [25]. Reproduced with permission from Elsevier.

In their study, the observed WORM effect was speculated to be related to the direct tunneling, TCLC, and Ohmic conduction transport mechanisms [25].

Due to the simplicity and ease of the spin-coating fabrication procedure, the NVM device structure can be designed efficiently to achieve multilevel switching states [32–34] for high-density data storage applications. For instance, two GQD layers can be embedded between three poly(methyl methacrylate) (PMMA) polymer layers to achieve tristable switching effects, as shown in Figure 4.5a [35]. The PMMA and GQDs thin films were spin-coated on the ITO-coated glass substrate in the sequence of PMMA/GQDs/PMMA/GQDs/PMMA. Again, the AgNWs TEs were spray-coated to complete the device construction. As a result, the device shows reprogrammable tristable switching states with a higher $10^5$ ON/OFF ratio, as can be seen in Figure 4.5b. The inset in Figure 4.5b shows the ignorable memory behavior of the reference samples without an embedded GQDs layer. The stability test of the device shows three different resistance states over 1000 endurance cycles and retention up to $1 \times 10^4$ seconds. Several conduction mechanisms, especially SE, PF emission, TCLC, and ohmic conduction, were proposed to be responsible for the tristable switching observation [35].

Notably, the tristable switching resistance states can also be achieved by introducing a carbon-related material to GQDs. Ooi et al. prepared the mixture of 2D $MoS_2$ flakes and GQDs in deionized water/ethanol [9]. The mixture was dispersed into polyvinylidene fluoride (PVDF) polymer host to create a GQD–$MoS_2$–PVDF nanocomposite (GMP-NC). The GMP-NC was spin-coated on the ITO-coated glass to form the charge-trapping layer. The TE contacts were deposited by the spray coating method. Figure 4.6a shows the fabricated device structure that consists of GMP-NC embedded between PVDF layers. Meanwhile, the SEM image in Figure 4.6b presents the cross-sectional structure of the fabricated tristable NVM device. The presence of GQD–$MoS_2$ heterostructure yielded tristable switching states and a high ON/OFF ratio of $10^7$, as shown in Figure 4.6b. During the SET process, the current switched from a low-conductivity (LC) state to an intermediate high-conductivity (IHC) state, followed by the highest conductivity (HC) when the applied voltage is swept from 0 to −1.0 V on top of the AgNWs electrode. When

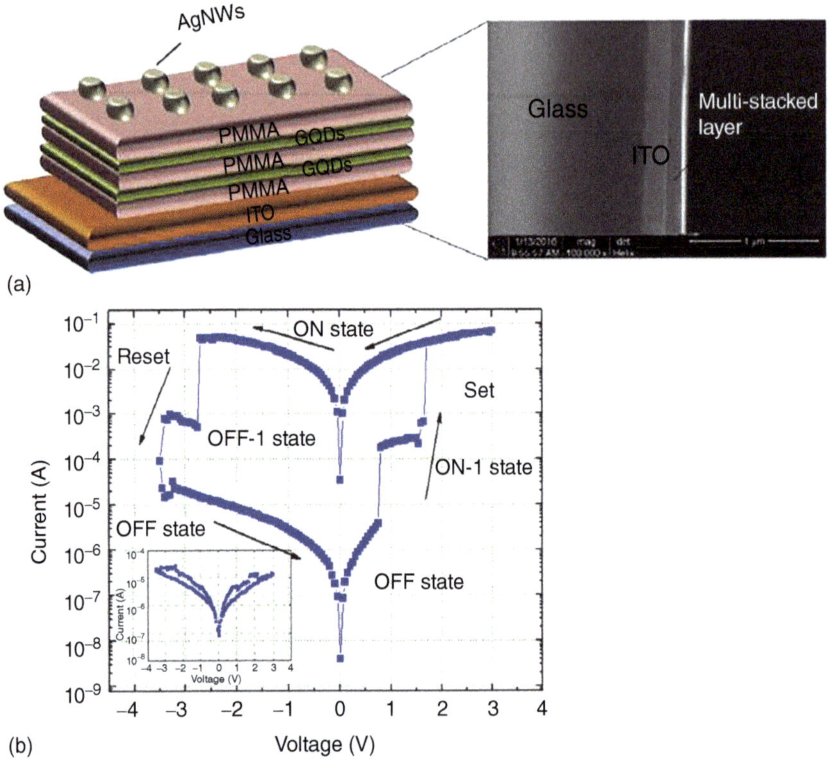


**Figure 4.5** (a) Schematic diagram and SEM cross-sectional image of the device. (b) Semi-log plot of the tristable switching *I–V* characteristics of the fabricated tristable NVM device. The inset presents the *I–V* characteristics of the reference device. Source: Ooi et al. [35]. Reproduced with permission from Elsevier.

the opposite polarity was applied from 0 to 1.5 V, the device can be RESET to intermediate low-conductivity (ILC) and eventually LC state. The fabricated device can be programmed and erased due to several transportation mechanisms, especially thermionic emission (TE), SE, PF conduction, TCLC conduction, direct tunneling conduction, and ohmic conduction were proposed. The three resistance states of the fabricated device were distinguishable and remained stable for $1 \times 10^4$ seconds [9].

### 4.5.2 Spray Coating

The spray coating technique could be considered a viable alternative approach to creating thin films of material that conform to the shape of the substrate and are suitable for a large area, high throughput, and industrially scalable process [26]. The principle of spray coating consists of the deposition of film from an atomized solution with typical droplet sizes in the micrometer range. The droplets can be produced using a nitrogen-filled nozzle or ultrasonic atomization to form a compact and closed thin film on the substrate surface. The most straightforward approach to producing an atomized spray can be created with conventional airbrush guns that are

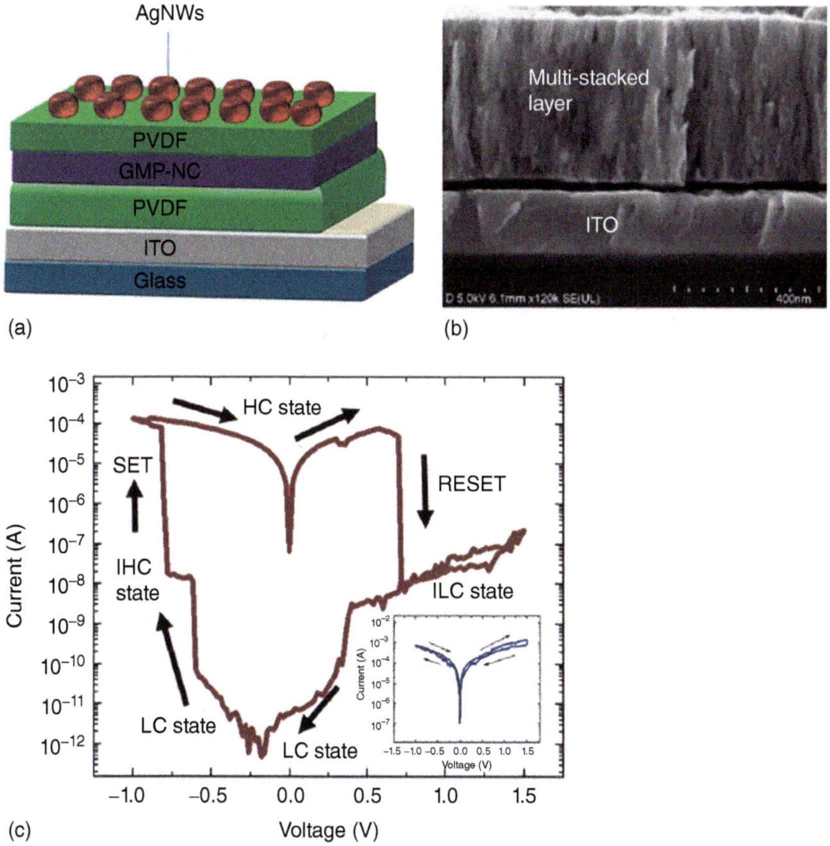


**Figure 4.6** (a) Schematic diagram and (b) SEM cross-sectional structure of the sandwiched tristable NVM device. (c) Tristable semi-log plot of *I*–*V* behavior of the fabricated NVM device. The inset presents the *I*–*V* characteristics of the control device. Source: Ooi et al. [9]. Springer Nature, CC BY 4.0.

relatively low cost. The droplet generation requires a low viscosity solution by diluting with a certain amount of solvent that could quickly evaporate from the droplets during in-flight between the spray nozzle and substrate. Nevertheless, pure nitrogen is highly recommended to be used in the system to prevent contamination of the solution with particles or atmospheric humidity on the low-pressure side of the spray nozzle [26]. For the progress of electronic device applications, spray coating of GO with an excess amount of hydrazine monohydrate was conducted by Pham et al. in 2010 on a quartz substrate to produce a high quality of rGO. The rGO film had a low sheet resistance of $2.2 \times 10^3\ \Omega\ \square^{-1}$ on a quartz substrate when spray treated at 240 °C, and it could be promising as a transparent electrode. The study observed high uniformity and continuous films due to the overlap of graphene sheets shown by the AFM image in Figure 4.7 [36].

For the fabrication of carbon-based NVM, the spray coating method has been used by Jaafar et al. for the deposition of AgNWs/GO/GQDs/GO structure on

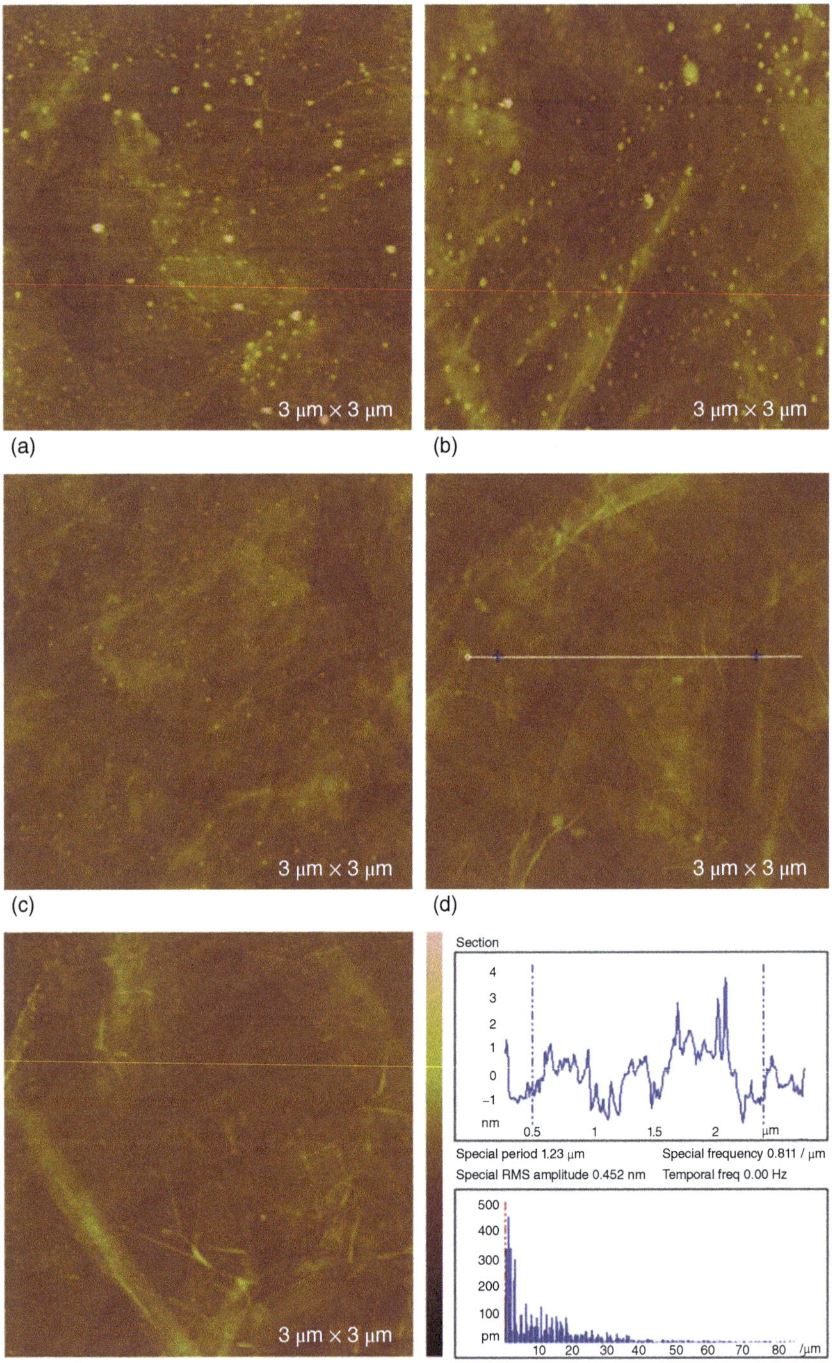


**Figure 4.7** (a–e) AFM images of the fabricated chemically converted graphene (CGC) films by spraying 2 mL of GO–hydrazine dispersion at various preheating temperatures: (a) 190 °C, (b) 210 °C, (c) 240 °C, (d) 270 °C, and (e) 295 °C. (f) Surface profile of the CCG film in (d). Source: Pham et al. [36]. Reproduced with permission from Elsevier.

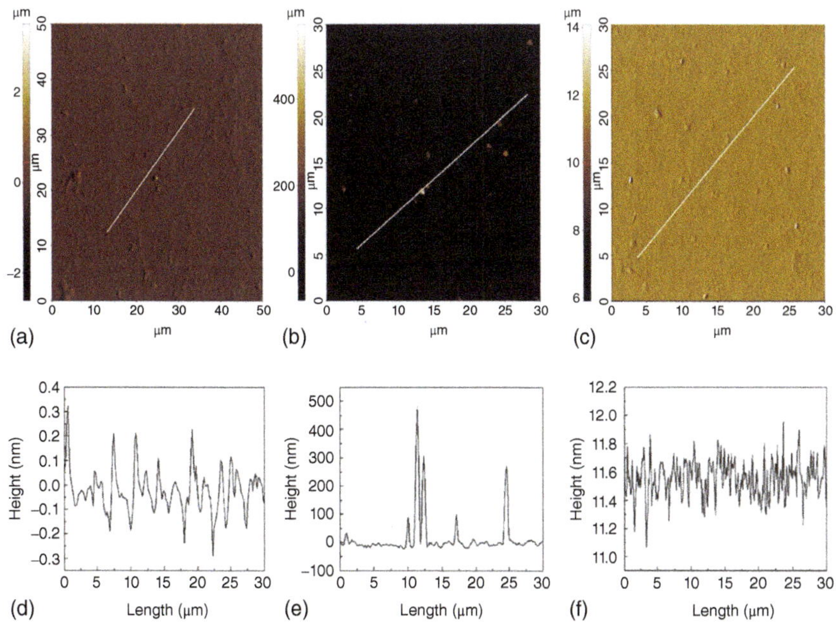

**Figure 4.8** AFM images of (a) first-layer 100-nm-thick GO film on PEDOT:PSS/PET substrate, (b) second-layer GQDs on GO/PEDOT:PSS/PET substrate, and (c) third-layer 140-nm-thick GO film on GQDs/GO/PEDOT:PSS/PET substrate by the spray-coating method. (d)–(f) Their corresponding height profile plots along with the selected white line as illustrated in (a)–(c). Source: Jaafar et al. [37]. Reproduced with permission from Springer Nature.

top of a spin-coated PEDOT:PSS deposited PET substrate. The multilayer deposition process was conducted on a hotplate at 100°C to evaporate the excessive solution. They found that the spray-coated GO/GQDs/GO attained was overall smooth, as shown in Figure 4.8, and it can be served as good contact with the top AgNWs electrode. Again, GQDs are acting as a charge-trapping layer embedded between the GO insulator layers. The constructed stacking structure shows bistable resistance states and reprogrammable. Several transportation mechanisms, typically SE, PF emission, TCLC, and filament formation, were speculated to be the dominant transportation mechanisms across the device. As a consequence, the fabricated device demonstrated bistable switching with a high ON/OFF ratio of $10^5$ was achieved [37]. In another recent study, Jo et al. fabricated the bistable resistive switching memory devices exclusively by the spray coating technique [38]. The fabricated device consists of liquid-exfoliated black phosphorus as the charge-trapping site and graphene layers to form the TE. The etched Si wafer substrate was utilized as the bottom conductive electrode. The charge-transfer mechanisms were dominated by TE, SCLC, and ohmic conduction. The electrical characterization of the fabricated device shows a high ON/OFF current ratio of $1.9 \times 10^3$ when read at 0.5 V with good retention stability up to $10^4$ seconds at room temperature [38].

### 4.5.3 Dip Coating

Another solution-based deposition technique known as dip coating has been used widely to fabricate carbon-based NVM. Briefly, the dip-coating method can be performed by immersing a substrate into the material in the solution form at a constant speed. Then, the substrate remains fully immersed and motionless for a specific time before its withdrawal. Due to its simplicity, this method can be automated in order to control the thickness of the coating, which is controlled mainly by the withdrawal speed and the viscosity of the solution. Nevertheless, it has limitations fundamentally related to the variation of the thickness from top to bottom attributed to the "wedge effect" [39]. Recently, Sun et al. have utilized the dip-coating method to construct the weavable all-carbon fibers NVM devices for the potential development of future smart e-textile [40]. In their study, the multi-walled carbon nanotube (MWCNT) fiber played the role of a conductive electrode, whereas the GO was used as the active layer material. The GO was coated twice on the outer surface of MWCNT to form the core–shell structure, denoted as MWCNT@GO fiber, as shown in Figure 4.9. The weavable MWCNT@GO fiber-based NVM devices can be fabricated by cross-stacking two MWCNT@GO fibers. The as-fabricated devices show bistable electrical behavior and are suitable for write-once-read-many-times (WORM) data storage applications. The observed WORM memory effect might be caused by the irreversible migration of oxygen in the GO layer as a result of the relatively inert MWCNT fiber electrodes. Nevertheless, the devices can attain a minimum ON/OFF state ratio of $10^3 : 1$ and stable retention for 500 seconds [40].

Meanwhile, Kim et al. demonstrated the feasibility of the sol–gel dip-coating method to fabricate the rGO-based transparent memory [41]. The rGO powder was dispersed in dichlorobenzene solution to prepare the rGO suspension. A 17 nm rGO layer was deposited on ITO during the dip-coating process. The rGO-based memory was fabricated in the ITO/rGO/ITO structure with device transmittance above 80% in the visible region, as shown in Figure 4.10. The high transparency feature is desired for the development of next-generation transparent electronic devices. The highly transparent memory device showed multilevel resistive switching behavior in the 00, 01, 10, and 11 states when the pulse height varied from 2 to 7 V. They concluded that the migration of oxygen ions within the rGO layer from the ITO electrode to form the GO layer is responsible for the observed switching behavior. The fabricated devices also show stable reliability at 85 °C with endurance and retention of over $10^5$ cycles and $10^5$ seconds, respectively [41].

In other respects, a two-terminal ZnO–rGO hybrid thin-film memristor was fabricated using the combination of dip-coating and inkjet-printing techniques by Khanal and coworkers [42, 43]. The introduction of GO into ZnO by a sol–gel synthesis approach demonstrates enhanced performance for electronics, optoelectronics, and photocatalytic applications. The prepared ZnO–GO solution was dip coated on the fluorine-doped tin oxide (FTO)-coated glass, and inkjet printed the top Ag electrodes. The insulating property of GO can be reduced to conductive rGO during the 3 hours of 500 °C treatment and create oxygen vacancies during the annealing process. The fabricated devices demonstrated bipolar switching and synaptic behaviors,

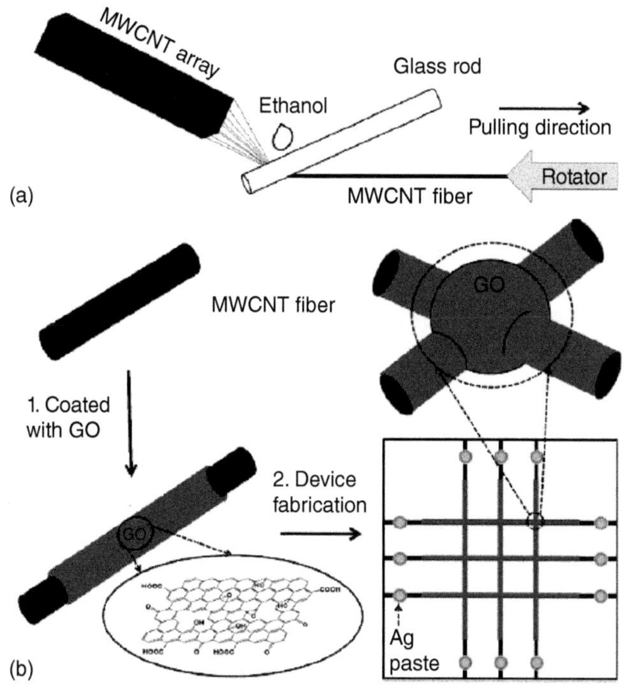


**Figure 4.9** (a) Schematic diagram of the spinning process for the formation of MWCNT fibers. (b) Schematic illustration of the fabrication of the MWCNT@GO fiber-based memory device. Source: Reproduced from Sun et al. [40] with the permission of John Wiley and Sons.

as shown in Figure 4.11. Figure 4.11a shows that the observed bipolar switching behavior was attributed to the forming and rupturing of conducting filament associated with the oxygen vacancies. While the device's synaptic signals, as shown in Figure 4.11b, were obtained from their reliable and realistic designed circuit on a printed circuit board. The constructed circuit can drive a memristor to imitate spike-time-dependence-plasticity behavior to generate better reliable and realistic data because it can accurately convert the time difference of arriving pulsed to a voltage level as compared to pulse mode or analog circuit measurements [42, 43].

### 4.5.4 Inkjet Printing

Inkjet printing is the conventional method deliberately developed for printing computer data in the form of text or image onto porous surfaces not limited to papers and transparencies but also cans and bottles. This deposition technique uses liquid-phase material, also known as inks consisting of a solute dissolved or otherwise dispersed in a solvent that will be ejected in a fixed quantity from a nozzle, usually through piezoelectric action. Piezoelectric printing is the preferred approach to thermal printing. Nonetheless, the thermal printing approach is restricted to vaporizable inks to form a bubble. Thus, the upraised operating temperature of thermal printing makes it unsuitable for developing polymer-based

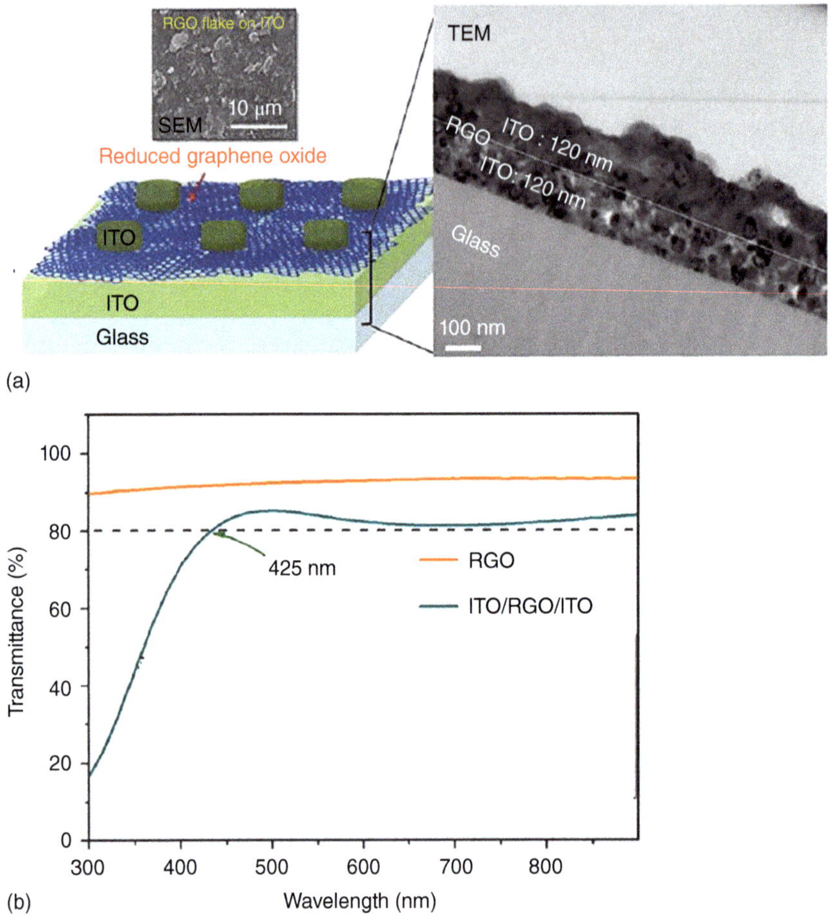


**Figure 4.10** (a) Schematic drawing and TEM image of the proposed ITO/RGO/ITO memory cell, with SEM image of the rGO flakes adsorbed on the ITO layer. (b) Optical transmission spectrum of the ITO/rGO/ITO/glass device. Source: Kim et al. [41]. Reproduced with permission from Springer Nature.

printing [44]. In recent years, inkjet printing has been developed as a free-form fabrication method to fabricate three-dimensional structures. Especially when it involves organic materials, inkjet printing enables the construction of electrical and optical devices, such as thin-film transistors, light-emitting diodes, solar cells, memory devices, and sensors. Besides, this technology is also capable of producing arrays of proteins and nucleic acids for biological applications. Furthermore, without the need for lithography and cleanroom processes, the printing process could significantly reduce production expenses by using low-cost plastic as a substrate. On the other hand, the highest resolution in the microscale can also be achieved in a single-step process performed at ambient temperature and pressure that can be established as a potentially promising route to low-cost, large-area, and flexible electronics.

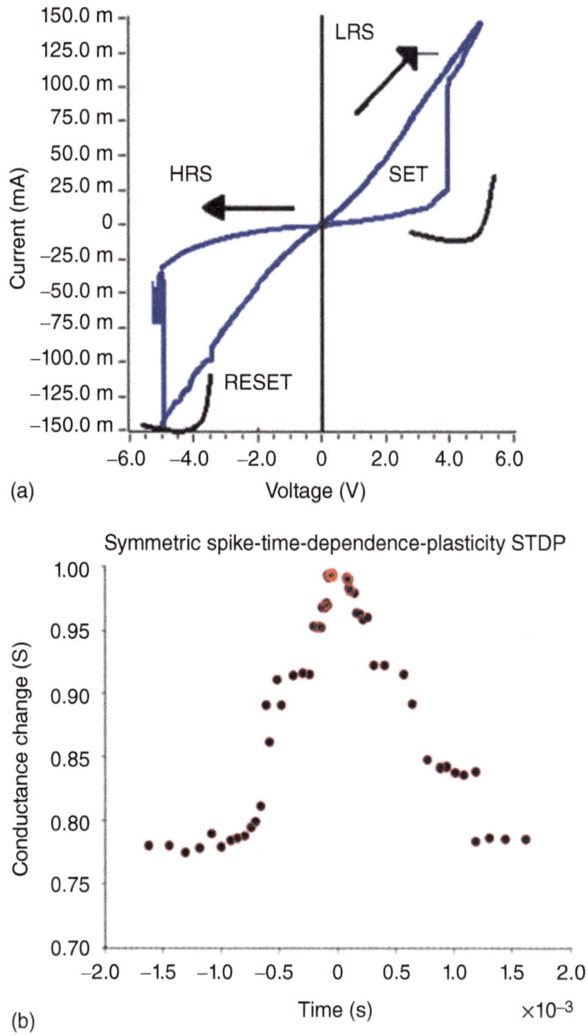


**Figure 4.11** (a) ZnO–rGO memristor *I*–*V* curve, and (b) STPD behavior. Source: Reproduced from Khanal et al. [43] with the permission of John Wiley and Sons.

Printing technology is also a promising approach for paper memory fabrication that is essential for the realization of paper-based electronic devices. This is because a memory device is the main component responsible for programming, data storage, and system setting to activate a paper-based device entirely. Without the use of printing techniques, manufacturing memory devices on rough and porous paper substrates is the current challenge. As a consequence, memory devices that require thin and uniform layers to be deposited on the fibrous paper surface would be inappropriate and difficult to employ the costly conventional techniques, such as chemical vapor deposition and sputtering techniques that usually require a vacuum and high-temperature conditions. The feasibility of all-printed paper memory

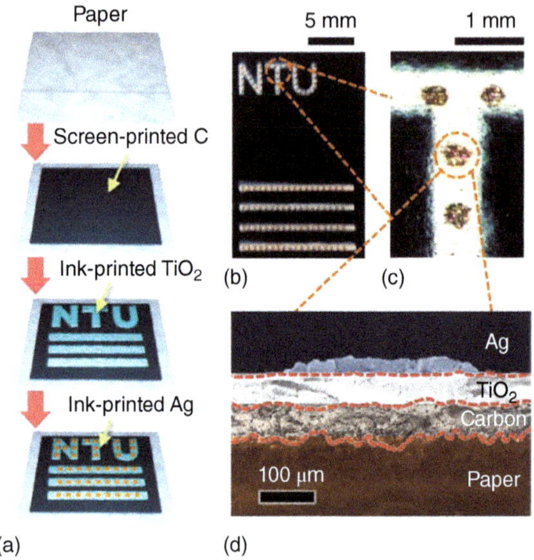


**Figure 4.12** Fabrication and geometry of a paper RRAM. (a) Schematic diagram of the fabrication process for the resistive paper memory device. (b) Photograph of the device taken by optical microscopy. The alphabetical letters and the line arrays composed of Ag and $TiO_2$ demonstrate the degree of freedom for the inkjet-printed patterns. (c) Zoom in an optical image from b. (d) Cross-sectional scanning electron microscopy image of the paper-based memory. Source: Lien et al. [45]. Reproduced with permission from American Chemical Society.

using a sequence of inkjet- and screen-printing techniques was demonstrated by Lien et al. [45]. Figure 4.12 shows the fabricated simple silver (Ag)/titanium oxide ($TiO_2$)/carbon (C) device structure on an A4 paper. The C bottom conductive layer was deposited using the screen-printing technique, whereas the resistive $TiO_2$ layer and Ag TE were formed by inkjet printing. Resultantly, the printed-paper-based memory devices (PPMDs) are able to store gigabyte capacities and exhibit reproducible switching endurance, reliable retention, tunable memory window, and the capability to operate under extreme bending conditions [45]. On the other hand, the inkjet-printing method is also suitable for depositing on the flexible plastic substrate.

Kim et al. fabricated all-solution-processed flexible organic transistor memory devices on a PES substrate utilizing inkjet printing and spray coating [46]. The electrode contacts (gate, source, and drain) and active n-type semiconductor layer were ink printed with PEDOT:PSS, and poly[[*N*,*N*9-bis(2-octyldodecyl)-naphthalene-1,4,5,8-bis(dicarboximide)-2,6-diyl]-alt-5,59-(2,29-bithiophene)] (P(NDI2OD-T2)), respectively. The patterns of the electrode contacts were formed without the aid of a shadow mask at the intended area, hence reducing the waste of chemical materials. Meanwhile, the charge trapping layer that consists of rGO and the gate PMMA dielectric layer were deposited by a spin-coating method. Figure 4.13 shows the fabrication process of flexible organic NVM. The fabricated all-solution-processed NVM devices demonstrated promising programmable memory behavior, and the rGO layer is responsible for electron trapping. The fabricated transistor memory

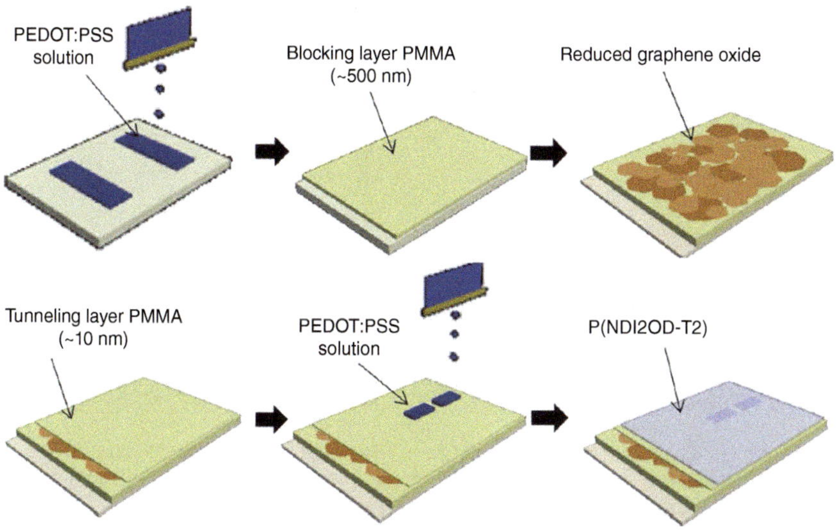


**Figure 4.13** Fabrication procedures for the n-type flexible organic transistor memory devices. Gate and source/drain electrodes (PEDOT:PSS) were deposited by inkjet printing. The blocking/tunneling dielectric layer (PMMA) and active layer (P(NDI2OD-T2)) were deposited by spin coating. The charge-trapping layer was composed of chemically synthesized reduced graphene oxide. All organic materials and solution processes were used to fabricate organic flexible memory devices. Source: Reproduced from Kim et al. [46] with the permission of IOI Publishing.

was mechanically stable and reliable as no significant difference was observed electrically when the device was bent and in a flat position. The obtained electrical data were retained and distinguishable for over $10^5$ seconds in the programmed and erased states [46].

Besides the spin-coating method to deposit GO layers, it is also feasible to deposit the GO using the inkjet-printing approach. Porro and Ricciardi fabricated GO-based memristor by means of inkjet printing, as shown in Figure 4.14 [47]. The GO dispersion in a water/ethanol was printed 20 passes on the ITO-coated glass substrate, and the TEs of the memristor devices were formed by thermal evaporation of Ag metal. Figure 4.14a shows the image of the fabricated 25 devices array of Ag/GO/ITO. GO flakes were distributed homogeneously after 20 passes of inkjet deposition, as shown in Figure 4.14b. They deduced that the observed memristive behavior could be related to the diffusion of carbon atoms from the GO layer into the Ag electrode during the electroforming occasion. However, the fabricated devices attained a relatively small ON/OFF ratio. Consequently, they might face limited usage for random access memories, but the low applied voltage would be suitable for adaptive and low-power applications [47].

### 4.5.5 Plasma Polymerization (PP) Deposition

PP deposition method can also be adopted for the fabrication of NVM devices. PP, also known as glow discharge polymerization, was discovered in 1960 by Yasuda,

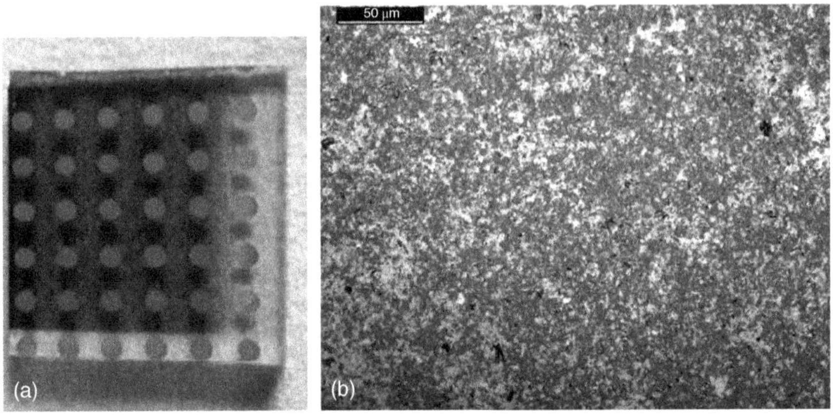


**Figure 4.14** (a) Photograph of the array of Ag/GO/ITO 25 devices. (b) Optical microscope image of 20 passes inkjet-printed GO on ITO. Source: Porro and Ricciardi [47]. Reproduced with permission from Royal Society of Chemistry.

where the deposition on the surface was found with the presence of organic vapor at low pressure when exposed to an electrical discharge [48]. This technique employs high-energy electrons excited by a plasma to dissociate, fragment, and ionize the monomer molecules present in a gaseous state. The polymerization process is induced due to the presence of a multitude of species in the plasma resulting from radicals of fragmented monomers, ions, metastable, excited electrons, atoms, and photons. The species would react between them and create chemically reactive species later deposited on the substrate's surface. The plasma polymerization process can occur at a low-temperature condition when thermal equilibrium is not maintained among electrons, ions, and neutral species. The low-temperature plasma created by electric glow discharge also can be referred to as cold plasma since it can be produced at room temperature [49].

In the nonequilibrium plasma, electrons acquire higher kinetic energy than ions and radicals, where the collision between them is relatively small to reach thermal equilibrium. The electrical discharge can be generated using direct current (DC), alternating current (AC), radio frequency (RF), and microwave (high-frequency source). The most common source used for PP is RF glow discharge operating at 13.56 MHz. This technique can be considered a green approach that can be utilized as a novel alternative to conventional wet-chemistry methods without the requirement for surface preparation. It can be performed on almost any type of substrate (metallic, plastic, or ceramic) as a result of its ambient temperature deposition with the ability to form a uniform deposition on a large area of the substrate. Furthermore, no solvent is required during the polymer preparation process, and thus almost no waste is produced compared to the aforementioned solution-processed deposition techniques. In this technique, a precursor liquid monomer is initially transformed into vapor and subsequently excited into the plasma under the influence of an electric field. The fragmented monomer species recombine onto any surface exposed to plasma and thus form a nano-thin layer.

**Table 4.1** Summary of various aspects of solution-processed and PP deposition techniques.

| Aspects | Spin coating | Spray coating | Dip coating | Inkjet Printing | Plasma polymerization |
|---|---|---|---|---|---|
| Control of thickness | >100 nm | >100 nm | >100 nm | >100 nm | Ultrathin (>10 nm) |
| Process time | Fast | Fast | Fast | Fast | Fast |
| *In situ* monitoring (thickness and composition) | No | No | No | No | Yes |
| Substrate independent | No | No | No | No | Yes |
| Control of functional group at surface | No | No | No | No | Yes |
| Temperature | Ambient | Ambient | Ambient | Ambient | Ambient |
| Crosslinking polymerization | Heat treatment required | No | No | No | Yes |
| Pressure | No | No | No | No | 10 millitorr |
| Usage of Solvent | Yes | Yes | No | Yes | No |
| Modification of composition | No | No | No | No | Yes |

In terms of stability, plasma polymer films are highly crosslinked with the presence of a specific functional group depending on the composition of the precursor [50].

Table 4.1 summarizes the comparison among the discussed deposition techniques from references [50–56]. Overall, plasma-based deposition has gained attention attributed to its versatility and flexibility in many aspects compared to the wet-based process. On occasion, which demands precision and consistency, such as in the biomedical and microelectronic industry, a vacuum-based process is preferable for industry practice compared to a solution process for a high-quality final product in terms of uniformity, purity, and flexibility. In addition, plasma-based deposition can be used to prepare nanocomposite, the same as the solution process method. Metal/polymer nanocomposite layers can be prepared by conducting PP and metal co-sputtering or evaporation simultaneously. Likewise, the metal oxide can be incorporated into the polymer matrix using RF co-sputtering from two magnetrons equipped with metal oxide and a polymer target. The absence of solvent during the deposition process will ensure the low-defect and low-impurities formation in the thin film. Hence, PP deposition is more suitable for producing a thin and smooth pin-hole-free surface polymer dielectric than solution-processed deposition. More importantly, PP deposited dielectric layer does not require heat treatment to achieve

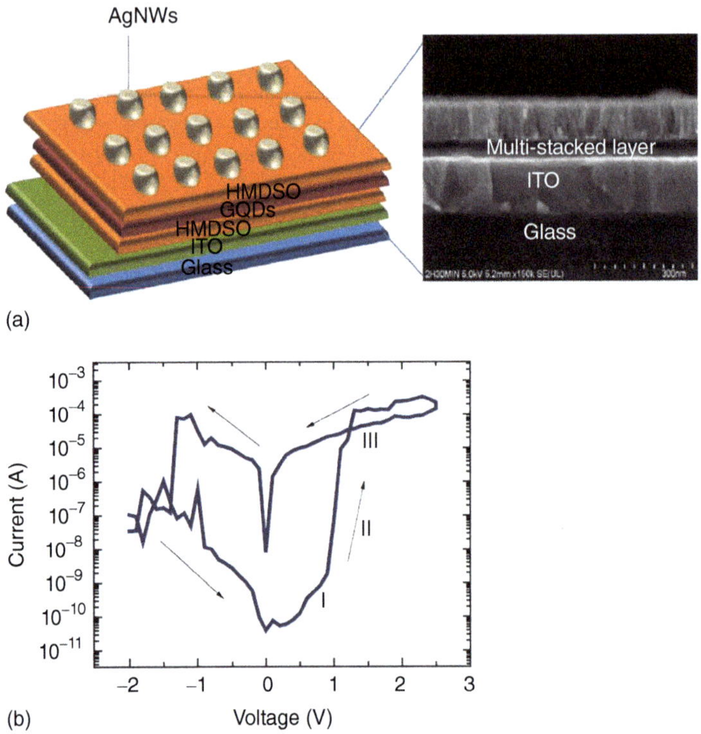


**Figure 4.15** (a) Schematic diagram and the SEM cross-sectional structure of the fabricated NVM devices. (b) *I–V* bistable switching behavior of the devices. Source: Ooi et al. [18]. Reproduced with permission from Elsevier.

structure crosslinking, unlike solution-processed deposition [57]. Note that heat treatment should be avoided or minimal for the development of plastic electronics.

For these reasons, Ooi et al. demonstrated the feasibility of using a combination of PP deposition and spin-coating methods to fabricate a bistable NVM device [18]. PP method was used to deposit hexamethyldisiloxane (HMDSO) dielectric layers, and the embedded GQD, charge-trapping layer, was deposited using the spin-coating approach. Figure 4.15a shows the schematic diagram and the SEM cross-sectional structure of the fabricated device bistable switching device. The device can be programmed/erased when the endurance test is performed and remains stable for up to $1 \times 10^4$ seconds cycles with a distinct ON/OFF ratio of $10^4$. Based on the measured *I–V* plot, they proposed that SE, PF emission, TCLC, and SCLC conduction mechanisms are responsible for the bistable switching behavior [18]. The exploration of using PP deposition to construct NVM devices is crucial to realize transparent and flexible electronic gadgets as no heat treatment is required to crosslink the deposited polymer layer. Hence, physical damage to the plastic substrate due to high-heat treatment can be avoided.

## 4.6 Conclusion

In brief, we have discussed the feasibility of fabricating the carbon-based NVM using solution-processing techniques, specifically spin coating, spray coating, dip coating, and inkjet printing. Besides, PP deposition also can be an alternate approach to depositing the dielectric layer without involving the high-temperature treatment. Hence, solution-processing techniques and the PP deposition method are principally suitable for producing mechanically flexible and optically transparent NVM devices. Overall, based on the reported literature, global researchers have explored the possible role of carbon nanomaterials as charge-trapping layers and conductive electrodes for developing low power, high-density storage, and reliable NVM. Thanks to the ease of solution-processing techniques, a low-cost yet high-density charge storage can be easily achieved by designing multi-stacking of charge-trapping and dielectric layers. Carbon nanomaterials and related 2D materials have also been explored over the past decade to improve carbon-based NVM devices. The blend of 2D carbon and related 2D material embedded in the polymer host suggests a new cross-point memory structure to offer high-density information storage. Various transportation mechanisms were proposed based on the obtained *I–V* behavior of the fabricated devices. Nevertheless, major challenges in the manufacturing and integration of carbon-based memory into the current silicon microfabrication platform, especially for large-scale integration, remain challenging hurdles to be solved by industrial and academic researchers before they can be widely introduced in the various mobile and wearable electronic gadgets.

## References


1 Ahn, E.C., Wong, H.-S.P., and Pop, E. (2018). Carbon nanomaterials for non-volatile memories. *Nature Reviews Materials* 3 (3): 1–15.

2 Lankhorst, M.H., Ketelaars, B.W., and Wolters, R.A. (2005). Low-cost and nanoscale non-volatile memory concept for future silicon chips. *Nature Materials* 4 (4): 347–352.

3 Scott, J.C. (2004). Is there an immortal memory? *Science* 304 (5667): 62–63.

4 Schaller, R.R. (1997). Moore's law: past, present and future. *IEEE Spectrum* 34 (6): 52–59.

5 Dehzangi, A., Larki, F., Wee, M.M.R. et al. (2017). Bollaert SJJoEM. Analog/RF study of self-aligned $In_{0.53}Ga_{0.47}As$ MOSFET with scaled gate length. *Journal of Electronic Materials* 46 (2): 782–789.

6 Shmagin, I., Muth, J., Kolbas, R. et al. (1997). Optical data storage in InGaN/GaN heterostructures. *Applied Physics Letters* 71 (10): 1382–1384.

7 Akasaki, I., Amano, H., Kito, M., and Hiramatsu, K. (1991). Photoluminescence of Mg-doped p-type GaN and electroluminescence of GaN pn junction LED. *Journal of Luminescence* 48: 666–670.

8 Bertolazzi, S., Bondavalli, P., Roche, S. et al. (2019). Nonvolatile memories based on graphene and related 2D materials. *Advanced Materials* 31 (10): 1806663.

9 Ooi, P.C., Mohammad Haniff, M.A.S., Mohd Razip Wee, M.F. et al. (2019). Electrical transportation mechanisms of molybdenum disulfide flakes-graphene quantum dots heterostructure embedded in polyvinylidene fluoride polymer. *Scientific Reports* 9 (1): 6761.

10 Chua, L. (1971). Memristor-the missing circuit element. *IEEE Transactions on Circuit Theory* 18 (5): 507–519.

11 Strukov, D.B., Snider, G.S., Stewart, D.R., and Williams, R.S. (2008). The missing memristor found. *Nature* 453 (7191): 80–83.

12 Kim, K.M., Kim, G.H., Song, S.J. et al. (2010). Electrically configurable electroforming and bipolar resistive switching in Pt/$TiO_2$/Pt structures. *Nanotechnology* 21 (30): 305203.

13 Sawa, A. (2008). Resistive switching in transition metal oxides. *Materials Today* 11 (6): 28–36.

14 Szot, K., Speier, W., Bihlmayer, G., and Waser, R. (2006). Switching the electrical resistance of individual dislocations in single-crystalline $SrTiO_3$. *Nature Materials* 5 (4): 312–320.

15 Chen, X., Zhang, H., Ruan, K., and Shi, W. (2012). Annealing effect on the bipolar resistive switching behaviors of $BiFeO_3$ thin films on $LaNiO_3$-buffered Si substrates. *Journal of Alloys and Compounds* 529: 108–112.

16 Kim, S. and Choi, Y.-K. (2008). Resistive switching of aluminum oxide for flexible memory. *Applied Physics Letters* 92 (22): 223508.

17 Bessonov, A.A., Kirikova, M.N., Petukhov, D.I. et al. (2015). Layered memristive and memcapacitive switches for printable electronics. *Nature Materials* 14 (2): 199–204.

18 Ooi, P.C., Wee, M.F.M.R., Dee, C.F. et al. (2018). Fabrication of transparent bistable switching memory device using plasmapolymerized hexamethyldisiloxane layers with embedded graphene quantum dots. *Thin Solid Films* 645: 45–50.

19 Chua, L. (2019). Resistance switching memories are memristors. In: *Handbook of Memristor Networks*, 197–230. Springer.

20 Borghetti, J., Snider, G.S., Kuekes, P.J. et al. (2010). 'Memristive'switches enable 'stateful'logic operations via material implication. *Nature* 464 (7290): 873–876.

21 Jo, S.H., Chang, T., Ebong, I. et al. (2010). Nanoscale memristor device as synapse in neuromorphic systems. *Nano Letters* 10 (4): 1297–1301.

22 Sun, L., Qin, G., Huang, H. et al. (2010). Flexible high-frequency microwave inductors and capacitors integrated on a polyethylene terephthalate substrate. *Applied Physics Letters* 96 (1): 013509.

23 Jang, J. (2016). Effect of electrode material on characteristics of non-volatile resistive memory consisting of $Ag_2S$ nanoparticles. *AIP Advances* 6 (7): 075006.

24 Liu, S., Dong, S., Jin, H. et al. (2018). Significant effects of electrode metal work function on resistive memory devices with gelatin biodielectric layer. *Journal of The Electrochemical Society* 165 (7): G90.

25 Ooi, P.C., Haniff, M.A.S.M., Wee, M.F.M.R. et al. (2017). Reduced graphene oxide preparation and its applications in solution-processed

write-once-read-many-times graphene-based memory device. *Carbon* 124: 547–554.

**26** Chaudhary, K.T. (2021, [cited 2022 Jul 06]. Available from: https://www.intechopen.com/chapters/75014). Thin film deposition: solution based approach. In: *Thin Films* [Internet] (ed. A.E. Ares). London: IntechOpen https://doi.org/10.5772/intechopen.94455.

**27** Jeong, H.Y., Kim, J.Y., Kim, J.W. et al. (2010). Graphene oxide thin films for flexible nonvolatile memory applications. *Nano Letters* 10 (11): 4381–4386.

**28** Liu, J., Yin, Z., Cao, X. et al. (2013). Fabrication of flexible, all-reduced graphene oxide non-volatile memory devices. *Advanced Materials* 25 (2): 233–238.

**29** Hui, F., Grustan-Gutierrez, E., Long, S. et al. (2017). Graphene and related materials for resistive random access memories. *Advanced Electronic Materials* 3 (8): 1600195.

**30** Lin, J., Ooi, P.C., Li, F. et al. (2015). Solution-processed, flexible, and transparent non-volatile memory with embedded graphene quantum dots in polymethylsilsesquioxane layers. *IEEE Electron Device Letters* 36 (11): 1212–1214.

**31** Ooi, P.C., Lin, J., Kim, T.W., and Li, F. (2016). Indium-tin-oxide, free, flexible, nonvolatile memory devices based on graphene quantum dots sandwiched between polymethylsilsesquioxane layers. *Organic Electronics* 32: 115–119.

**32** Wang, X., Xie, W., and Xu, J.B. (2014). Graphene based non-volatile memory devices. *Advanced Materials* 26 (31): 5496–5503.

**33** Lee, S.W., Park, S.J., Campbell, E.E., and Park, Y.W. (2011). A fast and low-power microelectromechanical system-based non-volatile memory device. *Nature Communications* 2 (1): 1–6.

**34** Shi, R., Wang, X., Wang, Z. et al. (2017). Fully solution-processed transparent nonvolatile and volatile multifunctional memory devices from conductive polymer and graphene oxide. *Advanced Electronic Materials* 3 (8): 1700135.

**35** Ooi, P.C., Lin, J., Kim, T.W., and Li, F. (2016). Tristable switching of the electrical conductivity through graphene quantum dots sandwiched in multi-stacked poly (methyl methacrylate) layers. *Organic Electronics* 38: 379–383.

**36** Pham, V.H., Cuong, T.V., Hur, S.H. et al. (2010). Fast and simple fabrication of a large transparent chemically-converted graphene film by spray-coating. *Carbon* 48 (7): 1945–1951.

**37** Jaafar, M.M., Ooi, P.C., Wee, M.M.R. et al. (2019). Electrical bistabilities behaviour of all-solution-processed non-volatile memories based on graphene quantum dots embedded in graphene oxide layers. *Journal of Materials Science: Materials in Electronics* 30 (17): 16415–16420.

**38** Jo, Y. and Dimitrakopoulos, C. (2020). Spray-coated, volatile and nonvolatile, two-terminal, resistive switching memory devices comprising liquid-exfoliated black phosphorus and graphene layers. *IEEE Transactions on Electron Devices* 67 (12): 5484–5489.

**39** Grosso, D. (2011). How to exploit the full potential of the dip-coating process to better control film formation. *Journal of Materials Chemistry* 21 (43): 17033–17038.

**40** Sun, G., Liu, J., Zheng, L. et al. (2013). Preparation of weavable, all-carbon fibers for non-volatile memory devices. *Angewandte Chemie* 125 (50): 13593–13597.

**41** Kim, H.-D., Yun, M.J., Lee, J.H. et al. (2014). Transparent multi-level resistive switching phenomena observed in ITO/RGO/ITO memory cells by the sol-gel dip-coating method. *Scientific Reports* 4 (1): 1–6.

**42** Khanal, G.M., Cardarilli, G., Chakraborty, A. et al. (ed.) (2016). A ZnO–rGO composite thin film discrete memristor. In: *2016 IEEE International Conference on Semiconductor Electronics (ICSE)*. IEEE.

**43** Khanal, G., Acciarito, S., Cardarilli, G. et al. (2017). Synaptic behaviour in ZnO–rGO composites thin film memristor. *Electronics Letters* 53 (5): 296–298.

**44** Zhan, Z., An, J., Wei, Y., and Du, H. (2017). Inkjet-printed optoelectronics. *Nanoscale* 9 (3): 965–993.

**45** Lien, D.-H., Kao, Z.-K., Huang, T.-H. et al. (2014). All-printed paper memory. *ACS Nano* 8 (8): 7613–7619.

**46** Kim, C., Song, J.-M., Lee, J.-S., and Lee, M.J. (2013). All-solution-processed nonvolatile flexible nano-floating gate memory devices. *Nanotechnology* 25 (1): 014016.

**47** Porro, S. and Ricciardi, C. (2015). Memristive behaviour in inkjet printed graphene oxide thin layers. *RSC Advances* 5 (84): 68565–68570.

**48** Yasuda, H.K. (2012). *Plasma Polymerization*. Academic Press.

**49** Yasuda, H. (1981). Glow discharge polymerization. *Journal of Polymer Science: Macromolecular Reviews* 16 (1): 199–293.

**50** Michelmore, A., Steele, D.A., Whittle, J.D. et al. (2013). Nanoscale deposition of chemically functionalised films via plasma polymerisation. *RSC Advances* 3 (33): 13540–13557.

**51** Cummins, G. and Desmulliez, M.P. (2012). Inkjet printing of conductive materials: a review. *Circuit World* 38 (4): 193–213.

**52** Eshkalak, S.K., Chinnappan, A., Jayathilaka, W. et al. (2017). A review on inkjet printing of CNT composites for smart applications. *Applied Materials Today* 9: 372–386.

**53** Aziz, F. and Ismail, A. (2015). Spray coating methods for polymer solar cells fabrication: a review. *Materials Science in Semiconductor Processing* 39: 416–425.

**54** Brinker, C.J., Frye, G., Hurd, A., and Ashley, C. (1991). Fundamentals of sol-gel dip coating. *Thin Solid Films* 201 (1): 97–108.

**55** Gu, J., Bullwinkel, M., and Campbell, G. (1995). Solvent concentration measurement for spin coating. *Journal of Applied Polymer Science* 57 (6): 717–725.

**56** Schneller, T., Waser, R., Kosec, M., and Payne, D. (2013). *Chemical Solution Deposition of Functional Oxide Thin Films*. Springer.

**57** Biederman, H., Kylian, O., Drabik, M. et al. (2012). Nanocomposite and nanostructured films with plasma polymer matrix. *Surface and Coating Technology* 211: 127–137.

# 5

# Carbonous-Based Optoelectronic Devices

*Nazia Nasr[1,2,3], Kashif Shahzad[3], Ishaq Ahmad[1,3], Chang Fu Dee[4], Badariah Bais[4,5], and Tingkai Zhao[1,2]*

[1] *Northwestern Polytechnical University, NPU-NCP Joint International Research Center on Advanced Nanomaterials and Defects Engineering, Youyi West Road, Xi'an 710072, China*
[2] *Northwestern Polytechnical University, School of Materials Science & Engineering, Youyi West Road, Xi'an 710072, China*
[3] *National Center for Physics (NCP), Quaid-i-Azam University Campus, Shahdra Valley Road, Islamabad, 44000, Pakistan*
[4] *Institute of Microengineering and Nanoelectronics (IMEN), Universiti Kebangsaan Malaysia, 43600 Bangi, Selangor, Malaysia*
[5] *Universiti Kebangsaan Malaysia (UKM), Department of Electrical, Electronic and Systems Engineering, Faculty of Engineering and Built Environment, 43600 Bangi, Selangor, Malaysia*

## 5.1 Introduction

Optoelectronics consists of two terms, optics and electronics. Optics includes the generation, detection, and transmission of light. When one does this electronically, it becomes optoelectronics. Optoelectronics has become a backbone of transmission technologies with high-quality transmission networks and low operational costs. Wide varieties of optoelectronic components are available nowadays. Many vital technologies, from medicine to entertainment to information processing, have benefited from optoelectronics. Semiconductor optoelectronics has a crucial role in processing information technologies, such as display, communication, and storage. The improvement and advancement are coming from low-dimensional systems. There is still tough competition for different materials available at low cost (silicon-based) and established fabrication facilities. However, new technologies are emerging by coupling other photonics with new and advanced materials. Various heterostructures have been widely used to confine electrons and holes to create low-dimensional electronic systems.

The discovery of graphene (GR) in 2004 from layered graphite has burst forth from internal energy, suitability in terms of various particular applications, popularity, and tremendous results [1, 2]. The bulk is like a stalk of sheets in-plane, bonded weakly through van der Waals forces. Each sheet consists of extraordinary individual properties. As a relatively young semiconductor (discovered in 2004), graphene is the first thermodynamically stable and atomically thin material obtained in human history [3]. Before its discovery, the existence of two-dimensional (2D) crystals was

*Enhanced Carbon-Based Materials and Their Applications*, First Edition.
Edited by Poh Choon Ooi, Mengying Xie, and Chang Fu Dee.

considered impossible due to thermodynamic considerations. With this landmark discovery, scientists have found a massive world with opportunities to study previously unexplored two-dimensional and layered materials. This discovery has triggered an exciting period of intense research in the scientific community, which has continued for nearly more than a decade. Its discovery and extraordinary properties have proven to be of great interest to scientific and industrial communities. Intensive research efforts on graphene have arisen worldwide, known as the graphene gold rush. The graphene gold rush has also inspired the exploration of other 2D electronic materials, such as phosphorene [4] and metal dichalcogenides [5, 6] that are potentially useful for research and various applications. The extraordinary properties of graphene stem from its simple crystal structure; however, the contact resistance issue needed to be resolved. The contact resistance at the metal/graphene interface is the key limiting factor affecting device performance. The transconductance, ON/OFF current ratio, and cut-off frequency do not make graphene suitable for digital electronics.

Two processes are involved in carrier transport at metal/graphene contact. The first process is the injection of charges from metal to graphene due to metal–graphene coupling. The second process deals with the transport through the pristine graphene sheet attributed to the width, shape, and height of the interface barrier [7]. The charge transfer during metallization induces doping in graphene [8]. Therefore, the work function difference, the quality of the interface, and sheet resistance of graphene under metal are three main factors contributing to the metal–graphene (M/G) interface's contact resistance. Thus, optimization of the above three factors is required, and failure could compromise M/G contact enhancement.

The high mobility of graphene and the Fermi velocity of electrons makes them a suitable candidate as a channel material in field-effect transistors [1]. The challenge of making powerful graphene field-effect transistors (GFETs) has begun. Moreover, the distinguished band structure and high carrier mobility make graphene an excellent alternative to silicon field-effect transistors (FETs). However, the graphene quality, which is hard to deal with, is the nonexistence bandgap, which prevents the complete shutdown of GFETs. The estimated bandgap energy of 0.4 eV is required to achieve a switching ratio of $10^4$ in FET [9]. Graphene is a kind of semimetallic material with zero bandgaps. Thus, this property makes GFETs not suitable for modern-day digital electronics.

On the other hand, electric-field application across bilayer graphene can create a bandgap of about 250 MeV [10], which is not enough for logic applications at room temperature. Nevertheless, the said bandgap creation is potentially suitable for GFETs and RF electronics. Another method introduced in the literature for bandgap opening uses one-dimensional graphene nanoribbons (GNRs), which produce a switching ratio of $10^6$ [3]. However, due to the difference in edge functionalization, crystal orientation, and edge roughness, the electrical properties of GNRs vary greatly, accompanied by the trade-off of bandgap mobility. Therefore, the FET fabricated on GNRs is worse than the silicon MOSFET. Despite small bandgap openings in GFETs, the maximum oscillation frequency $f_{max}$ is far inferior to the most advanced technology. The reason behind this is the poor current crowding at the interface and contact resistance of M/G [4]. Thus, to replace graphene as a mainstream channel material for FET, contact resistance problems need to be solved.

## 5.2 Graphene-Based Optoelectronics

Monolayers, such as graphene, have two surfaces with no volume in the middle. It presents the ultimate case of surface science. Graphene is the most studied material in two-dimensional crystals. It consists of a single layer of carbon atoms arranged in a two-dimensional honeycomb lattice. A sheet of graphene paper is one million times thinner than a sheet of paper. Graphene has attracted wide attention for many reasons, which can be summarized as having two discrete properties. First, the fundamental physics of graphene is elementary and clean. Scientists can produce graphene sheets with defect densities as low as 1 in $10^8$ [5]. Many exciting phenomena are also observed in graphenes; for example, electrons and holes are massless relativistic particles in graphene. They show integer and fractional quantum hall effects in magnetic fields [6].

The second reason for widespread interest in graphene is its potential. Graphene can transport electrons quickly, allowing fast switching in high-speed electronics. Recent research in graphene-based electronics has proved that transistors based on graphene run faster than today's silicon-based transistors. Graphene possesses unique properties, such as speed $v \sim 10^6\ \mathrm{m\,s^{-1}}$ [11], the optical absorption of 2.3% [12], mechanical breaking strength of ~130 GPa [13], and thermal conductivity of $K \sim 5 \times 10^3\ \mathrm{Wm^{-1}\ K^{-1}}$ [14] at room temperature and high melting point $T_\mathrm{m} \sim 4510\,°\mathrm{K}$ [15]. The extraordinary properties of graphene make it useful in many applications. Graphene has impacted the growth of modern-day technology for many applications, especially gas sensors [16], photodetectors [17], solar cells [18], FETs [19], transparent conductors for touch screens [20], interconnects [21], and flexible electronics [22]. These extraordinary properties lead to graphene penetration into many microelectronics applications.

The graphene properties make it an ideal material for the next generation of optoelectronic and optical communication systems. It has excellent electrical properties and broadband optical absorption characteristics. These characteristics made graphene very suitable for high-performance optoelectronic devices and easy to integrate with silicon photonic systems. Furthermore, the strong coupling between graphene and light makes it possible to observe single graphene with naked eyes under an optical microscope. The rich optical and electrical properties of graphene have aroused great interest. However, we believe its real potential lies in photonics and optoelectronics, fully using its unique optical and electronic characteristics even without a bandgap. The linear dispersion of Dirac electrons can achieve ultra-wideband tunability. These also contribute to the rise of graphene in photonics and optoelectronics, from solar cells and light-emitting devices to touch screens, photodetectors, and ultrafast lasers. The original graphene monolayer has a constant absorption of $\alpha\,\pi = 2.3\%$ in the infrared and visible light ranges, where $\alpha = \frac{e^2}{hc}$ is the fine structure constant, and $e$, $h$, and $c$ are charges, Plank's constant and speed of light, respectively. In addition, the broad optical absorption and optical transition can be controlled by electrical gating and by modifying the Fermi level of graphene. The strong electric absorption effect that has not been reported in bulk materials indicates that graphene has an active medium for photoelectric absorption modulators. However, the limited absorption of monolayers poses a challenge to graphene modulators. This limitation can be overcome by combining graphene with optical waveguides. These couplings increase the interaction length between impermanent waves and graphene.

Furthermore, the flexibility, robustness, and environmental stability of graphene make it suitable candidate material for recent multipurpose devices. The technological devices based on graphene have proven to be an essential part of the next generation of communications; the 5G high-performance optical communication systems via ultrahigh-speed and compact optoelectronic devices based on graphene. From lasers and optical switches to wireless communications and energy harvesting, graphene plays an essential role in optoelectronics.

Light and graphene are a perfect match. Owning the unique features of graphene, for instance, high carrier mobility and broadband absorption, a wide range of graphene-based optoelectronic and photonic devices has been developed till now. Optical detection, modulation, and switching are developing toward higher technology and manufacturing and have the potential to supplement traditional technologies. The exciting thing about graphene-based photonics and optoelectronics is not only to improve performance or reduce cost but also to introduce new technologies at the component level, e.g., graphene exhibits electrical absorption and refraction, so it can be used for ultrafast light modulation and optical detection. The graphene photodetectors detect light from ultraviolet to infrared and even terahertz. This detection was achieved by combining expensive materials that increase the complexity of the system. Data communications/telecommunications, optical spectroscopy, imaging devices, and consumer electronics, such as wearable devices, are just a few markets that benefit from graphene technology.

The complex system requires integrating different components to clearly understand the benefits of new technologies in semiconductor applications. Graphene-based applications are facing multiple constraints. Graphene handling and transfer, electrical interconnect, and interface engineering require utmost care and compatibility. Fortunately, graphene-based systems are emerging and are expected to meet these challenges in the next few years and demonstrate scalable processes, specifically those currently used in the microelectronics industry. At the prototype component level, modulators, detectors, and switches exist.

The zero-bandgap GFETs as ultrafast photodetectors have been explored. The mass and carrier concentration of graphene assist and support the process of light absorption. The external electric field that is applied to generate a photocurrent and electric field formed across the metal interconnect/graphene promotes ultrafast response in graphene. Because of the high carrier transmission speed, even in the medium electric field, there is no need to bias the source and drain directly to ensure ultrafast and high efficiency (the internal quantum efficiency in the photoelectric detection region is 6–16%). In the short-circuit state, the photocurrent depends mainly on the position of illumination and gate bias. Furthermore, the said efficiency (6–16%) is achieved if recombination of photogenerated carriers occurs outside the photogenerated region. In graphene-based photodetectors, photon interaction with graphene and photogenerated carriers' transport properties differs from those of the traditional group IV and III–V semiconductors. These unique properties of graphene enable it to achieve very high-bandwidth (potential > 500 GHz) light detection, detection of wide-wavelength regions, zero-dark-current operation, and good internal quantum efficiency.

GFETs are also used as a photodetectors. In GFETs photodetector [13], the built-in electric field liable to keep apart photogenerated carriers exists only in the narrow region (about 0.2 μm) adjacent to the metal/graphene (M/G) interface. This is because the band bending occurs at the metal–graphene interface. In the bulk graphene sheet, the generated charge carriers due to incident photon recombine without affecting external photocurrent. The enlarged high-electric-field light-detection area is achieved by multiple interdigitated metal fingers. Nonetheless, if the two electrodes are made of the same metal, the electric-field distribution inside the channel between the two adjacent fingers will be symmetrical, and the total photocurrent will disappear. An asymmetric metallization scheme can be used to break the channel's mirror symmetry so that the individual contributions can be added up to get the whole photocurrent.

The contact resistance generated at the M/G interface is the prominent factor affecting graphene utilization in modern-day electronic devices. The complex transmission phenomena occur while metal on graphene is deposited. The charge-carrier transport from three-dimensional metal electrodes into the 2D graphene layer starts. Two possibilities immediately follow when metal (M) and semiconductor (S) are contacted: (i) An ohmic contact, usually obtained by heavily doped S, in which the current can flow in two directions and can be easily adjusted by applying a voltage (the total resistance is the sum of the contact resistance R and the bulk resistance of the material); (ii) a rectifier Schottky diode, usually obtained on lightly doped semiconductors, thus allowing current to flow easily in one direction. The rectifying behavior stems from a Schottky barrier established at the interface. The former of the two possibilities arises during M/G contact. Despite the rapid growth in technology, the M/G interface's contact resistance is still needed to interpret bulk semiconductors correctly.

Due to the disappearance of the small density of states at the zero-energy gap and Dirac point, the interface between metal and graphene (G) is very complex. The absence of the energy gap poses an obstacle to the formation of the depletion layer. Thus, the conventional Schottky contact does not establish at the interface. The small density of states near the Dirac point restrains the current injection from metal to graphene. The graphene properties will be significantly affected when it contacts metals attributed to its 2D nature. So far, the above discussion mainly focuses on intrinsic graphene properties. External factors, such as charge impurities from supporting substrate, defects, and adatoms, on graphene influence graphene's electrical properties [23–25]. Carrier scattering in graphene is caused by impurities in support substrates. Consequently, the carrier mobility in graphene devices is reduced [23, 26]. The short-range scattering in graphene appears to happen between Dirac fermions and localized defects, along with long-range scattering occurring ascribed to charged impurities [26]. The significance of the scattering mechanism influences and varies significantly in different support substrates. The ambient temperatures and the presence of defects/impurities in graphene also affect the scattering mechanism. Moreover, the difference in work functions of metals used as contact and graphene causes a charge transfer through the interface leading to graphene's electrical doping. The density of states near the Dirac point in graphene is minimal. Even with limited charge transfer, the Fermi level will change immensely. There are two ways that the shift of Fermi level, $\Delta E_F$ can occur. It moves up when the electron moves from metal to graphene (N-doped) and down when the hole moves from metal to graphene (P-doped).

Compared with linear nanostructures, the photocurrent generated by single-layer graphene (SLG) has two-dimensional characteristics. It studies the dynamics of carriers (such as the collection and localization of photocurrent generation) and the overall electronic band structure of SLG related to spatial variation factors. The applicability of graphene as a photodetector is first demonstrated by local illumination near an electrical contact of a back gate graphene transistor [27–29]. Because of the band bending at the M/G interface, a detectable current is generated. The photocurrent is also obtained at the monolayer/bilayer graphene interface [30] and graphene p–n junction [31], attributed to the thermoelectric effect. Electronically controlled absorption is achieved by adjusting the Fermi level of the graphene sheet with grid electrodes [32, 33].

A broadband, high-speed, integrated waveguide electrode absorption modulator supported by SLG has been developed for the first time [34]. The graphene-based optical modulation process combines the advantages of small volume, low operating voltage, and ultrafast modulation speed in a wide wavelength range, providing a new structure for on-chip optical communication. Modulation is achieved by actively adjusting the Fermi level of SLG. The graphene-based modulator has the following advantages: (i) Strong optical graphene interaction; (ii) Broadband operation; (iii) High-speed operation; (iv) Compatibility with complementary metal–oxide–semiconductor (CMOS) process. A 50-nm-thick silicon layer is fabricated to deliver a waveguide to a 250-nm silicon bus and an electrode. A small amount of boron is added to the silicon layer and waveguide to reduce the resistance of the film. The atomic layer deposition method is used to deposit a 7-nm $Al_2O_3$ interlayer of 7 nm on the surface of the waveguide. Moreover, a graphene-based optical modulator is manufactured with a broad optical bandwidth (1.35–1.6 μm), small device size (25 μm), and high operating speed (1.2 GHz, 3 dBm) under ambient conditions, meeting the need for optical connectivity. The modulation efficiency of SLG film is comparable to that of other semiconductor materials, such as Si, GeSi, and InGaAs, and even lower than that of traditional materials with an order of magnitude larger.

The graphene-based photodetectors and high-efficiency wideband CMOS technology can be integrated [35]. The high-efficiency optical absorption ($\eta > 50\%$) can be attained with a small device length to achieve high-speed and intensive operation capacity. The device with the appropriate length completely absorbs the optical mode in a silicon waveguide. Nevertheless, the potential gradient across the metal/graphene (signal electrode) interface drives the photocurrent to the grounding (GND) conductor. In addition, the potential gradient can be increased by using the waveguide, as the back-grid electrode adjusts the potential across the graphene channel. The GND–S–GND structure is used to double the total photocurrent. Due to the zero bandgap of graphene, photogenerated carriers can almost pass through the GND electrode barrier unimpeded. Therefore, even without S-GND bias, high-bandwidth optical detection can be realized.

Another way to improve graphene photodetectors is by integrating graphene into silicon optical waveguide (SOI) materials on an insulator [36]. The graphene/silicon heterostructure waveguide photodetector on SOI works in the visible to mid-infrared spectrum range. The natural formation of graphene/silicon heterostructure shows a low dark current because of the junction barrier. Low-light responsivity is caused by the weak-light absorption of graphene. Therefore, a high-speed and wide spectral bandwidth waveguide integrated graphene photodetector with high responsivity is fabricated to overcome low light responsivity [37]. By integrating graphene photodetectors into SOI bus waveguides, the absorption of graphene and the corresponding optical detection efficiency can be significantly improved without compromising high-speed and wide spectral bandwidth. Spatial resolution photocurrent measurement assures the coherence of graphene junctions doped with metals. By deconvoluting the photocurrent and spot size of the exciting laser beam, the relative potential distribution on the graphene channel is obtained by numerical integration along with the straight line. Unlike traditional semiconductors, graphene has a very high mobility of electrons and holes. Therefore, the moderate internal electric field can attain ultrafast and efficient carrier separation. In addition, the device can have an ultrafast dynamic response under zero-bias voltage, thus reducing the power consumption of the chip.

Graphene is an excellent photoconductive material with a wide absorption spectrum and short response time. However, it is still challenging to attain a high response in graphene detectors due to their weak-light absorption and short photocarrier lifetime. The metal antenna structure can simultaneously improve the light absorption and optical carrier collection of graphene detectors [38]. The coupling antenna concentrates the free-space light, and the interaction between graphene and light is greatly enhanced by the ultrahigh electric-field intensity in the gap. At the same time, the metal antenna is designed as an electrode, which can collect the generated optical carriers very effectively. The photoconductive gain mechanism of the graphene detector shows that the responsivity of the mid-infrared antenna-assisted graphene detector at room temperature ($\lambda_0 = 4.45\,\mu\text{m} \sim 0.4\,\text{V}\,\text{W}^{-1}$) is 200 times higher than the device without antenna ($<2\,\text{MV}\,\text{W}^{-1}$).

High efficiency and patterned chemical doping technology can be used to fabricate photodetectors with large-area graphene film-forming controlled p–n junctions.

It is noticed that the ratio between the p–n junction region and the uniformly doped region of the two geometries is significantly different, which has a crucial impact on the optical response of the device. Because the p–n junction generated by chemical doping does not need a gate or dielectric layer, it is easy to fabricate ultra-thin, fully transparent flexible photodetectors [31]. In the wavelength range of 400–2000 nm, the transmittance of the device is more than 90%. The potential distribution and carrier polarity can be controlled by optical detection near the local electrostatic gate of graphene. When the p–n junction is formed, the intense light response appears, which allows the switch to control light detection. The photocurrent generated near the p–n junction is not biased and can be achieved with a submicron gate. Locally modulated photo response provides a new application range for graphene-based photodetectors, including pixel infrared imaging, to control the subwavelength size response. Therefore, graphene photodetectors grown by chemical doping CVD can be prepared on a large scale with broadband adsorption.

Graphene, being two-dimensional semiconductor material, is adapted to integrate with photonic devices built in a planar geometry. The planar geometry helps to improve the interaction length. Therefore, the device's performance is immensely improved [35, 37, 39]. The first optoelectronic device utilizing the tunable photoelectric properties of graphene was conceived through this important discovery that could act as both a modulator and a photodetector [40]. The electrostatic gate can be used to control the function of the device. The gate is also made of graphene segregated by a dielectric. The graphene gate is integrated on a planar silicon photonic waveguide. Graphene devices are integrated into silicon waveguides and can be tuned by a graphene grid. The near-infrared detection response and modulation depth of 64% are achieved in GHz bandwidth. The geometry is beneficial in terms of synchronous modulation of photocurrent and optical transmission. Thus, this geometry brings unprecedented optoelectronic applications.

Among various photoelectric detection schemes and structures proposed recently, the simplest photodetector structure is metal electrodes as drain and source upon graphene layer [27, 29, 36]. These photodetectors combined with metal nanostructures can increase local surface plasmon polaritons and enhance absorption and responsivity. However, the precise mechanism of light detection is still controversial. Based on this study, the light response of photodetectors can be quantified and controlled by the relative contribution of photo thermoelectric and photoelectric effects [41]. Metal–graphene (M/G)-interfaced photodetectors are easy to manufacture and play an important role because they do not rely on nanolithography technology. Since the interaction between light and matter is mainly determined by graphene, their working wavelength range is extensive. In addition, ultrahigh operating speed can be carried out successfully without using bandwidth-limiting materials. Each M/G photodetector consists of two graphene channels in contact with the same metal or two different metal electrodes. The work function difference across the metal M/G interface results in charge transfer. This leads to the shift of the Fermi level in graphene across the metal interconnect. At the M/G interface, the Fermi level gradually returns to the level of unconnected graphene. This distribution across the

M/G interface forms a p–p, p–n, or n–n junction at the interconnect. At the same time, the back gate controls the Fermi level at the channel region.

Graphene is an optical material with remarkable properties. It has a linear dispersion conduction band and valence band, as well as a strong interband transition. A novel ~1.5 μm (C-band optical communication) graphene-clad microfiber (GCM) all-optical modulator is conceived with a response time of ~2.2 ps. The fast response is determined only by the intrinsic response time of graphene [42]. The modulator is compatible with the existing high-speed optical fiber communication network, which opens the door to meeting future ultrafast optical signal processing requirements. The modulation is associated with the light–graphene interaction confined in the guided wave microfiber. The modulation depth can reach 38%. The weak infrared signal wave associated with the GCM is subjected to significant attenuation due to graphene absorption. When light falls on the graphene layer, it excites the carriers in graphene. The absorption threshold transfers to a higher frequency, making the attenuation of the signal wave lower. Therefore, the relaxation in excited carriers limits the response time. When the light intensity increases, the band-filling effect (Pauli blocking) of the excited carriers can significantly alter the absorption spectra of graphene. When the maximum power density is less than ~0.2 GW $cm^{-2}$, the linear range absorption of graphene is observed, resulting in a nearly constant transmittance of 15.5%. Nonetheless, when the density exceeds 1 GW $cm^{-2}$, the saturable absorption considerably increases the transmittance. When the density is close to ~2.5 GW $cm^{-2}$, it saturates and produces ~24% transmittance. The strong pumping effect of GCM absorption can be easily used in all-optical modulation.

The transmission of the signal depends on the pump's strength. The modulator operates at 1 GHz and has a high modulation depth (~0.16 dB $\mu m^{-1}$) at a medium driving voltage (5 V) [43]. The symmetry in the band structure of graphene across the Dirac point gives graphene an advantage as optical loss in silicon photonics exists primarily across the Dirac point. Hence, it benefits under the circumstances of graphene of low-energy consumption and low insertion loss [34]. The p–n junction is formed by injecting electrons and holes into the graphene layer, and the optical loss of silicon can be minimized [44]. The advantage of this device is that graphene has unique linear band dispersion and symmetric density of states near the Dirac point.

The modulation depth of the device is optimized by numerically analyzing different waveguide widths. Graphene can only absorb 2.3% of the radiation energy because of the short interaction layer. This weak-light absorption of graphene could make it ineffective for active optoelectronic devices [12]. Graphene microcavity photodetectors are used to overcome this problem. The light field in the cavity is significantly increased, thus increasing the absorption [45]. The remarkable discovery of SLG leads to an extraordinary research interest due to its exceptional properties. In recent years, in-depth research, especially in layered 2D materials and graphene, has opened a broad research field. Despite the rapid development of 2D materials, devices based on these materials still face many challenges and opportunities. Small size sets enormous obstacles to the realization of low contact resistance for electronic devices. The metal–graphene contact resistance elucidates the role of graphene in electronic applications.

## 5.3 Carbonous Materials in Photovoltaics

Carbonous materials have been regarded as highly attractive owing to their various features, including low cost, eco-friendly, long-term stability, high work function, charge-carrier mobility, and controlled properties. [46–48]. Figure 5.1 shows different carbonous structures.

A variety of carbonous materials, namely fullerene, carbon nanotubes (CNTs), paracrystalline carbon, namely, "carbon black (CB)," graphene (GR), and carbon quantum dots (CQDs), are widely used in many components of photovoltaic cells, mainly in 3rd-generation solar cells. For example, various carbonous materials are employed in multiple layers and perovskite solar cells (PSCs) structures and thus achieve diverse paybacks, including improved performance and stability. Fullerene and its derivatives have improved open-circuit voltage ($V_{oc}$) by reducing charge recombination in the device when employed as an electron transport layer (ETL) in PSC. Similarly, utilization of CQD and graphene quantum dots (GQDs) are proven to improve thermal stability when employed as an additive in the ETL of PSCs. Moreover, as far as CNTs are concerned, they are more beneficial and convenient, not only in an ETL and hole transport layer (HTL) but also as a counter electrode and in the perovskite layer as an additive. In DSSCs, CB, CNTs, and graphene are frequently employed as counter electrode materials as an alternative to Pt electrodes. Also, the high surface area and charge mobility, as well as the good conductivity and transparency of these carbonous materials, make them an excellent choice as a TCO layer [50–53].

This section will detail the role and performances of carbonous materials in various solar cells.

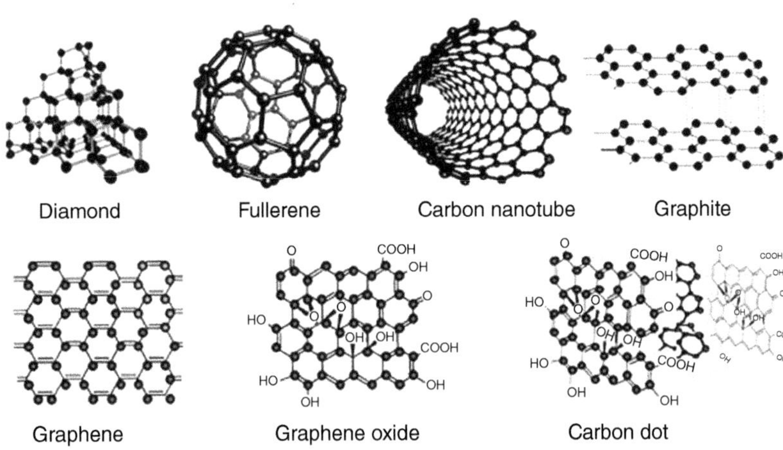


**Figure 5.1** Basic carbonous structures. Source: Yan et al. [49]/Public Domain CC BY 3.0.

## 5.4 Carbonous Materials in Dye-Sensitized Solar Cells

Dye-sensitized solar cells (DSSCs) were initially demonstrated by O'Regan and Grätzel in 1991. DSSC has been extensively researched due to its ease of fabrication, clean energy source, different colors, lowering cost, improving stability, and enhancing efficiency [54]. DSSC comprises a wide-bandgap semiconductor photoanode with a dye monolayer, a counter electrode, and an electrolyte solution inserted inside the sandwiched electrodes. When solar radiation falls on the surface of a DSSC, the absorption of photons leads to the excitation of sensitizer molecules. The sensitizer molecules inject electrons into the conduction band of the $TiO_2$, and these electrons will move to the FTO by diffusion. From there, the electrons reach the counter electrode via the external circuit. The sensitizing dye is oxidized by transferring the electron to the $TiO_2$. This oxidized dye becomes neutralized by getting an electron from the iodide ion present in the electrolyte. Upon oxidizing, the iodide ion is converted to a triiodide ion, which further moves toward the counter electrode by the diffusion mechanism. Hence, taking an electron from the counter electrode becomes the iodide ion, thus providing electrical energy to the external load [55] (Figure 5.2).

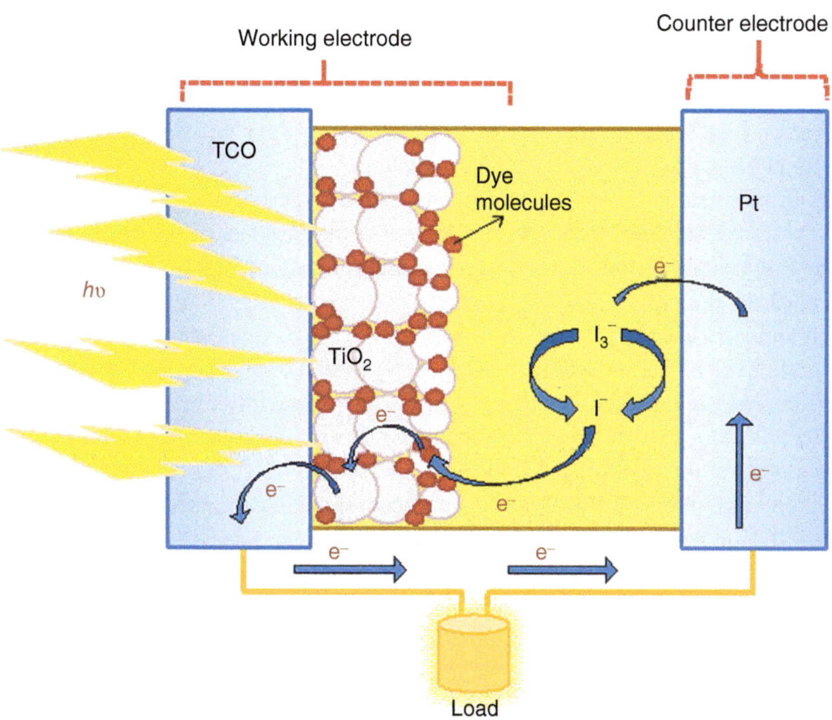


**Figure 5.2** Schematic representation of DSSC working. Source: Nazia Nasr.

Although platinum (Pt) has been the standard counter electrode until now, it works well by providing high catalytic activity in DSSCs. Still, the complexes it makes with the redox mediator and its high-cost challenge the stability of the DSSCs for commercialization purposes [56]. Similarly, the gold and silver in PSCs, used as back-electrodes BE, hinder the price and stability of PSCs. Consequently, substantial efforts have been accomplished for their alternates [57].

Till now, carbonous materials are an alternative for DSSCs and PSCs as a back electrode because of their high-charge mobility, photocatalytic ability, scalability, improved stability, and reduced price. In addition to this, the work function of carbon (5.0 eV) stands well to the work functions of gold (5.1 eV) and silver (4.7 eV). Grätzel and his coworker, for the first time in 1996, utilized carbon paste as a counter electrode in DSSC. This journey of the exploitation of carbonous materials in DSSC in different forms is continuing. A variety of atoms is incorporated into the carbon network to generate more active sites by perturbing its quantum spin and charge density by inserting heteroatoms. Thus, improving the catalytic ability of carbonous materials makes them more favorable for use in solar cells. Additionally, various metals (Fe, Co, Pt, etc.), metal oxides ($Fe_2O_3$, NiO, etc.), polymers, and metal alloys have been inserted into the carbon network to improve its catalytic behavior [58–65]. It has been reported that the use of carbon substrate provides additional oxygen vacancies for metal and metal oxides, thus improving their photocatalytic abilities.

CB and graphite possess a high surface-to-volume ratio, excellent conductivity, and electrocatalytic behavior, making them suitable for counter electrode material in DSSCs. Numerous reports are presented on the performance of CB counter electrodes alone. CB composites with diverse materials, such as polymer or transition metals, and their compounds are also up-to-date. High electrocatalytic activity and charge-transfer resistance are crucial for a better performance of the counter electrode. CB provides both of these features. These features are highly associated with the particle's dimensions and film thickness of CB, as reported by many researchers. With an increase in film thickness, a reduction in the charge-transfer resistance of the counter electrode is described. Studies showed that over 14 μm, the CB carries a large surface area, making the charge-transfer resistance lower and improving the device's PCE [66, 67]. The initial reports on CB as a counter electrode by Kim et al. also described the performance dependency of the device on the particle's dimensions and film thickness of CB. Keeping the particle's dimensions above 20 nm and below 100 nm while the film thicknesses up to 9 μm, they achieved high electrocatalytic and conductive counter electrodes [68]. Despite the high-specific-surface area and charge-carrier mobility, as well as optical transparency and excellent mechanical properties, pure graphene holds poor electrical conductivity as it possesses zero bandgaps. The bandgap of pure graphene has been altered; by modifying the surface of graphene and generating defects on its surface or by functionalization with heteroatoms. Such alteration in pure graphene opens its bandgap with an electrical conductivity as good as Pt [69–73].

Many researchers have exploited graphene nanoplatelets with catalytic edges in DSSCs as a counter electrode. They have demonstrated that catalytic edges, such as nanoplatelets, with low dimensions, have an equivalent electrocatalytic activity and device performance as that of Pt-based counter electrodes [74]. Like CB and graphene, CNTs with different sizes and dimensions have been utilized differently in DSSCs since their first demonstration by Suzuki et al. in 2003. These CNTs include SWCNTs, DWCNTs, TWCNTs, and MWCNTs [75–77]. The electrical properties of CNTs are subjected to the delocalized electrons and the curvature effects in the nanotubes. The sidewall imperfections, impurities, and misorientation of $\pi$-orbitals are responsible for instigating the catalytic activity in CNTs. Among all types of CNTS, SWCNTs possess low catalytic activities and electrical properties. However, through sidewall functionalization, their properties and solvent dispersibility could be significantly enhanced. The better catalytic activity in other CNTs, such as double or triple-walled CNTs, is that the outer walls defect and turn themselves into catalytic sites, and the inner walls behave as metallic electron conductors [78, 79]. Zhang et al. also demonstrated the role of different CNTs in DSSCs. They also reported better performance of DWCNTs in comparison. They suggested that the insertion of metallic NPs into the CNTs' surface could enhance the electrocatalytic behavior of CNTs [75]. More recently, NiS, $ZnFe_2O_3$, ZIF-8-derived N/transition metal, CoSe@NPC/CoSe@CNT, and N-doped C@$\gamma$-MoC/Ni nanoparticles have been analyzed as co-dopants with carbon as efficient CEs for DSSCs [80–85].

Besides other carbonous structures, CQD has been studied to be potentially exploited as an alternative to semiconductive sensitizers in solar cells. These quantum-confined nanoparticles have a variable bandgap, thus possessing absorption spectra (from near-IR to UV region) and high absorption coefficient, which can absorb light and generate an electron–hole pair. An exciting spectacle of CQDs is that they are electron sinkers and can reduce the electrons and holes recombination, thus increasing the electron's lifetime [50, 86]. As reported by Zhang et al., the functionalized CQDs efficiently attach to the semiconducting photoanode, such as $TiO_2$. Their bands are well adjustable for efficient electron transfer in the device. Their studies demonstrated an improvement in the PCE of the CQDs sensitized photoanode DSSCs. They accredited this improvement of PCE to the hopping effect of electrons at the CQDs' surface and fast charge separation, which increases the electron's lifetime [87]. A group of researchers also demonstrated using CQDs mixed with ZnO film-based photoanodes and N719 dye as a sensitizer in DSSCs and achieved an enhancement in the PCE by sevenfold. A reduction in charge-transfer resistance was observed due to CQDs incorporation in ZnO film. A unique perspective of biomass-generated CQDs is electricity generation in both light and dark. The CQDs produced from biomass have shown photoluminescent behavior. Such phosphors CQDs built photosensitive loading in the photoanode and became excited and brightened by green light in the dark. An efficiency of around 8% has been achieved from phosphors CQDs-based DSSCs [88, 89]. Similarly, a group of researchers has demonstrated the use of fluorescent phosphors with ruthenium and N-CQDs co-sensitization as an EDS layer in DSSC. They were able to increase the absorption range in the green wavelength. The efficient charge extraction by CQDs utilization was observed. They reported a PCE enhancement of more than 9% for their CQDs modified DSSC [90].

## 5.5 Carbonous Materials in Perovskite Solar Cells (PSCs)

A PSC typically consists of a perovskite layer sandwiched between an ETL and HTL. The ETL is a thin film comprising an electron-transporting material called an n-type semiconductor anode, usually $TiO_2$ or $ZnO_2$. In contrast, the HTL is based on the hole-transporting material and is responsible for transporting holes in the device.

In PSCs, an electron–hole pair is formed in the perovskite layer upon light absorption. Electrons are collected by the electron-transport material (ETM) from the perovskite layer and then further transported toward the anode (n-type). While holes are migrated toward the hole-transport material (HTM) from the perovskite layer, they move to the back electrode (p-type) [91]. Figure 5.3 shows different structures of PSCs.

The back-electrode material has an essential role in the PSCs. It is responsible for the extraction and collection of holes from the HTM. Suppose the material of the back electrode is morphologically good, as well as its electrical conductivity. In that case, it will make better contact with HTL and help reduce series resistances. Also, the work function of the back electrode should match well with all other layers in the device. In this way, the migration of holes will be smooth in the device, and hence the device will perform well. Ku et al. were the first to demonstrate carbon paste as a back electrode in an HTM-free PSC, with an insulating layer of $ZrO_2$ on the $TiO_2$ layer. They reported the use of morphologically suitable spheroidal graphite for improved conductivity and pore filling in the carbon composite electrode with an improvement in the PCE from 4.08% to 6.64%. Their devices not only showed an improvement in PCE but were also more stable. This discovery proved the constructive role of the carbon layer in the device [92, 93].

Till now, carbonous materials, including C60, graphite, graphene, and CNT, have been exploited in different PSC device structures as additives, charge-transport layers, front transparent electrodes, and/or back electrodes. At present, inverted planar heterojunction PSCs with ETL-containing C60 layers have a certified PCE of 20.9% [94]. The use of zero-dimensional carbon material was initiated by Jeng et al. By employing C60 as an ETL in PSCs, a recorded efficiency of 3.9% was achieved, followed by Sun et al. with a conversion efficiency of more than 7% using [6,6]-phenyl-C61-butyric acid methyl ester (PCBM) as an ETL [95, 96]. In an n–i–p PSC architecture (FTO/$TiO_2$/bis[$C_{60}$]PCBM:DMC/$CH_3NH_3PbI_3$/Spiro-OMeTAD/

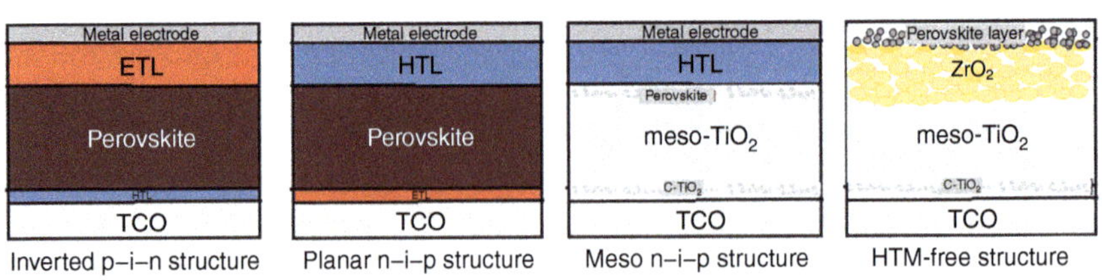


**Figure 5.3** Schematic representation of different perovskite solar cell structures. Source: Nazia Nasr.

$MoO_3$/Ag), the role of n-type doped PCBM were investigated by a group of researcher from Zhejiang University in 2017. They reported an efficient electron extraction through bandgap tailoring. Their carbonous-based film possesses efficient electron mobility and good conductivity, achieving an overall PSC efficiency of 20.14% with a DMC-doped bis-PCBM layer [97]. Likewise, the use of CQDs has become more pronounced in different PSC architectures. Kim et al. have recently reported the incorporation of CQDs in NiO band structure in p–i–n PSC. They proposed a different approach for engineering solution-processable metal oxide MO interlayers. With the use of CQDs, they altered the NiO band energy and work function, making it more aligned with the ITO and perovskite layer. Their modified PSCs showed an enhancement of more than 10% photoelectrochemical (PEC) and more stability than PSCs without incorporated CQDs, even under room conditions. They revealed a reduced charge recombination rate and hysteresis decline from 4.5% to below 1% in their modified devices [98]. Hao Li et al. in 2017 were able to achieve 19% PEC by incorporating CQD with $TiO_2$ in ETL in n–i–p heterojunction PSC. They demonstrated an enhancement in charge collection and mobility with well-aligned energy levels [99]. Most recently, CQDs have also been used as additives to the methylammonium iodide solution. CQDs were found to improve the crystallinity and particle size of the MAPBI3 film and help in the passivation of crystal defects and imperfections, which results in reducing recombination rate and improved efficiency and stability of the devices [100, 101]. Also, the use of GQDs in PSCs by many researchers has been reported. In this regard, Zhuo et al. synthesized ETL based on GQDs/$SnO_2$ composites with better charge mobility and band alignment with the perovskite and other layers in the device compared to the $SnO_2$-based ETL. They achieved a conversion efficiency of 19.6% for the flexible GQDs/$SnO_2$ composite ETL-based PSC with extraordinary mechanical strength and stability, preserving above 90% of the original PCE value after 500 bending cycles with a bending radius of 7 mm [102].

Low optical density, good transparency, and high conductivity make CNTs an alternative preference for PSCs. It was demonstrated initially by Zhen et al. in 2014 by applying layered CNTs by substituting metal counter electrodes in PSCs. Afterward, many researchers exploited its use in PSCs primarily as an HTL or back electrode. Li et al. demonstrated the use of both kinds of single-walled and double-walled CNTs with a dimension of up to 2-nm range. They utilize CNTs networks of both metallic and conductive nature. They were able to obtain the power conversion efficiency of PSCs by 9.9% with the addition of Spiro-OMeTAD [103–106]. Similarly, Xiaoli et al. utilize boron-doped MWCNTs instead of HTL and metal electrodes. The doping of CNTs helped to improve charge collection and extraction by bandgap alignment with the perovskite layer. They reported a stable PSC with about 14% power conversion efficiency even after 80 days of placement under room conditions. They credited the excellent stability of the device to the hydrophobic nature of the MWCNTs. Likewise, Saemon et al. utilized a CNTs/PEDOT:PSS hybrid hole collector for the efficiency and stability improvement of inverted planar PSCs. Their hybrid HTL showed only minor parasitic absorption loss, confirmed by the UV–Vis spectroscopy studies [107–109].

Aitola et al. used the SWCNTs and Spiro-OMeTAD mixture as an HTL in n–i–p PSC architecture for rapid extraction and hole collection. Meanwhile Batmunkh et al. employed the SWCNTs@$TiO_2$ nanofiber network as an ETL for electron collection and transportation [110, 111]. Likewise, Jeon et al. used SWCNTs and DWCNTs as transparent conductive electrodes in indium-free inverted planar p–i–n PSCs. They reported an increase in PCE by 25% with TFMS-doped DWCNTs inserted into inverted planar p–i–n structured PSCs as transparent conductive electrodes [112]. Besides this, the CNTs have been utilized as an interlayer in the PSC structure. In this regard, Siva et al. validated the reduction in charge recombination of PSCs by integrating an interlayer of SWCNTs to the interface between $TiO_2$ and perovskite for mesoscopic n–i–p structured PSCs. This interfacial modification endorsed an enhancement in PCE to more than 37% due to the improvement in short-circuit current density ($J_{sc}$) and open-circuit voltage ($V_{oc}$)values [113]. In short, all the above data showed that CNTs had been established to be adequate substitutes and supplements to PSC device architectures for extended stability and excellent performances. However, we encounter many challenges still in the way to exploiting its benefits in PSCs fully.

The high-charge mobility and zero bandgaps of graphene make it a reasonable substitute for the ITO and FTO substrates. Not only as an electrode, like other carbonous materials, but graphene has also been employed in various other layers of PSCs as an additive, including ETL and HTL [114–117]. For example, the GO and r-GO have been used as effective HTM with improved performance, replacing spiro-OMeTAD. Initially, GO was employed with PEDOT: PSS in PSCs. This approach was useful in improving the conductivity of PEDOT: PSS [118–120]. Similarly, Yoon et al. fabricated graphene/PEN-based flexible electrodes. They demonstrated that graphene/PEN-based electrodes are flexible and stable, improving PSCs' efficiency due to minimal charge interface recombination. Moreover, the graphene/PEN electrode-based PSCs showed a stable performance of their starting PCE with a minimal loss, even after 1000 bending cycles [121]. Similarly, Heo et al. utilized $AuCl_3$/graphene to improve the stacking of the graphene with a PET substrate using an interlayer of 3-aminopropyltriethoxysilane [122]. Their modified electrodes showed similar behavior as that of graphene/PEN-based electrodes; demonstrated by Yoon et al. In short, from the above discussion and evidence, it is our opinion that with the use of graphene and its derivatives, stable, efficient, and flexible PSCs is the future of the solar market.

Carbon pastes (graphite and CB) are primarily utilized in PSCs as back electrodes in two device configurations. One architecture is mesoporous PSCs with an insulating spacer layer (known as meso-PSC). In contrast, the other device configuration is a conventional HTM-free PSC. In this device configuration, the low-temperature processed carbon-film-containing electrode is employed on the perovskite film. For the mesoscopic PSC architecture, a PCE of more than 15% has been surpassed [123, 124]. While for the planar structure, the efficiency has been reached up to 17.46% based on Ni-doped $TiO_2$ [125]. Besides utilizing any single form of carbon, these carbonous materials have been utilized in mixed forms to take the combined benefits. For example, Ryu et al. used MWCNTs in chlorobenzene mixed with perovskite precursor solution on a carbon counter electrode. In this way, they witnessed that the CNTs were implanted in the perovskite film with an upward orientation. They demonstrated an improvement in the charge-transfer process. As a result, contact between perovskite and carbon interface was improved, thus improving overall device performance parameters [126]. One interesting study was performed by Mi et al. and Luo et al. They applied variable forms of carbon in different layers in a single device [127, 128]. Luo et al. demonstrated graphene/PEN as a front electrode and cross-stacked CNTs as a back electrode, showcasing the benefits of combining mechanical and thermal steadiness of flexible PSCs with an 11.9% conversion efficiency. The modified devices were incredibly stable and retained more than 85% PCE even after 2000 bending cycles. Zhuo et al. and Meng et al. demonstrated an "all-carbon" solar cell. Zhao et al. reported HTM-free all-carbon PSC by employing hexamethonium bromide (HMB)-doped C60 as an ETL. They reported an improvement in their devices' various parameters, including improved charge transportation, electron mobility, reduction in electron trap density, and extraordinary stability. While Meng et al. studies were centered on a comparison between utilizing $TiO_2$ and C60 as ETL in an FTO/ETL/$MAPbI_3$/carbon structure, accomplishing a remarkable conversion efficiency of 15.38% for C60 ETL-based PSC. They also substitute the traditional conductive electrode with the monolayer graphene-coated quartz, demonstrating high stability and performance [129, 130].

From the above discussion, though, the performance of carbonous materials in PSCs has been evidenced. However, progress is still needed to eradicate the limitations and flaws in its way to the commercial level. Many factors limit the production of carbon-based PSCs on a large scale. Among them are the exfoliation of graphene at larger intensity, organic solvents' requirement at greater volume, high temperature for the epitaxial growth, and film thickness control. Also, CNTs are comprised of graphene layers rolled up in a cylindrical shape. As a result, an elevated charge-transfer resistance is built up in them that needs to be minimized. Additionally, at the perovskite/carbon interface, the contacts are not perfect. Work needs to be done to overcome these limitations.

## References


1 Novoselov, K.S., Jiang, D., Schedin, F. et al. (2005). Two-dimensional atomic crystals. *Proceedings of the National Academy of Sciences of the United States of America* 102 (30): 10451.

2 Novoselov, K.S., Geim, A.K., Morozov, S.V. et al. (2005). Two-dimensional gas of massless Dirac fermions in graphene. *Nature* 438 (7065): 197–200.

3 Li, X., Wang, X., Zhang, L. et al. (2008). Chemically derived, ultrasmooth graphene nanoribbon semiconductors. *Science* 319 (5867): 1229.

4 Schwierz, F. (2013). Graphene transistors: status, prospects, and problems. *Proceedings of the IEEE* 101 (7): 1567–1584.

5 Ni, Z.H., Ponomarenko, L.A., Nair, R.R. et al. (2010). On resonant scatterers as a factor limiting carrier mobility in graphene. *Nano Letters* 10: 3868–3872.

6 Zhang, Y., Tan, Y.W., Stormer, H.L., and Kim, P. et al. (2005). Experimental observation of the quantum Hall effect and Berry's phase in graphene. *Nature* 438 (7065): 201–204.
7 Malec, C.E., Elkus, B., and Davidović, D. (2011). Vacuum-annealed Cu contacts for graphene electronics. *Solid State Communications* 151 (23): 1791–1793.
8 Giovannetti, G., Khomyakov, P.A., Brocks, G. et al. (2008). Doping graphene with metal contacts. *Physical Review Letters* 101: 026803.
9 Schwierz, F., Granzner, R., and Pezoldt, J. (2015). Two-dimensional materials and their prospects in transistor electronics. *Nanoscale* 7: 8261–8283.
10 Castro, E., Novoselov, K.S., Morozov, S.V. et al. (2007). Biased bilayer graphene: semiconductor with a gap tunable by the electric field effect. *Physical Review Letters* 99: 216802.
11 Novoselov, K.S., Geim, A.K., Morozov, S.V. et al. (2004). Electric field effect in atomically thin carbon films. *Science* 306 (5696): 666–669.
12 Nair, R.R., Blake, P., Grigorenko, A.N. et al. (2008). Fine structure constant defines visual transparency of graphene. *Science* 320 (5881): 1308.
13 Lee, C., Wei, X., Kysar, J.W., and Hone, J. et al. (2008). Measurement of the elastic properties and intrinsic strength of monolayer graphene. *Science* 321 (5887): 385–388.
14 Balandin, A.A., Ghosh, S., Bao, W. et al. (2008). Superior thermal conductivity of single-layer graphene. *Nano Letters* 8 (3): 902–907.
15 Los, J.H., Zakharchenko, K.V., Katsnelson, M.I., and Fasolino, A. et al. (2015). Melting temperature of graphene. *Physical Review B* 91 (4): 045415.
16 Schedin, F., Geim, A.K., Morozov, S.V. et al. (2007). Detection of individual gas molecules adsorbed on graphene. *Nature Materials* 6: 652–655.
17 Xia, F., Mueller, T., Lin, Y. et al. (2009). Ultrafast graphene photodetector. *Nature Nanotechnology* 4: 839–843.
18 Li, X.M., Zhu, H., Wang, K. et al. (2010). Graphene-on-silicon Schottky junction solar cells. *Advanced Materials* 22: 2743–2748.
19 Schwierz, F. (2010). Graphene transistors. *Nature Nanotechnology* 5: 487.
20 Bae, S.K., Kim, H., Lee, Y. et al. (2011). Roll-to-roll production of 30-in. graphene films for transparent electrodes. *Nature Nanotechnology* 5: 574–578.
21 Murali, R., Brenner, K., Yang, Y. et al. (2009). Resistivity of graphene nanoribbon interconnects. *IEEE Electron Device Letters* 30: 611–613.
22 Eda, G., Fanchini, G., and Chhowalla, M. (2008). Large-area ultrathin films of reduced graphene oxide as a transparent and flexible electronic material. *Nature Nanotechnology* 3: 270–274.
23 Chen, J.H., Jang, C., Adam, S. et al. (2008). Charged-impurity scattering in graphene. *Nature Physics* 4 (5): 377–381.
24 Cançado, L.G., Jorio, A., Ferreira, E.H.M. et al. (2011). Quantifying defects in graphene via Raman spectroscopy at different excitation energies. *Nano Letters* 11 (8): 3190–3196.
25 Meyer, J.C., Geim, A.K., Katsnelson, M.I. et al. (2007). The structure of suspended graphene sheets. *Nature* 446 (7131): 60–63.

26 Tan, Y.-W., Zhang, Y., Bolotin, K. et al. (2008). Measurement of scattering rate and minimum conductivity in graphene. *Physical Review Letters* 99: 246803.

27 Lee, E.J.H., Balasubramanian, K., Weitz, R.T. et al. (2008). Contact and edge effects in graphene devices. *Nature Nanotechnology* 3: 486.

28 Xia, F., Mueller, T., Golizadeh-Mojarad, R. et al. (2009). Photocurrent imaging and efficient photon detection in a graphene transistor. *Nano Letters* 9 (3): 1039–1044.

29 Park, J., Ahn, Y.H., and Ruiz-Vargas, C. (2009). Imaging of photocurrent generation and collection in single-layer graphene. *Nano Letters* 9 (5): 1742–1746.

30 Xu, X., Gabor, N.M., Alden, J.S. et al. (2010). Photo-thermoelectric effect at a graphene interface junction. *Nano Letters* 10 (2): 562–566.

31 Lemme, M.C., Koppens, F.H.L., Falk, A.L. et al. (2011). Gate-activated photoresponse in a graphene p–n junction. *Nano Letters* 11 (10): 4134–4137.

32 Li, Z.Q., Henriksen, E.A., Jiang, Z. et al. (2008). Dirac charge dynamics in graphene by infrared spectroscopy. *Nature Physics* 4 (7): 532–535.

33 Wang, F., Zhang, Y., Tian, C. et al. (2008). Gate-variable optical transitions in graphene. *Science* 320 (5873): 206.

34 Liu, M., Yin, X., Ulin-Avila, E. et al. (2011). A graphene-based broadband optical modulator. *Nature* 474 (7349): 64–67.

35 Wang, X., Cheng, Z., Xu, K. et al. (2013). High-responsivity graphene/silicon-heterostructure waveguide photodetectors. *Nature Photonics* 7 (11): 888–891.

36 Mueller, T., Xia, F., Freitag, M. et al. (2009). Role of contacts in graphene transistors: a scanning photocurrent study. *Physical Review B* 79 (24): 245430.

37 Gan, X., Shiue, R.J., Gao, Y. et al. (2013). Chip-integrated ultrafast graphene photodetector with high responsivity. *Nature Photonics* 7 (11): 883–887.

38 Yao, Y., Shankar, R., Rauter, P. et al. (2014). High-responsivity mid-infrared graphene detectors with antenna-enhanced photocarrier generation and collection. *Nano Letters* 14 (7): 3749–3754.

39 Pospischil, A., Humer, M., Furchi, M.M. et al. (2013). CMOS-compatible graphene photodetector covering all optical communication bands. *Nature Photonics* 7 (11): 892–896.

40 Youngblood, N., Anugrah, Y., Ma, R. et al. (2014). Multifunctional graphene optical modulator and photodetector integrated on silicon waveguides. *Nano Letters* 14 (5): 2741–2746.

41 Echtermeyer, T.J., Nene, P.S., Trushin, M. et al. (2014). Photothermoelectric and photoelectric contributions to light detection in metal–graphene–metal photodetectors. *Nano Letters* 14 (7): 3733–3742.

42 Li, W., Chen, B., Meng, C. et al. (2014). Ultrafast all-optical graphene modulator. *Nano Letters* 14 (2): 955–959.

43 Liu, M., Yin, X., and Zhang, X. (2012). Double-layer graphene optical modulator. *Nano Letters* 12 (3): 1482–1485.

44 Reed, G.T., Mashanovich, G., Gardes, F.Y., and Thomson, D.J. (2010). Silicon optical modulators. *Nature Photonics* 4 (8): 518–526.

**45** Furchi, M., Urich, A., Pospischil, A. et al. (2012). Microcavity-integrated graphene photodetector. *Nano Letters* 12 (6): 2773–2777.

**46** Almkhelfe, H., Li, X., Thapa, P. et al. (2018). Carbon nanotube-supported catalysts prepared by a modified photo-Fenton process for Fischer–Tropsch synthesis. *Journal of Catalysis* 361: 278–289.

**47** Jiang, P., Xiong, Y., Xu, M. et al. (2018). The influence of the work function of hybrid carbon electrodes on printable mesoscopic perovskite solar cells. *The Journal of Physical Chemistry C* 122 (29): 16481–16487.

**48** Sun, P.-P., Bai, L., Kripalani, D.R., and Zhou, K. et al. (2019). A new carbon phase with direct bandgap and high carrier mobility as electron transport material for perovskite solar cells. *Computational Materials* 5 (1): 1–7.

**49** Yan, Q.-L., Gozin, M., Zhao, F.Q. et al. (2016). Highly energetic compositions based on functionalized carbon nanomaterials. *Nanoscale* 8 (9): 4799–4851.

**50** Riaz, R., Ali, M., Maiyalagan, T. et al. (2019). Dye-sensitized solar cell (DSSC) coated with energy down shift layer of nitrogen-doped carbon quantum dots (N-CQDs) for enhanced current density and stability. *Applied Surface Science* 483: 425–431.

**51** Baker, J., Hooper, K., Meroni, S. et al. (2017). High throughput fabrication of mesoporous carbon perovskite solar cells. *Journal of Materials Chemistry A* 5 (35): 18643–18650.

**52** Yang, J., Tang, Q., Meng, Q. et al. (2017). Photoelectric conversion beyond sunny days: all-weather carbon quantum dot solar cells. *Journal of Materials Chemistry A* 5 (5): 2143–2150.

**53** Mugadza, K., Nyamori, V.O., Mola, G.T. et al. (2017). Low temperature synthesis of multiwalled carbon nanotubes and incorporation into an organic solar cell. *Journal of Experimental Nanoscience* 12 (1): 363–383.

**54** O'regan, B. and Grätzel, M. (1991). A low-cost, high-efficiency solar cell based on dye-sensitized colloidal $TiO_2$ films. *Nature* 353 (6346): 737.

**55** Nasr, N., Sayyad, M.H., Elbohy, H., and Qiao, Q. (2018). Efficiency enhancement in plasmonic dye-sensitized solar cell employing high performance $TiO_2$ photoanode doped with silver nanoparticles. *International Journal of Sustainable Energy and Environmental Research* 7 (2): 44–52.

**56** Denizalti, S., Ali, A.K., Ela, Ç., Ekmekci, M. et al. (2018). Dye-sensitized solar cells using ionic liquids as redox mediator. *Chemical Physics Letters* 691: 373–378.

**57** Rong, Y., Hu, Y., Mei, A. et al. (2018). Challenges for commercializing perovskite solar cells. *Science* 361 (6408): eaat8235.

**58** Jaafar, H., Ahmad, Z.A., and Ain, M.F. (2018). The use of carbon black-$TiO_2$ composite prepared using solid state method as counter electrode and E. conferta as sensitizer for dye-sensitized solar cell (DSSC) applications. *Optical Materials* 79: 366–371.

**59** Liao, Y., Tian, Y., Ma, X. et al. (2020). Screen-printed high-performance flexible electrothermal films based on three-dimensional intercalation graphene nanosheets/MWCNT/carbon black composite. *ACS Applied Materials & Interfaces* 12 (42): 48077–48083.

**60** Igman, E., Bayram, O., Mavi, A. et al. (2020). Photovoltaic performance of non-covalent functionalized single-layer graphene in dye-sensitized solar cells (DSSCs). *Journal of Materials Science* 56 (6): 4184–4196.

**61** Durairasan, M., Karthik, P.S., Balaji, J. et al. (2021). Design and fabrication of WSe2/CNTs hybrid network: a highly efficient and stable electrodes for dye sensitized solar cells (DSSCs). *Diamond and Related Materials* 111: 108174.

**62** Aseena, S., Abraham, N., and Babu, V.S. (2020). Synthesis of CNT based nanocomposites and their application as photoanode material for improved efficiency in DSSC. *Ceramics International* 46 (18): 28355–28362.

**63** Kilic, B., Simsek, E.B., Turkdogan, S. et al. (2020). Carbon nanofiber based CuO nanorod counter electrode for enhanced solar cell performance and adsorptive photocatalytic activity. *Journal of Nanoparticle Research* 22: 1–11.

**64** Murakami, T.N., Ito, S., Wang, Q. et al. (2006). Highly efficient dye-sensitized solar cells based on carbon black counter electrodes. *Journal of the Electrochemical Society* 153 (12): A2255.

**65** Khan, M.W., Zuo, X., Yang, Q. et al. (2020). Quantum dot embedded N-doped functionalized multiwall carbon nanotubes boost the short-circuit current of Ru(II) based dye-sensitized solar cells. *Nanoscale* 12 (2): 1046–1060.

**66** Jaafar, H., Ain, M.F., and Ahmad, Z.A. (2020). Performance of dye-sensitized solar cell (DSSC) using carbon black-TiO2 composite as counter electrode subjected to different annealing temperatures. *Optical and Quantum Electronics* 52 (4): 221.

**67** Solehudin, S., Diantoro, M., and Hidayat, A. (2020). Analysis of nanoporous semiconductor film thickness effect on short circuit current density of β-carotene dye based DSSC: theoretical and experimental approach. *JPSE (Journal of Physical Science and Engineering)* 4 (1): 1–7.

**68** Kim, H.-N. and Moon, J.H. (2012). Enhanced photovoltaic properties of $Nb_2O_5$-coated $TiO_2$ 3D ordered porous electrodes in dye-sensitized solar cells. *ACS Applied Materials & Interfaces* 4 (11): 5821–5825.

**69** Bhaumik, A., Haque, A., Taufique, M.F.N. et al. (2017). Reduced graphene oxide thin films with very large charge carrier mobility using pulsed laser deposition. *Journal of Materials Science and Engineering* 6 (4): 1–11.

**70** Wang, Y., Li, Z., Tang, C. et al. (2019). Few-layered mesoporous graphene for high-performance toluene adsorption and regeneration. *Environmental Science: Nano* 6 (10): 3113–3122.

**71** Wang, Y., Wang, H., Liu, F. et al. (2020). Flexible printed circuit board based on graphene/
polyimide composites with excellent thermal conductivity and sandwich structure. *Composites Part A: Applied Science and Manufacturing* 138: 106075.

**72** Young, R.J., Liu, M., Kinloch, I.A. et al. (2018). The mechanics of reinforcement of polymers by graphene nanoplatelets. *Composites Science and Technology* 154: 110–116.

**73** Ghany, N.A.A., Elsherif, S.A., and Handal, H.T. (2017). Revolution of graphene for different applications: state-of-the-art. *Surfaces and Interfaces* 9: 93–106.

**74** Yang, W., Xu, X., Gao, Y. et al. (2016). High-surface-area nanomesh graphene with enriched edge sites as efficient metal-free cathodes for dye-sensitized solar cells. *Nanoscale* 8 (26): 13059–13066.

**75** Suzuki, K., Yamaguchi, M., Kumagai, M., and Yanagida, S. (2003). Application of carbon nanotubes to counter electrodes of dye-sensitized solar cells. *Chemistry Letters* 32 (1): 28–29.

**76** Bumika, M., Mallick, M.K., Mohanty, S. et al. (2020). One-pot electrodeposition of polyaniline/SWCNT/ZnO film and its positive influence on photovoltaic performance as counter electrode material. *Materials Letters* 279: 128473.

**77** Mohammadnezhad, M., Selopal, G.S., Cavuslar, O. et al. (2021). Gold nanoparticle decorated carbon nanotube nanocomposite for dye-sensitized solar cell performance and stability enhancement. *Chemical Engineering Journal* 421: 127756.

**78** Saifuddin, N., Raziah, A.Z., and Junizah, A.R. (2013). Carbon nanotubes: a review on structure and their interaction with proteins. *Journal of Chemistry* 2013: 676815.

**79** Roch, A., Greifzu, M., Talens, E.R. et al. (2015). Ambient effects on the electrical conductivity of carbon nanotubes. *Carbon* 95: 347–353.

**80** Chen, X., Zhang, Y., Pang, Y. et al. (2021). NiS modified porous cotton-carbon fiber cloth with high conductivity and excellent electrocatalytic performance for DSSCs. *Materials Letters* 284: 128991.

**81** Yang, J., Zhang, C., Niu, Y. et al. (2021). N-doped C-CoS2@ CoS2/MoS2 nano polyhedrons with hierarchical yolk-shelled structures as bifunctional catalysts for enhanced photovoltaics and hydrogen evolution. *Chemical Engineering Journal* 409: 128293.

**82** Li, Z., Liu, L., Li, L. et al. (2020). In situ synthesis of ZnFe2O4 rough nanospheres on carbon nanofibers as an efficient titanium mesh substrate counter electrode for triiodide reduction in dye-sensitized solar cells. *Applied Surface Science* 148429.

**83** 袁双喜 and 云斯宁 (2020). ZIF-8 derived nitrogen/transition metal co-doped porous carbon for counter electrode of DSSCs. *Proceedings of 7th Symposium on Novel Solar Cell Materials Science and Technology.*

**84** Wang, T., Li, Y., Li, H. et al. (2020). Rational design of hierarchical structural CoSe@ NPC/CoSe@ CNT nanocomposites derived from metal–organic frameworks as a robust Pt-free electrocatalyst for dye-sensitized solar cells. *ACS Omega* 5 (40): 26253–26261.

**85** Zhang, X., Wang, T., Li, X. et al. (2020). Nitrogen-doped carbon encapsulating γ-MoC/Ni nanoparticles as efficient counter electrodes for dye-sensitized solar cells. *Journal of Coordination Chemistry* 73: 2621–2631.

**86** Zhu, W., Zhao, Y., Duan, J. et al. (2017). Carbon quantum dot tailored counter electrode for 7.01%-rear efficiency in a bifacial dye-sensitized solar cell. *Chemical Communications* 53 (71): 9894–9897.

**87** Zhang, Y.-Q., Ma, D.K., Zhang, Y.G. et al. (2013). N-doped carbon quantum dots for $TiO_2$-based photocatalysts and dye-sensitized solar cells. *Nano Energy* 2 (5): 545–552.

**88** Essner, J.B. and Baker, G.A. (2017). The emerging roles of carbon dots in solar photovoltaics: a critical review. *Environmental Science: Nano* 4 (6): 1216–1263.

**89** Zamiri, G. and Bagheri, S. (2018). Fabrication of green dye-sensitized solar cell based on ZnO nanoparticles as a photoanode and graphene quantum dots as a photo-sensitizer. *Journal of Colloid and Interface Science* 511: 318–324.

**90** Zhao, Y., Duan, J., He, B. et al. (2018). Improved charge extraction with N-doped carbon quantum dots in dye-sensitized solar cells. *Electrochimica Acta* 282: 255–262.

**91** Assadi, M.K., Bakhoda, S., Saidur, R., and Hanaei, H. (2018). Recent progress in perovskite solar cells. *Renewable and Sustainable Energy Reviews* 81: 2812–2822.

**92** Mei, A., Li, X., Liu, L. et al. (2014). A hole-conductor–free, fully printable mesoscopic perovskite solar cell with high stability. *Science* 345 (6194): 295–298.

**93** Ku, Z., Rong, Y., Xu, M. et al. (2013). Full printable processed mesoscopic $CH_3$ $NH_3PbI_3/TiO_2$ heterojunction solar cells with carbon counter electrode. *Scientific Reports* 3: 3132.

**94** Luo, D., Yang, W., Wang, Z. et al. (2018). Enhanced photovoltage for inverted planar heterojunction perovskite solar cells. *Science* 360 (6396): 1442–1446.

**95** Liu, D., Wang, Q., Traverse, C.J. et al. (2018). Impact of ultrathin C60 on perovskite photovoltaic devices. *ACS Nano* 12 (1): 876–883.

**96** Sun, S., Salim, T., Mathews, N. et al. (2014). The origin of high efficiency in low-temperature solution-processable bilayer organometal halide hybrid solar cells. *Energy & Environmental Science* 7 (1): 399–407.

**97** Yang, Z., Xie, J., Arivazhagan, V. et al. (2017). Efficient and highly light stable planar perovskite solar cells with graphene quantum dots doped PCBM electron transport layer. *Nano Energy* 40: 345–351.

**98** Kim, J.K., Nguyen, D.N., Lee, J.H. et al. (2020). Carbon quantum dot-incorporated nickel oxide for planar pin type perovskite solar cells with enhanced efficiency and stability. *Journal of Alloys and Compounds* 818: 152887.

**99** Li, H., Shi, W., Huang, W. et al. (2017). Carbon quantum dots/$TiO_x$ electron transport layer boosts efficiency of planar heterojunction perovskite solar cells to 19%. *Nano Letters* 17 (4): 2328–2335.

**100** Wen, Y., Zhu, G., and Shao, Y. (2020). Improving the power conversion efficiency of perovskite solar cells by adding carbon quantum dots. *Journal of Materials Science* 55 (7): 2937–2946.

**101** Ma, Y., Zhang, H., Zhang, Y. et al. (2018). Enhancing the performance of inverted perovskite solar cells via grain boundary passivation with carbon quantum dots. *ACS Applied Materials & Interfaces* 11 (3): 3044–3052.

**102** Zhou, Y., Yang, S., Yin, X. et al. (2019). Enhancing electron transport via graphene quantum dot/$SnO_2$ composites for efficient and durable flexible perovskite photovoltaics. *Journal of Materials Chemistry A* 7 (4): 1878–1888.

**103** Jeon, I., Xiang, R., Shawky, A. et al. (2019). Solar cells: single-walled carbon nanotubes in emerging solar cells: synthesis and electrode applications. *Advanced Energy Materials* 9 (23): 1970091.

**104** Li, Z., Kandel, H.R., Dervishi, E. et al. (2008). Comparative study on different carbon nanotube materials in terms of transparent conductive coatings. *Langmuir* 24 (6): 2655–2662.

**105** Shen, C., Brozena, A.H., and Wang, Y. (2011). Double-walled carbon nanotubes: challenges and opportunities. *Nanoscale* 3 (2): 503–518.

**106** Li, Z., Kulkarni, S.A., Boix, P.P. et al. (2014). Laminated carbon nanotube networks for metal electrode-free efficient perovskite solar cells. *ACS Nano* 8 (7): 6797–6804.

**107** Habisreutinger, S.N., Leijtens, T., Eperon, G.E. et al. (2014). Carbon nanotube/polymer composites as a highly stable hole collection layer in perovskite solar cells. *Nano Letters* 14 (10): 5561–5568.

**108** Zheng, X., Chen, H., Li, Q. et al. (2017). Boron doping of multiwalled carbon nanotubes significantly enhances hole extraction in carbon-based perovskite solar cells. *Nano Letters* 17 (4): 2496–2505.

**109** Yoon, S., Ha, T.-J., and Kang, D.-W. (2017). Improving the performance and reliability of inverted planar perovskite solar cells with a carbon nanotubes/PEDOT: PSS hybrid hole collector. *Nanoscale* 9 (27): 9754–9761.

**110** Aitola, K., Domanski, K., Correa–Baena, J.P. et al. (2017). High temperature-stable perovskite solar cell based on low-cost carbon nanotube hole contact. *Advanced Materials* 29 (17): 1606398.

**111** Batmunkh, M., Shearer, C.J., Bat-Erdene, M. et al. (2017). Single-walled carbon nanotubes enhance the efficiency and stability of mesoscopic perovskite solar cells. *ACS Applied Materials & Interfaces* 9 (23): 19945–19954.

**112** Jeon, I., Chiba, T., Delacou, C. et al. (2015). Single-walled carbon nanotube film as electrode in indium-free planar heterojunction perovskite solar cells: investigation of electron-blocking layers and dopants. *Nano Letters* 15 (10): 6665–6671.

**113** Siva, U., Murugathas, T., Yohi, S. et al. (2020). Single walled carbon nanotube incorporated titanium dioxide and Poly (3-hexylthiophene) as electron and hole transport materials for perovskite solar cells. *Materials Letters* 276: 128174.

**114** Guo, B., Fang, L., Zhang, B., and Gong, J.R. (2011). Graphene doping: a review. *Insciences Journal* 1 (2): 80–89.

**115** Agresti, A., Pescetelli, S., Taheri, B. et al. (2016). Graphene–perovskite solar cells exceed 18% efficiency: a stability study. *ChemSusChem* 9 (18): 2609–2619.

**116** You, P., Liu, Z., Tai, Q. et al. (2015). Efficient semitransparent perovskite solar cells with graphene electrodes. *Advanced Materials* 27 (24): 3632–3638.

**117** Gan, X., Yang, S., Zhang, J. et al. (2019). Graphite-N doped graphene quantum dots as semiconductor additive in perovskite solar cells. *ACS Applied Materials & Interfaces* 11 (41): 37796–37803.

**118** Yan, W., Ye, S., Li, Y. et al. (2016). Hole-transporting materials in inverted planar perovskite solar cells. *Advanced Energy Materials* 6 (17): 1600474.

**119** Palma, A.L., Cinà, L., Pescetelli, S. et al. (2016). Reduced graphene oxide as efficient and stable hole transporting material in mesoscopic perovskite solar cells. *Nano Energy* 22: 349–360.

**120** Suragtkhuu, S., Tserendavag, O., Vandandoo, U. et al. (2020). Efficiency and stability enhancement of perovskite solar cells using reduced graphene oxide derived from earth-abundant natural graphite. *RSC Advances* 10 (15): 9133–9139.

**121** Yoon, J., Sung, H., Lee, G. et al. (2017). Superflexible, high-efficiency perovskite solar cells utilizing graphene electrodes: towards future foldable power sources. *Energy & Environmental Science* 10 (1): 337–345.

**122** Heo, J.H., Shin, D.H., Jang, M.H. et al. (2017). Highly flexible, high-performance perovskite solar cells with adhesion promoted $AuCl_3$-doped graphene electrodes. *Journal of Materials Chemistry A* 5 (40): 21146–21152.

**123** Tsai, C.-M., Wu, G.W., Narra, S. et al. (2017). Control of preferred orientation with slow crystallization for carbon-based mesoscopic perovskite solar cells attaining efficiency 15%. *Journal of Materials Chemistry A* 5 (2): 739–747.

**124** Tian, C., Mei, A., Zhang, S. et al. (2018). Oxygen management in carbon electrode for high-performance printable perovskite solar cells. *Nano Energy* 53: 160–167.

**125** Liu, X., Liu, Z., Sun, B. et al. (2018). 17.46% efficient and highly stable carbon-based planar perovskite solar cells employing Ni-doped rutile $TiO_2$ as electron transport layer. *Nano Energy* 50: 201–211.

**126** Ryu, J., Lee, K., Yun, J. et al. (2017). Paintable carbon-based perovskite solar cells with engineered perovskite/carbon interface using carbon nanotubes dripping method. *Small* 13 (38): 1701225.

**127** Mi, L., Zhang, Y., Chen, T. et al. (2020). Carbon electrode engineering for high efficiency all-inorganic perovskite solar cells. *RSC Advances* 10 (21): 12298–12303.

**128** Luo, Q., Ma, H., Hou, Q. et al. (2018). All-carbon-electrode-based endurable flexible perovskite solar cells. *Advanced Functional Materials* 28 (11): 1706777.

**129** Meng, X., Zhou, J., Hou, J. et al. (2018). Versatility of carbon enables all carbon based perovskite solar cells to achieve high efficiency and high stability. *Advanced Materials* 30 (21): 1706975.

**130** Zhou, J., Hou, J., Tao, X. et al. (2019). Solution-processed electron transport layer of n-doped fullerene for efficient and stable all carbon based perovskite solar cells. *Journal of Materials Chemistry A* 7 (13): 7710–7716.

# 6

# Thermoelectric Energy Harvesters and Applications

*Ravi A. Kishore and Kerry Rippy*

*National Renewable Energy Laboratory, 15013 Denver West Parkway, Golden, CO 80401, USA*

## 6.1 Introduction

Thermoelectric generators, often referred to as TEGs, are currently one of the most reliable and advanced solid-state thermal-to-electric energy conversion technologies, providing promising clean energy solution for generating electricity from heat [1, 2]. TEGs, in principle, are based on the Seebeck effect, where two dissimilar thermoelectric (TE) materials joined at their ends generate a voltage difference in response to a temperature difference applied across the two ends of the materials. Figure 6.1a shows a TEG module comprising only one set of legs—thus called a thermocouple—illustrating the flow of charge carriers in the TE materials. TE materials, typically semiconductors, can either be p-type or n-type. The p-type materials contain an excess of positive charge carriers (holes), whereas the n-type materials possess an excess of negative charge carriers (electrons). Thus, when a temperature difference is applied across the p- and n-type TE materials, holes in p-type legs and electrons in n-type materials flow from the hot side to the cold side, causing an electric current, termed thermoelectricity.

TEGs are typically composed of several thermocouples, arranged electrically in series and thermally in parallel using copper electrodes, as shown in Figure 6.1b. This configuration allows induced voltage to increase in proportion to the number of thermocouples. TEGs are also integrated with thin ceramic substrates on both sides to electrically insulate the module while minimizing thermal resistance to the heat flow. Figure 6.1c depicts an actual TEG attached to a hot pipe, generating electricity using the heat losses from the hot pipe, which would otherwise just be wasted in the atmosphere [3]. Note that, unlike Rankine cycle-based power plants, TEGs require no working fluid, involve no chemical reactions, and produce no emissions [4, 5]. They are noiseless, lightweight, and scalable, making them ideal for power generation from distributed thermal energy sources [6, 7], such as waste heat [8] or natural heat [9]. Additionally, TEGs are completely solid-state systems with no moving parts; therefore, they require minimal maintenance and provide long-term and reliable operation. For instance, TE modules used in multi-mission

*Enhanced Carbon-Based Materials and Their Applications*, First Edition.
Edited by Poh Choon Ooi, Mengying Xie, and Chang Fu Dee.

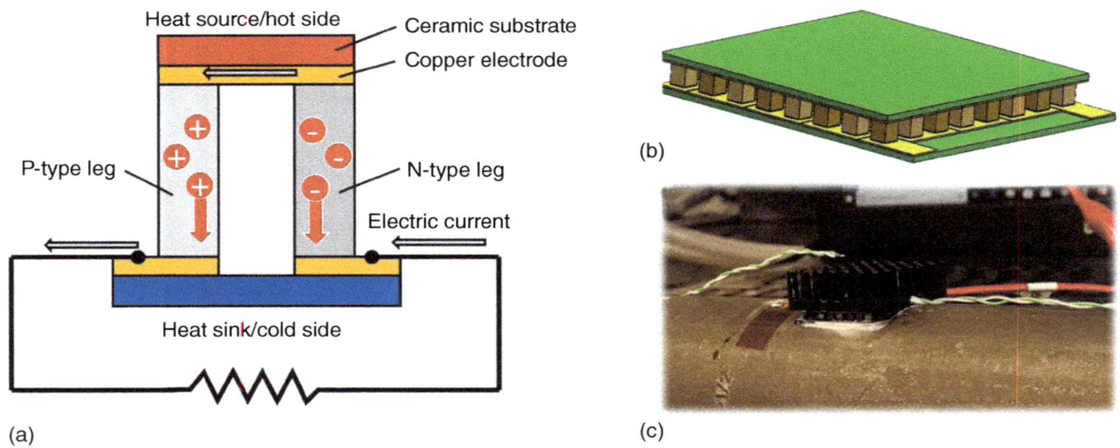


**Figure 6.1** Thermoelectric energy generator: (a) schematic illustrating charge flow in a thermocouple comprising a p-type and an n-type leg; (b) a complete thermoelectric module comprising several p- and n-type legs connected electrically in series and thermally in parallel; (c) a thermoelectric generator deployed on a hot-water pipe for low-grade waste heat recovery. Source: Adapted from Kishore et al. [3], with permission of American Chemical Society.

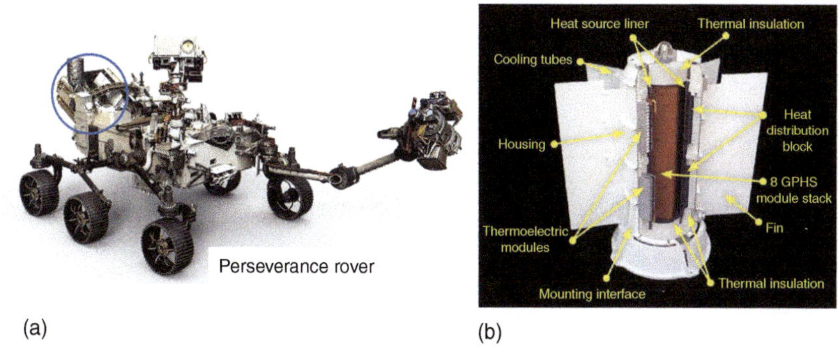


**Figure 6.2** Multi-mission radioisotope thermoelectric generator (MMRTG) used on Curiosity and Perseverance Mars rovers. (a) Perseverance Mars rovers (b) MMRTG. Source: Taken from NASA [10, 11]/National Aeronautics and Space Administration / Public Domain.

radioisotope generators for space exploration missions are designed for a minimum lifetime of 14 years while generating continuous power and ensuring a high degree of safety [10]. Figure 6.2 shows a multi-mission radioisotope thermoelectric generator (MMRTG) used on Curiosity and Perseverance Mars rovers. MMRTG generates electricity using the decay of Pu-238 to heat the hot side of the thermoelectric modules and the cold of space or a planetary atmosphere to produce a low temperature at the cold-side of the thermoelectric modules [10, 11].

In recent years, TEGs have been a key thermal-to-electrical conversion technology, gaining tremendous attention from researchers as well as an industry worldwide. According to a recent report, the global TEG market has been estimated to be about US$ 250 million in 2016, and it is projected to grow to over US$ 550 million by 2022 [12]. TEGs are used for a wide range of applications, including direct power generation from solar and geothermal heat [9, 13–15], waste heat recovery in automotive and aerospace industries [16–19], cogeneration in power plants [18, 20], and energy harvesting for self-powered sensors and wearables [21–24].

As the application areas of TEGs are growing, there is an increased emphasis on developing new TE materials that are high performing, cost-effective, and readily available. One such recent TEG application is power generation from low-grade heat sources, such as the human body for self-powered sensors and wearable health-monitoring devices. These applications present several new challenges in TEG design due to the small thermal gradient (<50 °C) and low-power generation (~mW) requirement. Additionally, TEGs for such applications need to be lightweight, compact, flexible/conformal, and cheap. While there are numerous TE materials developed for various applications, bismuth telluride ($Be_2Te_3$) and its related compounds are currently the most common TE material available for commercialization. Bismuth telluride-related compounds have relatively low earth abundance and are fabricated through highly complex vacuum processing routes [25]. Organic thermoelectric materials, especially conductive polymers, have been proposed as a promising alternative, particularly for low-temperature (<100 °C) energy-harvesting applications. Organic TE materials are nontoxic, flexible as they can be printed on thin foils, easy to manufacture, scalable to large areas, and potentially available at low cost [25]. In this chapter, we will discuss various aspects of the TEGs with a specific focus on organic TE materials and organic TEGs as power generation devices.

## 6.2 Thermoelectric Effect and Properties

The thermoelectric effect results from three different phenomena: the Seebeck effect, Peltier effect, and Thomson effect; each of these phenomena are described in Sections (6.2.1–6.2.3).

### 6.2.1 Seebeck Effect

Seebeck effect is described as the phenomenon that results in an induced voltage across an electrical conductor or semiconductor when a temperature difference is applied between its two ends (Figure 6.3). This induced voltage, often called Seebeck voltage, is directly proportional to the temperature difference and an intrinsic property of the material called the Seebeck coefficient.

Seebeck voltage ($V_p$) in the p-type material is described as [26, 27]:

$$V_p = S_p(T_h - T_c) \tag{6.1}$$

where $S_p$ is the Seebeck coefficient of the p-type TE material, and $T_h$ and $T_c$ are hot- and cold-side temperatures.

In the n-type semiconductor, the induced voltage ($V_n$) is given as:

$$V_n = -S_n(T_h - T_c) \tag{6.2}$$

where $S_n$ is the Seebeck coefficient of the n-type TE material.

The total voltage induced in a thermocouple, comprising a p-type and an n-type leg, can thus be expressed as:

$$V = V_p + V_n = (S_p - S_n)(T_h - T_c) \tag{6.3}$$

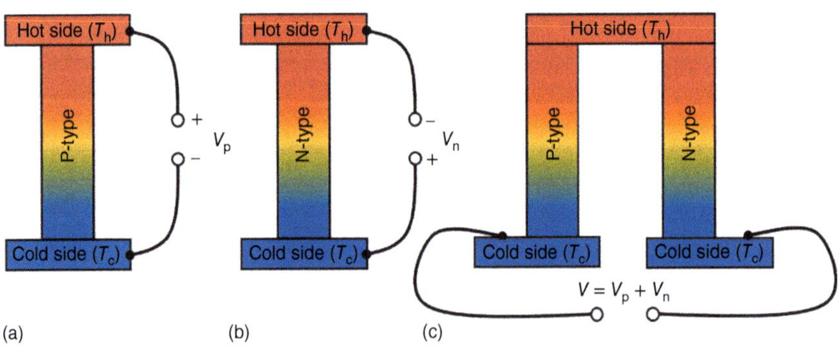


**Figure 6.3** Seebeck effect causes an induced voltage in response to the applied temperature difference across the two ends of TE materials: (a) p-type, (b) n-type, and (c) thermocouple.

In most cases, the Seebeck coefficient is not constant but varies with temperature; therefore, the induced voltage needs to be defined as a function of temperature:

$$V = \int_{T_c}^{T_h} (S_p(T) - S_n(T))\mathrm{d}T \quad (6.4)$$

Lastly, the open-circuit voltage ($V_{OC}$) induced in a thermoelectric module comprising "$n$" thermocouples is given as [28]:

$$V_{OC} = n \int_{T_c}^{T_h} (S_p(T) - S_n(T))\mathrm{d}T \quad (6.5)$$

### 6.2.2 Peltier Effect

The Peltier effect is defined as the reverse of the Seebeck effect. It results in cooling or heating at the ends of a TE material when an electric current is passed through it. As shown in Figure 6.4, the rate of Peltier heating or cooling in the p-type and n-type TE materials is given as:

$$Q_p = \pi_p I \quad (6.6)$$

$$Q_n = -\pi_n I \quad (6.7)$$

where $I$ denotes the electric current, and $\pi_p$ and $\pi_n$ are the Peltier coefficient of the p-type and n-type TE materials.

The rate of Peltier cooling in a thermocouple, comprising a p-type and an n-type leg, is expressed as:

$$Q = (\pi_p - \pi_n)I \quad (6.8)$$

Seebeck and Peltier coefficients of a TE material are found to be correlated using Kelvin relation, which is expressed as [27]:

$$\pi = ST \quad (6.9)$$

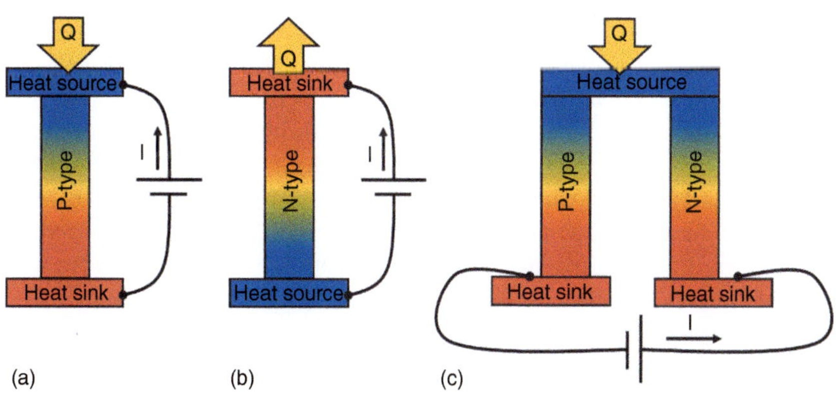


**Figure 6.4** Peltier effect causes cooling or heating in response to an applied electric current in TE materials: (a) p-type, (b) n-type, and (c) thermocouple.

Because Seebeck and Peltier coefficients are typically temperature dependent, Peltier heating or cooling in a TE module comprising n thermocouples is expressed as:

$$Q = nI \int_{T_c}^{T_h} (S_p(T) - S_n(T))dT \tag{6.10}$$

### 6.2.3 Thomson Effect

As described in Sections 6.2.1 and 6.2.2, for most TE materials, Seebeck and Peltier coefficients are temperature dependent; therefore, when a spatial temperature gradient exists in a TE material, and a current is passed through it, a continuous version of Seebeck and Peltier effects occur, which results in Thomson heat. The rate of Thomson heating/cooling per unit length of a TE leg is directly proportional to electric current and temperature gradient, which can be expressed as [29]:

$$\frac{dQ}{dx} = \tau I \frac{dT}{dx} \tag{6.11}$$

where $\tau$ is defined as the Thomson coefficient and $x$ is the spatial coordinate in the current direction.

Thomson coefficient is also related to Seebeck coefficient using Kelvin relation, and it is given as [30]:

$$\tau = T\frac{dS}{dT} \tag{6.12}$$

#### 6.2.3.1 Figure-of-Merit and Power Factor

The performance of thermoelectric material is normally expressed using two parameters: first, a dimensionless number called figure-of-merit, and second, a parameter called power factor.

The figure-of-merit ($zT$) and power factor (pf) are given as [31, 32]:

$$zT = \frac{S^2}{\rho k}T \tag{6.13}$$

$$\text{pf} = \frac{S^2}{\rho} \tag{6.14}$$

where $S$ is the Seebeck coefficient, $\rho$ is electrical resistivity, $k$ is thermal conductivity, and $T$ is the temperature in Kelvin.

Higher $zT$ value of TE materials typically implies higher energy conversion efficiency, whereas higher pf of TE materials suggests higher electric power generation by a TEG. To maximize TEG performance, TE materials should have a large Seebeck coefficient but small electrical resistivity and thermal conductivity. However, these transport parameters are not independent and depend on various factors, such as band structure and carrier concentration. Increasing $zT$ for thermoelectric materials much beyond 1.0 has been a key research challenge for several decades.

Figures 6.5 and 6.6 depict the temperature-dependent material properties of p-type and n-type bismuth telluride, which is one of the most common TE materials used for commercial TEGs. Note that $zT$ of bismuth telluride is high (~1.2 for p-type

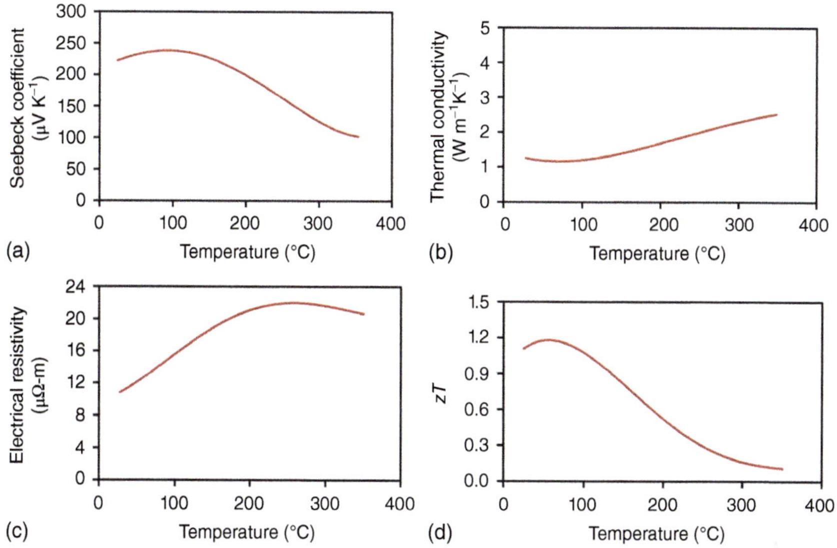


**Figure 6.5** Material properties of commercial p-type bismuth telluride. (a) Seebeck coefficient, (b) Thermal conductivity, (c) Electrical resistivity, and (d) $zT$. Source: Data are taken from Hu et al. [33].

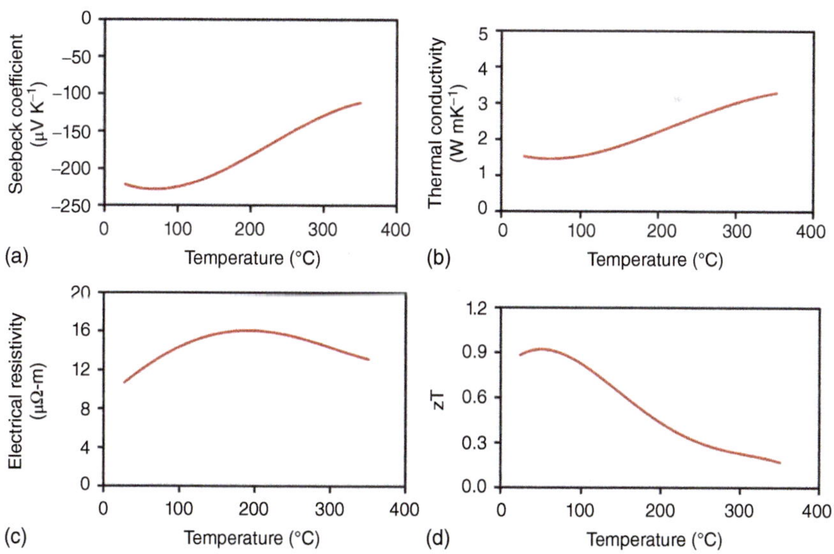


**Figure 6.6** Material properties of commercial n-type bismuth telluride. (a) Seebeck coefficient, (b) Thermal conductivity, (c) Electrical resistivity, and (d) $zT$. Source: Data are taken from Hu et al. [33].

and ~1.0 for n-type) at temperatures below 100 °C; however, it reduces dramatically when temperature increases. Bismuth telluride, therefore, is primarily used to fabricate TEGs for low-temperature applications. Section 6.4 will discuss other material options for mid- and high-temperature applications.

## 6.3 Thermoelectric Power and Efficiency

TEG power and efficiency can either be calculated using a simplified one-dimensional model or a complex three-dimensional model. The simplified one-dimensional analysis is typically performed using a decoupled model where thermal and electrical physics are solved separately. A one-dimensional model is easy to implement and provides fairly accurate results (error up to 10% [34, 35]) when the thermal gradient is small, material properties are temperature independent, and contact resistances are small [36]. However, for more robust analysis, a three-dimensional model is required to account for temperature-dependent material properties and various kinds of losses. The coupled three-dimensional model, however, requires numerical methods to obtain the solutions.

### 6.3.1 Simplified One-Dimensional Decoupled Model

For the simplified analysis, we assume that TEG is in steady-state condition, and the heat losses to the environment are negligible. Also, material properties are assumed to be temperature independent. Because the TE material properties typically vary with temperature, effective material properties can be obtained using the equations below [37]:

$$S_{\mathrm{eff}} = \frac{1}{T_{\mathrm{h}} - T_{\mathrm{c}}} \int_{T_{\mathrm{c}}}^{T_{\mathrm{h}}} S(T)\,\mathrm{d}T \tag{6.15}$$

$$k_{\mathrm{eff}} = \frac{1}{T_{\mathrm{h}} - T_{\mathrm{c}}} \int_{T_{\mathrm{c}}}^{T_{\mathrm{h}}} k(T)\,\mathrm{d}T \tag{6.16}$$

$$\rho_{\mathrm{eff}} = \frac{1}{T_{\mathrm{h}} - T_{\mathrm{c}}} \int_{T_{\mathrm{c}}}^{T_{\mathrm{h}}} \rho(T)\,\mathrm{d}T \tag{6.17}$$

Note that because the TE properties are assumed to be constant, the Thomson coefficient becomes zero. In other words, the Thomson effect is automatically ignored.

Figure 6.7 depicts a one-dimensional thermal circuit showing the heat flow from a heat source to a heat sink through a TEG [38]. Energy balance equations at the cold and hot sides of the TEG can be given as [35]:

$$Q_{\mathrm{h}} + \frac{Q_{\mathrm{j}}}{2} - Q_{\mathrm{ph}} - Q_{\mathrm{TEG}} = 0 \tag{6.18}$$

$$-Q_{\mathrm{c}} + \frac{Q_{\mathrm{j}}}{2} + Q_{\mathrm{pc}} + Q_{\mathrm{TEG}} = 0 \tag{6.19}$$

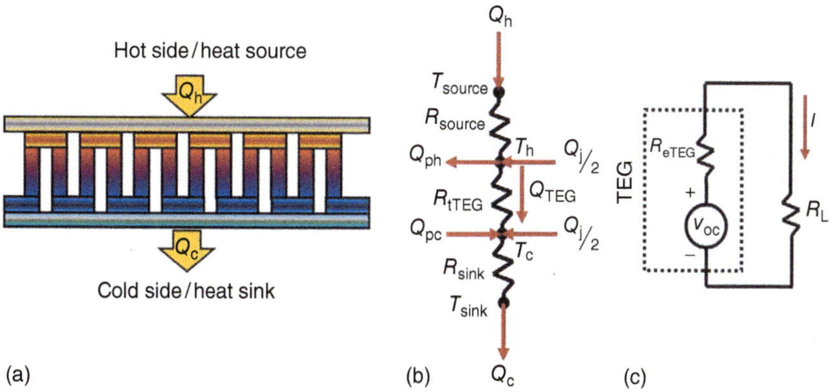


**Figure 6.7** Simplified one-dimensional energy equilibrium model for TEG. (a) A TEG is placed between a heat source and a heat sink. Heat flows from the heat source to the heat sink along with charge carriers resulting in an electrical current. (b) The one-dimensional thermal circuit illustrates the heat flow from a heat source to a heat sink via TEG. (c) The electrical circuit illustrates TEG, Seebeck voltage $V_{OC}$, and thermoelectric current I through the electrical load $R_L$. Source: Adapted from Kishore et al. [3].

where $Q_h$, $Q_c$, and $Q_{TEG}$ are the heat flow related to the heat source, heat sink, and TEG because of the internal temperature difference. $Q_{ph}$ is Peltier cooling generated at the hot side of TEG, whereas $Q_{pc}$ is Peltier heating generated at the cold side of TEG. Lastly, $Q_j$ is the Joule heating, which can be assumed to be distributed equally between the nodes at hot and cold sides.

The heat flow terms can be mathematically expressed as [39]:

$$Q_h = \frac{T_{source} - T_h}{R_{source}} \tag{6.20}$$

$$Q_c = \frac{T_c - T_{sink}}{R_{sink}} \tag{6.21}$$

$$Q_{TEG} = \frac{T_h - T_c}{R_{tTEG}} \tag{6.22}$$

$$Q_{pc} - nST_cI \tag{6.23}$$

$$Q_{ph} = nST_hI \tag{6.24}$$

$$Q_j = I^2R_{eTEC}, \tag{6.25}$$

where $T_{source}$ and $T_{sink}$ are the temperatures of the heat source and heat sink, $T_h$ and $T_c$ are the temperatures of the hot and cold sides of the TEG, $R_{source}$ and $R_{sink}$ are the thermal resistances of the heat source and sink, $n$ is the total number of pairs of p- and n-type legs, and $S$ is the net Seebeck coefficient of a uni-thermocouple given as:

$$S = S_p - S_n \tag{6.26}$$

where $S_p$ and $S_n$ are the effective Seebeck coefficient of p- and n-type legs, respectively.

The thermal and electrical resistances of the TEG are expressed as [40]:

$$R_{tTEG} = \frac{1}{n}\frac{x}{a}\frac{1}{(\kappa_p + \kappa_p)} \tag{6.27}$$

$$R_{eTEG} = n\frac{x}{a}(\rho_p + \rho_n) \tag{6.28}$$

where $\kappa_p$ and $\kappa_n$ are the thermal conductivity of p- and n-type legs, $\rho_p$ and $\rho_n$ are the electrical resistivity of p- and n-type legs, $x$ is the leg height, and $a$ is the cross-sectional area of the legs.

As shown in Figure 6.7c, if $I$ is the current in the electrical circuit containing the electrical load $R_L$, the heat inflow into TEG can be expressed as:

$$Q_h = \frac{T_h - T_c}{R_{tTEC}} + nST_hI - \frac{I^2R_{eTEG}}{2} \tag{6.29}$$

The open-circuit Seebeck voltage $V_{OC}$ is given as [41]:

$$V_{OC} = nS(T_h - T_c) \tag{6.30}$$

The current $I$ in the electrical circuit can be determined using:

$$I = \frac{V_{OC}}{R_{eTEG} + R_L} = \frac{nS(T_h - T_c)}{R_{eTEG} + R_L} \tag{6.31}$$

The voltage across the resistive load is given as:

$$V_L = IR_L = \frac{nS(T_h - T_c)}{R_{eTEG} + R_L}R_L \tag{6.32}$$

The output electrical power $P_{out}$ by the TEG is given by [42]:

$$P_{out} = I^2R_L = \left(\frac{nS(T_h - T_c)}{R_{eTEG} + R_L}\right)^2 R_L \tag{6.33}$$

Using differential calculus, it can be shown that $P_{out}$ is maximum when $R_{eTEG} = R_L$; therefore,

$$P_{max} = \frac{(nS(T_h - T_c))^2}{4R_{eTEG}} \tag{6.34}$$

The efficiency, $\eta$, of TEG is measured using relations:

$$\eta = \frac{P_{out}}{Q_h} \tag{6.35}$$

Using Eqs. (6.23) and (6.33), $\eta$ can be expressed as [40]:

$$\eta = \eta_c \frac{\beta}{(1+\beta) + (1+\beta)^2(ZT_h)^{-1} - \eta_c/2} \tag{6.36}$$

where Carnot efficiency $\eta_c = \frac{(T_h - T_c)}{T_h}$, $\beta = \frac{R_L}{R_{eTEG}}$, and $Z = \frac{S^2}{R_{eTEG}} R_{tTEG}$.

η is theoretically maximum when:

$$\beta = \sqrt{1 + Z\overline{T}}, \text{or } R_L = R_{eTEG}\sqrt{1 + Z\overline{T}} \tag{6.37}$$

where $\overline{T} = {}^{(T_h + T_c)}/_2$.

The maximum theoretical efficiency of a TEG can be expressed as:

$$\eta_{max} = \frac{(T_h - T_c)}{T_h}\left[\frac{\sqrt{1 + Z\overline{T}} - 1}{\sqrt{1 + Z\overline{T}} + \frac{T_c}{T_h}}\right] \tag{6.38}$$

$Z\overline{T}$ is termed as device figure-of-merit. This is different from the figure-of-merit expressed as $zT$.

TEG power at the optimal load, $R_L = R_{eTEG}\sqrt{1 + Z\overline{T}}$, can be expressed as:

$$P_{opt} = \frac{(nS(T_h - T_c))^2}{R_{eTEG}} \frac{\sqrt{1 + Z\overline{T}}}{(1 + \sqrt{1 + Z\overline{T}})^2} \tag{6.39}$$

### 6.3.2 Three-Dimensional Coupled Multiphysics Model

For the three-dimensional analysis of TEGs, we can obtain equations of thermoelectricity by applying the principle of conservation of energy and continuity of electric charge, which is given as [1]:

$$\nabla(\kappa\nabla T) + \rho\mathbf{J}^2 - T\mathbf{J}.\left[\left(\frac{\partial S}{\partial T}\right)\nabla T + (\nabla S)_T\right] = 0 \tag{6.40}$$

$$\nabla.\mathbf{J} = 0 \tag{6.41}$$

where, vector **J** is current density in three dimensions, which is given as [1]:

$$\mathbf{J} = -\frac{1}{\rho}(\nabla V + S\nabla T), \tag{6.42}$$

where $V$ is the electrostatic potential, $T$ is the absolute temperature, and $\kappa$, $\rho$, and $S$ denote the temperature-dependent thermal conductivity, electrical resistivity, and Seebeck coefficient of the TE materials, respectively.

The terms on the left-hand side in Eq. (6.40) represent conduction heat-transfer rate, Joule heat-generation rate, Thomson heat, and Peltier heat, respectively. Equations (6.40), (6.41), and (6.42) are coupled and thus need to be solved numerically.

## 6.4 Thermoelectric Materials

### 6.4.1 Inorganic Thermoelectric Materials

While there are various families of TE materials, they are broadly divided into three categories based on their operating temperature range: low temperature (<200 °C), medium temperature (200–600 °C), and high temperature (>600 °C). Some of these materials are described below.

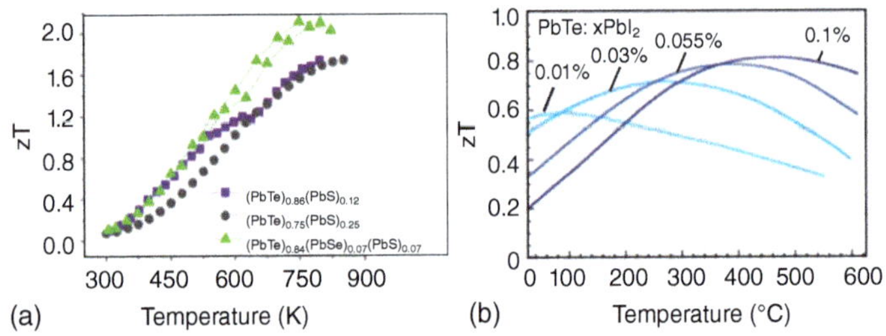


**Figure 6.8** Figure-of-merit, $zT$ of (a) p-type. Source: Reproduced from Korkosz et al. [50]/American Chemical Society, and (b) n-type. Source: Reproduced from Snyder and Toberer [7]/Springer Nature Limited, PbTe alloys.

**Low-temperature TE materials (<200 °C)**: For applications with operating temperatures less than 200 °C, bismuth telluride ($Bi_2Te_3$), antimony telluride ($Sb_2Te_3$), and their alloys are widely used. The most common TE material, bismuth telluride, is typically alloyed with antimony or selenium to optimize its $zT$ by altering the carrier concentration through material composition in p-type alloy to form $Bi_{2-x}Sb_xTe_3$ and n-type alloy to form $Bi_2Te_{3-y}Se_y$. p-type bismuth telluride with compositions near $(Sb_{0.8}Bi_{0.2})_2Te_3$ provides the highest $zT$ of ~1.1, and n-type bismuth telluride with compositions close to $Bi_2(Te_{0.8}Se_{0.2})_3$ provides the highest $zT$ of ~0.8 [7]. Figures 6.5 and 6.6 depict $zT$ of p-type and n-type bismuth telluride alloys.

**Medium-Temperature TE materials (200–600 °C)**: Medium-temperature TE materials typically include group-IV tellurides, such as PbTe [43], GeTe [44–46], and SnTe [47–49]. Within this group, PbTe and its alloys are the most widely used TE material in this temperature range. Studies have reported $zT$ of 0.8–1.8 for p-type and around 1.0 for n-type PbTe alloys. Figure 6.8 shows a few PbTe alloys with appropriate dopants tuned for enhanced $zT$ for mid-temperature applications [50]. A few other mid-temperature TE materials are skutterudite alloys of $CoSb_3$ and $CeFe_4Sb_{12}$ with maximum $zT \sim 0.8$, TAGS $((AgSbTe_2)_{0.15}(GeTe)_{0.85})$ with peak $zT \sim 1.2$ [51], $Zn_4Sb_3$ with peak $zT \sim 1.3$ [52], $Ag(Pb_{1-y}Sn_y)_mSbTe_{2+m}$ series exhibiting peak $zT \sim 1.45$ [53], and $AgPb_mSbTe_{2+m}$ series with peak $zT \sim 2.2$ [54].

**High-Temperature TE Materials (>600 °C)**: There are not many TE materials for high-temperature applications. Silicon–germanium alloys (SiGe) are mostly used for temperatures above 600 °C. Due to high lattice thermal conductivity, the $zT$ value of these TE materials is less than unity.

Figure 6.9 depicts $zT$ plots of some additional thermoelectric materials for applications in a wide temperature range between 0 and 1000 °C.

### 6.4.2 Organic Thermoelectric Materials

Organic TE materials are ideally suited for low-temperature applications, such as room temperature cooling and power generation [25, 55]. They can be efficiently and cost-effectively synthesized on a large scale [56]. In some cases, organic TE materials

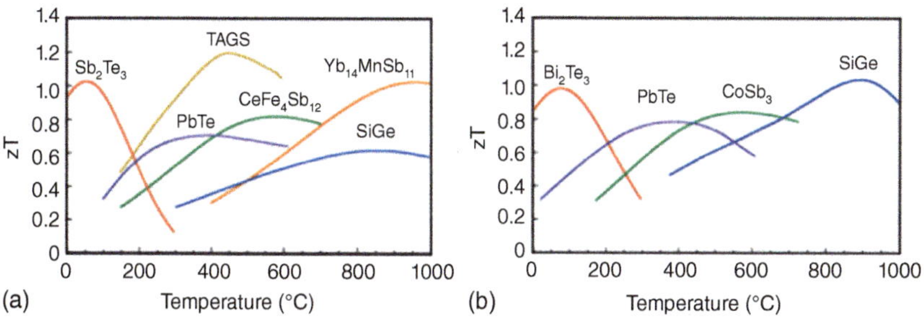


**Figure 6.9** Some of the widely used TE materials with their operating temperature and figure-of-merit, $zT$. (a) P-type TE materials, (b) N-type TE materials. Source: Reproduced from Snyder and Toberer [7]/Springer Nature Limited.

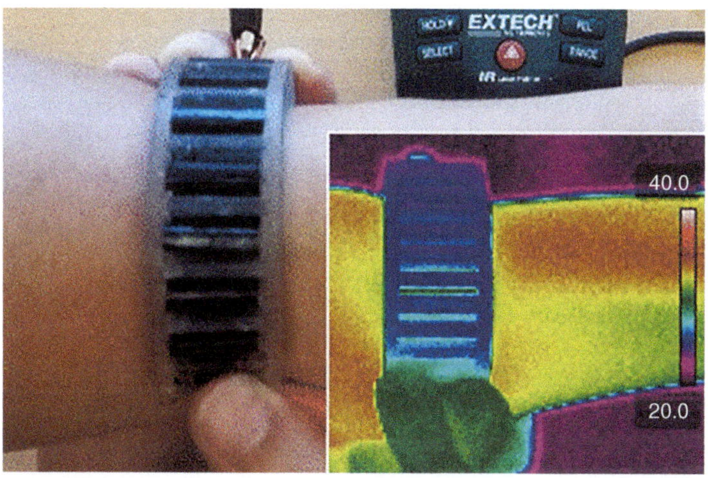


**Figure 6.10** A wearable organic TE generator based on carbon nanotubes could be used to power a wearable electronic device. Source: Choi et al. [57], with permission of American Chemical Society.

offer advantages over inorganic TE materials. For example, organic TE materials are generally lightweight, flexible, nontoxic, and often low cost [56]. This makes them ideal for wearable and portable devices. For example, a wearable TEG is shown in Figure 6.10 [57].

One advantage of organic TE materials is that, in general, they demonstrate low thermal conductivity [56]. It must be noted that low thermal conductivity is associated with disordered materials, and the influence of morphology on thermal conductivity in organic TE materials is an active area of investigation, which will be discussed further below. However, in general, the thermal conductivity for organic TE materials is lower than for inorganic TE materials.

Despite their low thermal conductivity, the figure of merit ($zT$, Eq. (6.13)) and power factor (pf, Eq. (6.14) for organic TE materials are lower than those of inorganic materials. It remains difficult to obtain organic TE materials with a $zT$

over 0.5, mainly due to their relatively high electrical resistivity and low Seebeck coefficient. Correlations among the Seebeck coefficient, thermal conductivity, and electrical resistivity make improvements challenging. Nevertheless, significant strides have been made in the last decade using methods, such as nanostructure formation, solvent treatment, and electrochemical doping [56]. A selection of high-performance TE materials is given below, and fundamental properties that can be leveraged to improve organic TE material performance are discussed.

Most high-performance organic thermoelectric materials that have been identified to date are polymers. A few notable exceptions are small molecules, including pentacene, C60, and thiophene derivatives, with Seebeck coefficients in the range of $mV\,K^{-1}$ [58]. However, conjugated polymers are generally better candidates for TE materials due to the delocalized π-bonds along with the conjugated backbone, which lead to favorable organic semiconductor characteristics [59]. Selected high-performing p-type organic polymer TE materials are depicted in Table 6.1, and selected n-type organic polymers are depicted in Table 6.2.

As reflected in these tables, there are currently more high-performance p-type TE materials available than organic n-type TE materials, and the properties of the organic p-type TE materials are generally more favorable than the properties of organic n-type TE materials. The poorer performance of n-type TE materials relative to p-type TE materials matches the general trend among organic semiconductors for all applications [70]. Due to the relatively small electron affinity of many organic materials, it is challenging to create a chemically stable negatively charged (n-type) organic molecule. Often, reactions to the environment (for example, water or oxygen) will occur, degrading the material. Stable n-type materials must have high electron affinities (as high as ~4 eV by some estimates) to avoid this degradation. Therefore, stable n-type materials are based on units with large electron affinity [71].

For both p- and n-type TE materials, charge carrier mobility and transport remain a significant hurdle that must be overcome to optimize carrier concentration for high-performance organic TE materials [56]. In inorganic TE materials, charge carriers move readily along valence and conduction bands. In organic TE materials, the frontier molecular orbitals are analogous to these bands; the highest-occupied molecular orbital (HOMO) behaves like the valence band, and the lowest-unoccupied molecular orbital (LUMO) behaves like the conduction band. However, charge transport through the HOMO and LUMO of organic materials is generally orders of magnitude lower than charge transport through valence and conduction bands of inorganic materials. While the exact mechanism of charge transport in organic semiconductors, including organic TE materials, remains controversial, there is evidence that it proceeds via hopping and quantum tunneling [72]. Thus, charge carrier mobility within disordered organic TE materials relates to the electronic coupling between molecules, interchain transport in polymers within crystalline domains, and the ability of these charge carriers to cross-domain boundaries. Therefore, optimization of parameters, such as crystalline packing, has been demonstrated as a route to improving organic TE materials [73].

**Table 6.1** Selected high-performance p-type organic TE materials and their Seebeck coefficient (*S*), electrical conductivity ($\sigma$), and power factor (pf).

| Selected p-type organic TE materials | | $S$ (μV K$^{-1}$) | $\sigma$ (S cm$^{-1}$) | pf | References |
|---|---|---|---|---|---|
| PEDOT-tosylate | 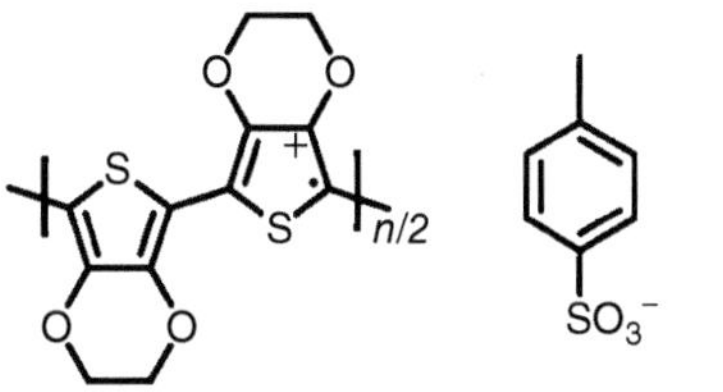 | 117 | 923 | 1270 | [61] |
| PEDOT:PSS | 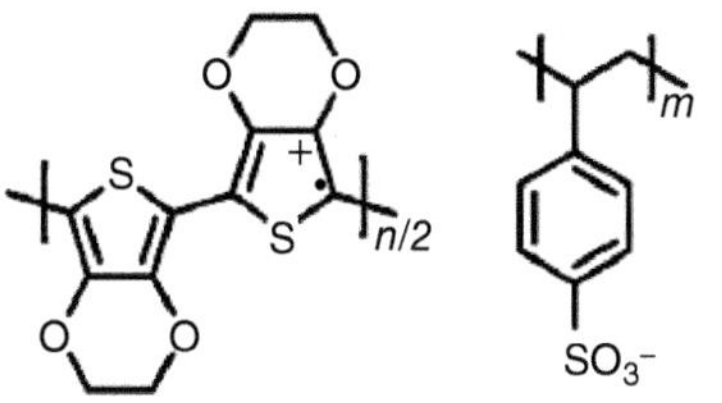 | 72.6 | 890 | 469 | [62] |
| P3HT | 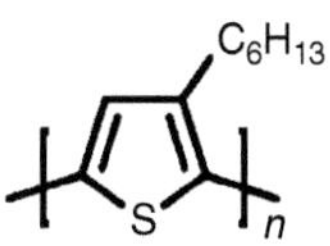 | 39.5 | 2 | 31 | [63] |
| P3HT:TFSI$^-$ | 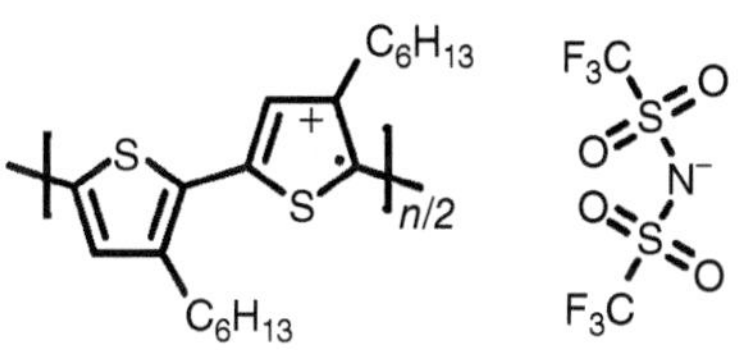 | 51 | 92 | 23.9 | [63] |
| PDPP3T | 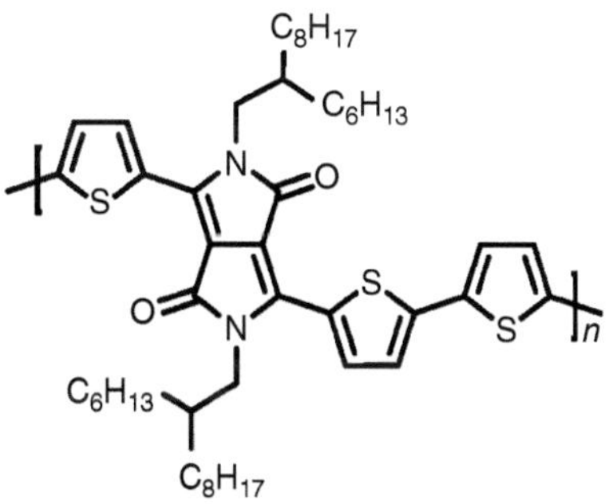 | 70 | 0.5 | 25 | [64] |
| PANI:CSA | 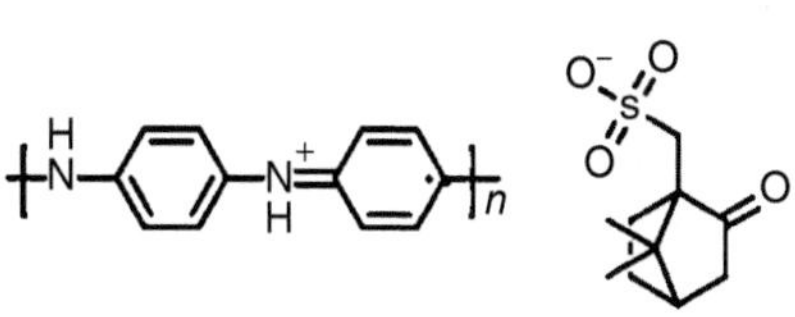 | 34 | 169 | 19 | [65] |

**Table 6.1** (Continued)

| Selected p-type organic TE materials | | $S$ ($\mu V\,K^{-1}$) | $\sigma$ ($S\,cm^{-1}$) | pf | References |
|---|---|---|---|---|---|
| PBTTT-C12 | 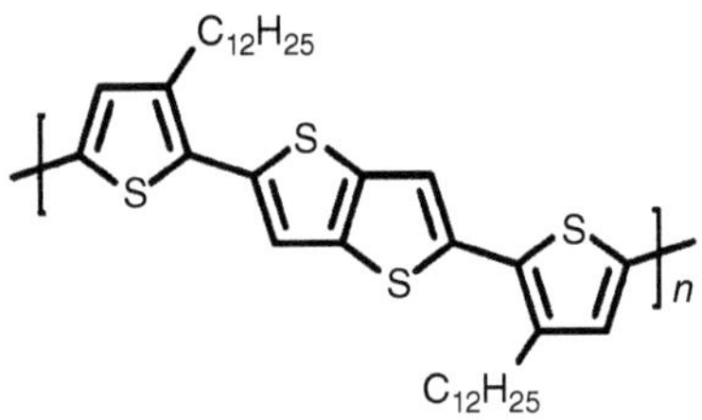 | 27 | 1.9 | 14 | [64] |
| PBTTT-C14 | 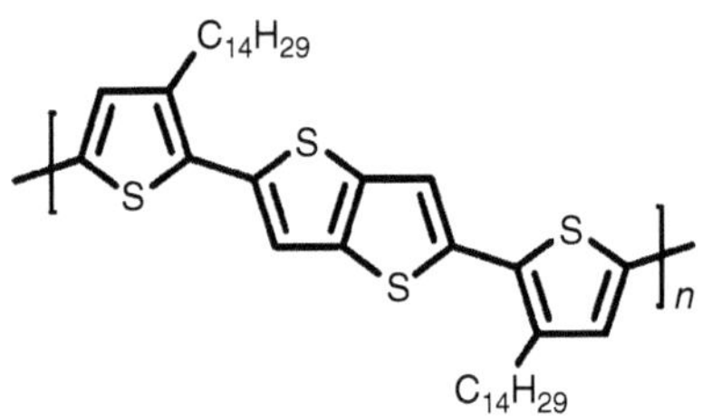 | 32.7 | 1000 | 109 | [66] |

Source: Adapted from Lindorf et al. [60].

Another common route to improving the performance of organic TE materials is doping. Doping (Figure 6.11) improves the power factor by raising the conductivity of the TE material (Figure 6.12). For lightly doped organic semiconductors, thermopower is generally independent of temperature, consistent with hopping behavior for a homogeneous system. However, the morphology of heavily doped polymers is poorly understood, but it is known that doping can be inhomogeneous and that the thermopower often increases with temperature [56].

To expand the number of stably doped materials, efforts have focused on the molecular design of dopants because efficient charge transfer relies on the favorable alignment of the frontier electronic levels of the dopant relative to those of the active semiconducting organic material. Lower ionization energies are necessary for n-type dopants, and higher electron affinities are required for p-type dopants. Consistent with the prevalence of p-type organic TE materials compared to n-type, doping to form p-type materials is more common. Electrical doping of organic materials relies on the formation of a material that is chemically stable on addition or removal of an electron by a dopant. Therefore, neutral or unlopped conjugated polymers are most often doped by oxidation, forming p-type materials, which are more stable. Note that while organic semiconductors can be ambipolar and, in some cases, the hole and electron mobilities are comparable, one sign of carrier is often preferred owing to the ability to inject or stably dope the material with it. The stability of electrically doped materials is determined in part by the ionization energy and electron affinity.

**Table 6.2** Selected high-performance n-type organic TE materials and their Seebeck coefficient ($S$), electrical conductivity ($\sigma$), and power factor (pf).

| Selected n-type organic TE materials | | $S$ ($\mu V K^{-1}$) | $\sigma$ ($S cm^{-1}$) | pf | References |
|---|---|---|---|---|---|
| P(NDI2OD-T2) | 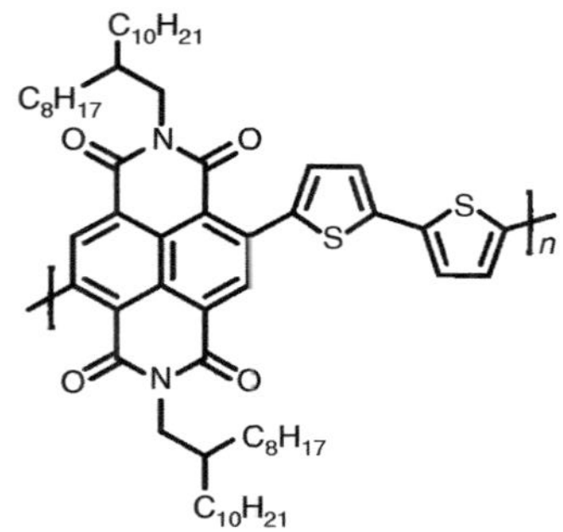 | −850 | 0.008 | 0.6 | [67] |
| BDPPV | 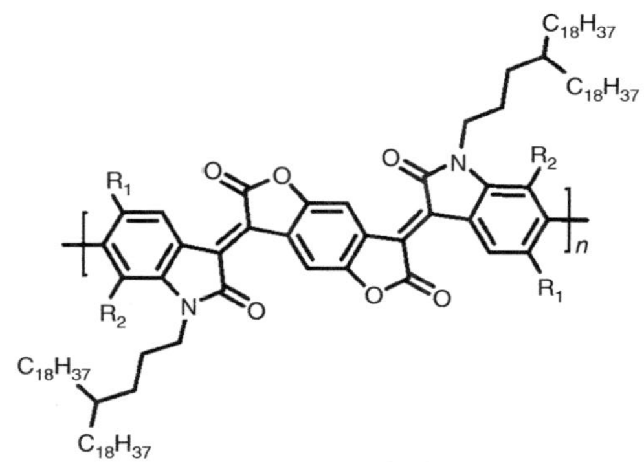 | −141 | 14 | 28 | [68] |
| BBL | 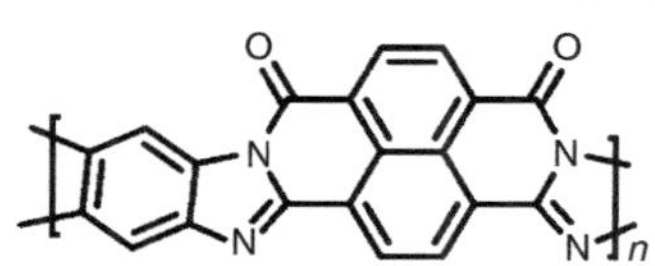 | −60 | 1 | 0.43 | [69] |

Source: Adapted from Lindorf et al. [60].

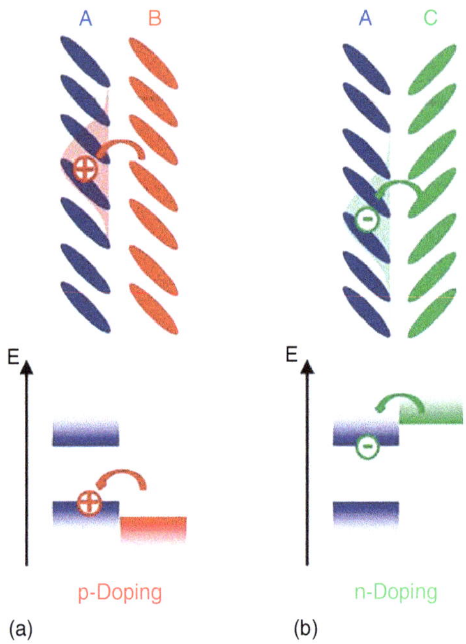


**Figure 6.11** Schematic illustration of the integer charge transfer (CT) within a CT complex leading to an effective p-doping (a) or n-doping (b). Shaded areas indicate the possible delocalization of the transferred charge along with the crystalline molecular stack. Source: Reproduced from Lindorf et al. [60]/The Royal Society of Chemistry.

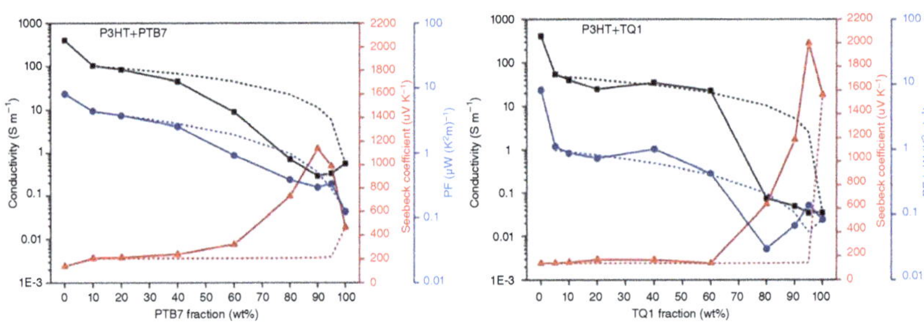


**Figure 6.12** Thermoelectric properties of P3HT (which has a relatively shallow HOMO level) and its binary mixtures PTB7 (left, graph a) and TQ1 (right, graph b) both of which are polymers containing deeper HOMO levels. The large increases in the Seebeck coefficient are attributed to control over the density of states due to the differences in the ionization potentials of the polymers. Source: Reproduced from Lindorf et al. [60]/with permission of John Wiley & sons.

### 6.4.3 Hybrid Organic–Inorganic Thermoelectric Materials

Because material properties, such as electronic structure, charge carrier mobility, Seebeck coefficient, thermal conductivity, and electrical resistivity, are interrelated, it is difficult to optimize them all in a single TE material. To circumvent this challenge, hybrid organic–inorganic TE materials have been developed. These hybrid materials combine favorable properties of inorganic components with favorable properties of organic materials. For example, hybrid materials demonstrate lower thermal conductivity than inorganic TEs because the thermal conductivity of organic TEs is about an order of magnitude lower than that of inorganic TEs [74]. Furthermore, due to the inorganic components, hybrid materials do not suffer from the high electrical resistivity, low Seebeck coefficient, and low-power factor associated with organic TE materials [75]. Among the many advantages offered by this approach, hybrid materials can demonstrate high $zT$ values, flexibility, transparency, low cost, and simplified fabrication processes.

Various architectures of hybrid TE materials have been explored. These include hybrid polymer–inorganic TE materials, metal–organic coordination polymers, organic–inorganic perovskites, and organometallic charge-transfer complexes. Optimization of these materials often involves controlling the nanostructure to increase the Seebeck coefficient through quantum confinement effects while increasing the thermal scattering of phonons at interfaces in order to decrease thermal conductivity.

Hybrid polymer–organic TE materials are fabricated using straightforward methods, such as physical mixing, solution mixing, and *in situ* polymerization. A variety of polymers, such as polyaniline (PANI), polythiophene (PTH), and poly(3,4-ethylenedioxythiophene), are often used in combination with inorganic TE materials, as discussed above. PANI is a typical conducting polymer with high stability and has demonstrated electrical conductivity as high as $10^5\, S\, m^{-1}$. It has been mixed with $V_2O_5$, $BaTiO_3$, BiCuSeO, $Ca_3Co_4O_9$, Bi, $Bi_2Te_3$ and its alloys, $NaFe_4P_{12}$, carbon nanotubes, and PbTe. These materials showed an improved Seebeck coefficient compared to the polymer alone, even though their power factor was only on the order of $10^{-6}\, W\, mK^{-1}$. Similarly, PTH has poor solubility and low electrical conductivity, but when used with $Bi_2TE_3$, they were found to achieve an electrical conductivity of up to $8\, S\, cm^{-1}$ under optimal conditions. This is a seven-fold improvement compared to pure PTH. Finally, PEDOT has also attracted much attention due to its favorable properties, such as high electrical conductivity, optical transparency, stability, low density, and flexibility. However, it has limited solubility and is thus often emulsified with polystyrene sulfonate (PSS). This PEDOT:PSS blends have been used in conjunction with inorganic components, such as carbon nanotubes and Te nanorods, $Bi_2Te_3$, $Ca_3CO_4O_9$ compounds, and Au nanoparticles, yielding high-performance TE materials [76].

Metal–organic coordination polymers, complexes consisting of metal ions and organic ligands linked together to form an organometallic species, have also recently begun to receive attention for their TE properties [60]. In these complexes, metal ions connect ligand linkers. Several examples of this class of material are given in Table 6.3. This class of materials also includes metal–organic frameworks (MOFs). For MOF-based TE materials, power factors of up to $10\, mW\, (mK)^{-2}$ and $zT$ values up to 104 have been recorded [79]. Both n-type and p-type coordination polymers have been demonstrated.

Recently, hybrid perovskites have attracted attention as hybrid TE materials. Perovskites spontaneously crystallize, and thus, morphology optimization is straightforward. While the figure-of-merit as TE material is relatively low for these materials, some materials, such as $Ch_2NH_3PbI_3$ and $CH_3NH_3SNI_3$, have

**Table 6.3** Selected high-performance organometallic TE materials and their Seebeck coefficient (*S*), electrical conductivity ($\sigma$), and power factor (*pf*).

| Selected coordination polymer and charge-transfer complex TE materials | | $S$ ($\mu V K^{-1}$) | $\sigma$ ($S cm^{-1}$) | *pf* | References |
|---|---|---|---|---|---|
| Poly[$Cu_x$(Cu-ett)] | 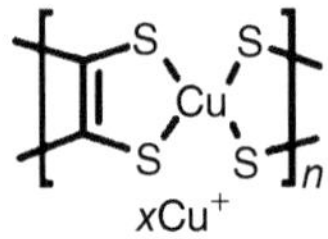 | 83 | 9.5 | 6.5 | [77] |
| $TTT_2I_3$ | 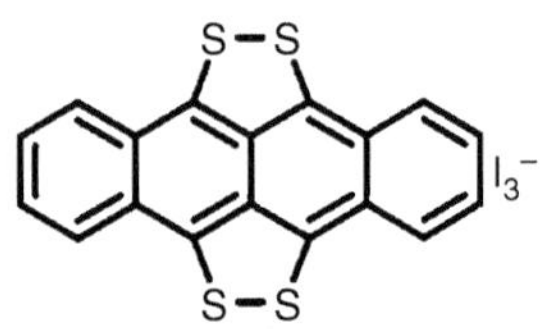 | 42 | 2150 | 387 | [78] |
| Poly[$K_x$(Ni-ett)] | 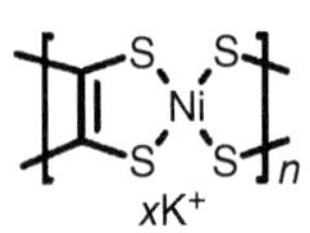 | −122 | 44 | 66 | [77] |
| $Cu(DCNQI)_2$ | 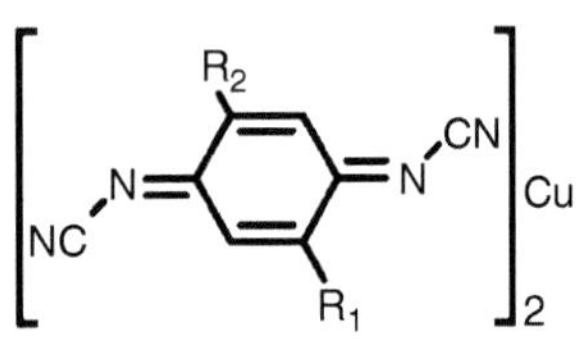 | −34 | 1000 | 110 | [78] |

Source: Adapted from Lindorf et al. [60].

shown promise. Furthermore, optimization of phase and temperature can lead to improvements [76].

Finally, organometallic charge-transfer complexes have begun to attract interest as TE materials [25]. In these complexes, an integer charge transfer results from strong electron affinity or ionization potential of one or more molecule(s) in the complex, leading to p- or n-type doping, respectively. Delocalization of this charge, which is dependent upon the crystalline packing of the charge-transfer complex, allows it to behave like a quasi-free charge carrier. This behavior has been correlated to favorable TE properties, such as higher charge carrier mobility, and thus, higher conductivity. Several of these complexes are given in Table 6.3.

## 6.5 Application of Organic Thermoelectric Generators

New wearable and bio-integrated devices, used for various applications, such as real-time health monitoring, point-of-care diagnostics, and biological actuators, require a small amount (i.e. micro- to milli-watts) of continuous power. TEGs can harness body heat and provide a nonstop power supply without any concern for battery charging or replacement. Researchers have proposed several types of wearable

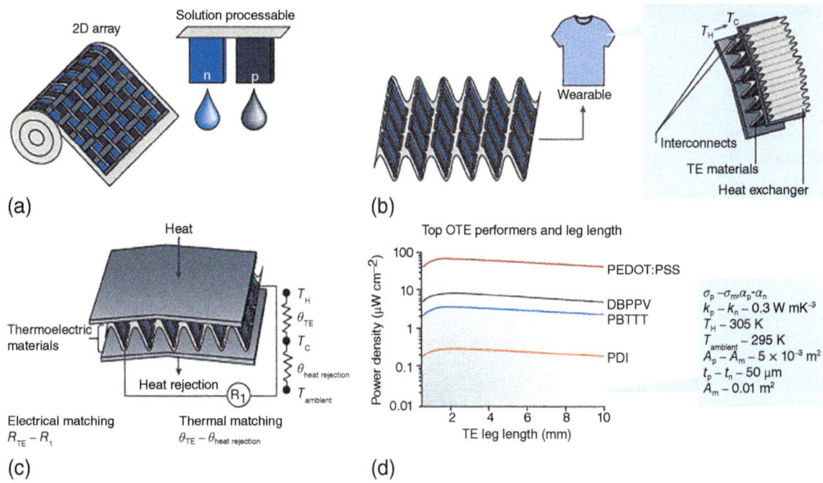


**Figure 6.13** Schematic illustrating OTEG using a roll-to-roll printing process, its application as wearables, and estimated power output. (a) 2D array, (b) customizable 3D architecture, (c) transport matching condition, and (d) top OTE performers and leg length. Source: Reproduced from Russ et al. [25]/Springer Nature.

TEGs based on inorganic TE materials; however, flexibility and weight remain vital concerns. Organic TE material-based TEGs (typically referred to as OTEGs) are more suitable for this application, thanks to their flexible and lightweight nature.

OTEG is currently in the proof-of-concept stage; therefore, most of the devices present in the literature are miniature-scale laboratory prototypes. Figure 6.13 depicts a schematic illustrating a possible method of printing OTEGs where numerous p- and n-type legs can be deposited on a thermally conductive polymer to generate the required voltage and power necessary for the operation of wearable devices. These OTEGs can either be deployed directly on the human body or integrated into clothing or other wearables. Depending on the $zT$ of TE materials and other TEG-related parameters, such as leg height and cross-sectional area, OTEGs can provide power density on the order of a few $\mu W\ cm^{-2}$.

Figure 6.14 shows an actual OTEG based on an organic TE film, poly(3,4-ethylene dioxythiophene) or PEDOT, that generates electricity at the touch of a fingertip. The PEDOT film is synthesized using a mixture of pyridine and poly(ethylene glycol)-block-poly(propylene glycol)-block-poly(ethylene glycol) triblock copolymer (PEPG), thus called PP-PEDOT. The PP-PEDOT OTEG is thin and flexible so that it can be easily bent, twisted, or even cut by scissors. Upon the touch of a fingertip, the OTEG generates an average voltage of about 592 μV for a temperature gradient of around 5.3 K.

Figure 6.15 illustrates a single-couple OTEG with poly(3,4-ethylene dioxythiophene) polystyrene sulfonate (PEDOT: PSS) as the active materials printed onto a glass substrate [55]. Maximum power outputs of $5.17 \pm 0.5$ and $4.08 \pm 0.5$ nW were reported for the OTEGs with different characteristic lengths of 30 and 40 mm at a temperature gradient of 120 °C. The OTEG was designed to be used on building windows for power generation using solar thermal energy harvesting.

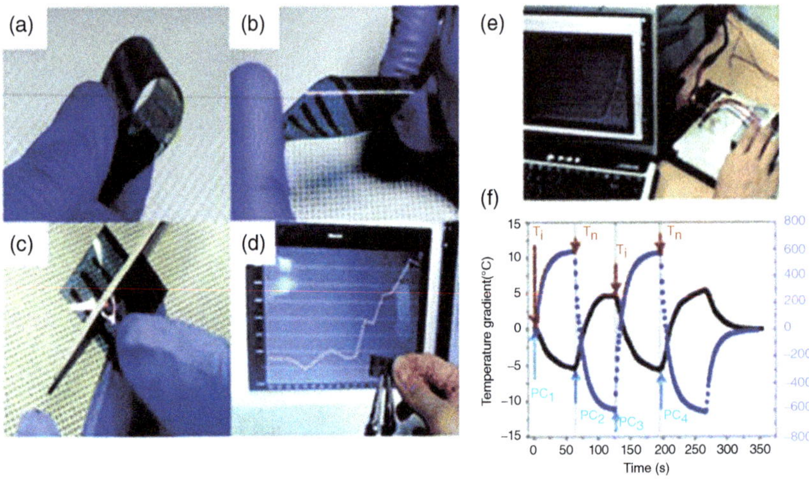


**Figure 6.14** Photographic images of electricity generation by the flexible PP-PEDOT thermoelectric film due to the touch by fingertips. (a) Bending, (b) twisting, and (c) cutting, (d) electricity generation by touching, (e) a Peltier module, and (f) electricity generation (blue square) and the corresponding temperature gradient (black square). Source: Reproduced from Park et al. [61]/Royal Society of Chemistry.

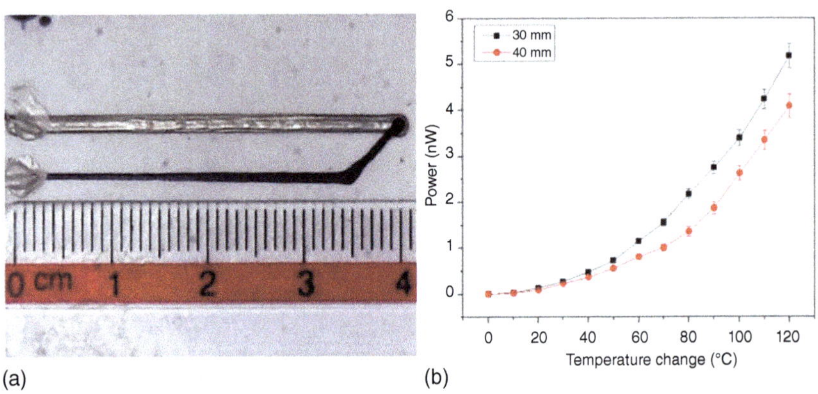


**Figure 6.15** Single-couple OTEG printed on a glass substrate for power generation using solar thermal energy harvesting. (a) Image of the printed OTEG (b) Power versus temperature. Source: Shakeel et al. [55], with permission of Elsevier.

Figure 6.16 depicts another flexible OTEG composed of only p-type carbon nanotube/poly-3-hexylthiophene (CNT/P3HT) nanocomposite patterns spray printed on a polyimide substrate [80]. The OTEG has 41 p-type active lines, each having a width of 1 mm and a length of 15 mm. The legs are connected in series by dispenser-printed silver electrodes. At a temperature difference of about 10 °C, the OTEG was reported to generate a maximum open-circuit voltage of ~41.8 mV and a maximum power of 32.7 nW at the optimal load of 13.5 kΩ. Because output power is small, the device performance needs to be improved by optimizing the device design and adding compatible n-type legs.

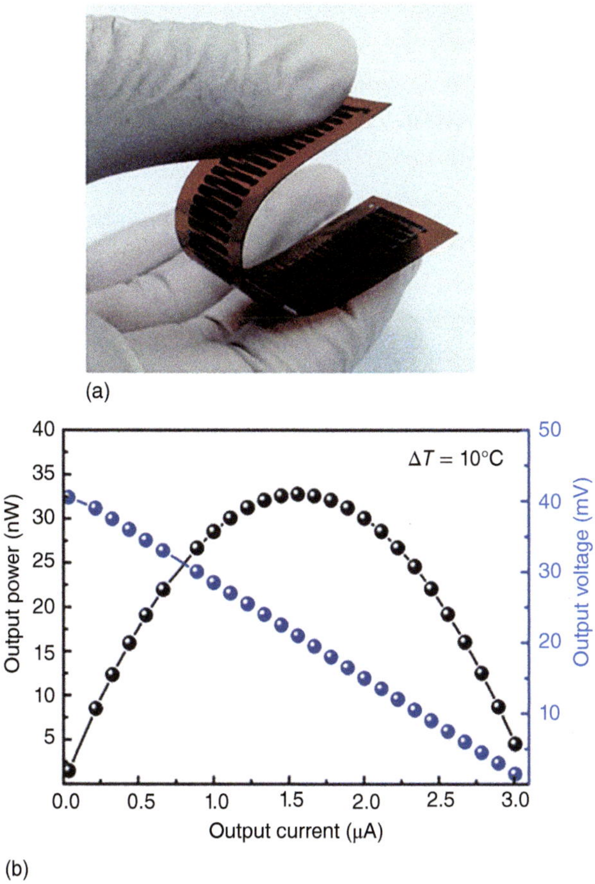


**Figure 6.16** (a) Photograph and (b) output power and voltage versus current of the spray-printed flexible CNT/P3HT OTEG. Source: Hong et al. [80], Taken with permission from Royal Society of Chemistry.

## 6.6 Summary/Future Perspective

In summary, this chapter discussed thermoelectric energy harvesting, including TE materials, working principles, TE devices, and their applications. While we briefly covered traditional inorganic TE materials, we focused mostly on organic and hybrid TE materials, because organic TE materials are nontoxic, flexible, easy to manufacture and scale, and potentially available at low cost. They present a promising alternative for lightweight and conformal thermoelectric power generation and cooling. Organic TE materials, however, have relatively low figure-of-merit ($zT < 0.5$); therefore, organic TEGs are primarily suitable for low-power generation (order of micro-watts to milli-watts) applications. Although organic TEGs are currently at a low-technology readiness level, recent developments and results are highly encouraging.

## References


1 Zhang, Q., Liao, J., Tang, Y. et al. (2017). Realizing a thermoelectric conversion efficiency of 12% in bismuth telluride/skutterudite segmented modules through full-parameter optimization and energy-loss minimized integration. *Energy & Environmental Science* 10: 956–963.

2 Shi, X. and Chen, L. (2016). Thermoelectric materials step up. *Nature Materials* 15: 691.

3 Kishore, R.A., Nozariasbmarz, A., Poudel, B., and Priya, S. (2020). High-performance thermoelectric generators for field deployments. *ACS Applied Materials & Interfaces* 12: 10389–10401.

4 Anant Kishore, R., Kumar, P., Sanghadasa, M., and Priya, S. (2017). Taguchi optimization of Bismuth–Telluride based thermoelectric cooler. *Journal of Applied Physics* 122: 025109.

5 Minnich, A., Dresselhaus, M., Ren, Z., and Chen, G. (2009). Bulk nanostructured thermoelectric materials: current research and future prospects. *Energy & Environmental Science* 2: 466–479.

6 Rowe, D.M. (1995). *CRC Handbook of Thermoelectrics*. CRC Press.

7 Snyder, G.J. and Toberer, E.S. (2008). Complex thermoelectric materials. *Nature Materials* 7: 105.

8 Bell, L.E. (2008). Cooling, heating, generating power, and recovering waste heat with thermoelectric systems. *Science* 321: 1457–1461.

9 Baranowski, L.L., Snyder, G.J., and Toberer, E.S. (2012). Concentrated solar thermoelectric generators. *Energy & Environmental Science* 5: 9055–9067.

10 NASA. Multi-Mission Radioisotope Thermoelectric Generator (MMRTG): https://mars.nasa.gov/internal_resources/788/ (accessed on 15 May 2021).

11 NASA. Perseverance rover: https://mars.nasa.gov/mars2020/spacecraft/rover/electrical-power/ (accessed on 15 May 2021).

12 Stratistics-MRC. Thermoelectric Generator – Global Market Outlook (2016–2022): https://wwwprnewswirecom/news-releases/thermoelectric-generator---global-market-outlook-2016-2022-300490467html (accessed on 15 May 2021).

13 Chen, T., Guai, G.H., Gong, C. et al. (2012). Thermoelectric $Bi_2Te_3$-improved charge collection for high-performance dye-sensitized solar cells. *Energy & Environmental Science* 5: 6294–6298.

14 Sundarraj, P., Maity, D., Roy, S.S., and Taylor, R.A. (2014). Recent advances in thermoelectric materials and solar thermoelectric generators–a critical review. *RSC Advances* 4: 46860–46874.

15 Shi, B., Su, H., Li, J. et al. (2017). Clean power generation from the intractable natural coalfield fires: turn harm into benefit. *Scientific Reports* 7: 5302.

16 Tuley, R., Placha, K., Robbins, M. et al. (2016). Automotive power harvesting/thermoelectric applications. *Thermoelectric Materials and Devices* 230–251. https://doi.org/10.1039/9781782624042-00230.

17 Schierning, G. (2018). Bring on the heat. *Nature Energy* 3: 92.

**18** Zebarjadi, M., Esfarjani, K., Dresselhaus, M. et al. (2012). Perspectives on thermoelectrics: from fundamentals to device applications. *Energy & Environmental Science* 5: 5147–5162.

**19** Kishore, R.A., Kumar, P., and Priya, S. (2018). A comprehensive optimization study on $Bi_2Te_3$-based thermoelectric generators using the Taguchi method. *Sustainable Energy & Fuels* 2: 175–190.

**20** Takahashi, K., Kanno, T., Sakai, A. et al. (2013). Bifunctional thermoelectric tube made of tilted multilayer material as an alternative to standard heat exchangers. *Scientific Reports* 3: 1501.

**21** Bahk, J.-H., Fang, H., Yazawa, K., and Shakouri, A. (2015). Flexible thermoelectric materials and device optimization for wearable energy harvesting. *Journal of Materials Chemistry C* 3: 10362–10374.

**22** Du, Y., Cai, K., Chen, S. et al. (2015). Thermoelectric fabrics: toward power generating clothing. *Scientific Reports* 5: 6411.

**23** Ito, M., Koizumi, T., Kojima, H. et al. (2017). From materials to device design of a thermoelectric fabric for wearable energy harvesters. *Journal of Materials Chemistry A* 5: 12068–12072.

**24** Nozariasbmarz, A., Kishore, R.A., Poudel, B. et al. (2019). High power density body heat energy harvesting. *ACS Applied Materials & Interfaces* 11: 40107–40113.

**25** Russ, B., Glaudell, A., Urban, J.J. et al. (2016). Organic thermoelectric materials for energy harvesting and temperature control. *Nature Reviews Materials* 1: 1–14.

**26** Hsu, C.-T., Huang, G.-Y., Chu, H.-S. et al. (2011). An effective Seebeck coefficient obtained by experimental results of a thermoelectric generator module. *Applied Energy* 88: 5173–5179.

**27** Goldsmid HJ. (2017) Physics of Thermoelectric Energy Conversion, Online ISBN: 978-1-6817-4641-8, Print ISBN: 978-1-6817-4640-1, https://iopscience.iop.org/book/978-1-6817-4641-8.

**28** Kishore, R.A., Mahajan, R.L., and Priya, S. (2018). Combinatory finite element and artificial neural network model for predicting performance of thermoelectric generator. *Energies* 11: 2216.

**29** Nolas, G.S., Sharp, J., and Goldsmid, J. (2013). *Thermoelectrics: Basic Principles and New Materials Developments*. Springer Science & Business Media.

**30** Kishore, R.A. and Priya, S. (2018). A review on low-grade thermal energy harvesting: materials, methods and devices. *Materials* 11: 1433.

**31** Elsheikh, M.H., Shnawah, D.A., Sabri, M.F.M. et al. (2014). A review on thermoelectric renewable energy: Principle parameters that affect their performance. *Renewable and Sustainable Energy Reviews* 30: 337–355.

**32** Kishore, R.A., Nozariasbmarz, A., Poudel, B. et al. (2019). Ultra-high performance wearable thermoelectric coolers with less materials. *Nature Communications* 10: 1–13.

**33** Hu, X., Jood, P., Ohta, M. et al. (2016). Power generation from nanostructured PbTe-based thermoelectrics: comprehensive development from materials to modules. *Energy & Environmental Science* 9: 517–529.

34 Zaitsev, V., Fedorov, M., Eremin, I. et al. (2006). *Thermoelectrics Handbook: Macro to Nano*. Boca Raton: CRC Press, Taylor & Francis.

35 Zhao, D. and Tan, G. (2014). A review of thermoelectric cooling: materials, modeling and applications. *Applied Thermal Engineering* 66: 15–24.

36 Kishore, R.A., Sanghadasa, M., and Priya, S. (2017). Optimization of segmented thermoelectric generator using Taguchi and ANOVA techniques. *Scientific Reports* 7: 16746.

37 Lee, H., Sharp, J., Stokes, D. et al. (2018). Modeling and analysis of the effect of thermal losses on thermoelectric generator performance using effective properties. *Applied Energy* 211: 987–996.

38 Dunham, M.T., Barako, M.T., LeBlanc, S. et al. (2015). Power density optimization for micro thermoelectric generators. *Energy* 93: 2006–2017.

39 Jeong, E.S. (2014). A new approach to optimize thermoelectric cooling modules. *Cryogenics* 59: 38–43.

40 Ouyang, Z. and Li, D. (2016). Modelling of segmented high-performance thermoelectric generators with effects of thermal radiation, electrical and thermal contact resistances. *Scientific Reports* 6: 24123.

41 Riffat, S. and Ma, X. (2004). Improving the coefficient of performance of thermoelectric cooling systems: a review. *International Journal of Energy Research* 28: 753–768.

42 Fraisse, G., Ramousse, J., Sgorlon, D., and Goupil, C. (2013). Comparison of different modeling approaches for thermoelectric elements. *Energy Conversion and Management* 65: 351–356.

43 Gelbstein, Y., Dashevsky, Z., and Dariel, M. (2005). High performance n-type PbTe-based materials for thermoelectric applications. *Physica B: Condensed Matter* 363: 196–205.

44 Wu, D., Zhao, L.-D., Hao, S. et al. (2014). Origin of the high performance in GeTe-based thermoelectric materials upon $Bi_2Te_3$ doping. *Journal of the American Chemical Society* 136: 11412–11419.

45 Perumal, S., Roychowdhury, S., and Biswas, K. (2016). High performance thermoelectric materials and devices based on GeTe. *Journal of Materials Chemistry C* 4: 7520–7536.

46 Li, J., Zhang, X., Lin, S. et al. (2016). Realizing the high thermoelectric performance of GeTe by Sb-doping and Se-alloying. *Chemistry of Materials* 29: 605–611.

47 Zhao, L.-D., Zhang, X., Wu, H. et al. (2016). Enhanced thermoelectric properties in the counter-doped SnTe system with strained endotaxial SrTe. *Journal of the American Chemical Society* 138: 2366–2373.

48 Banik, A., Shenoy, U.S., Saha, S. et al. (2016). High power factor and enhanced thermoelectric performance of SnTe–$AgInTe_2$: synergistic effect of resonance level and valence band convergence. *Journal of the American Chemical Society* 138: 13068–13075.

49 Zhou, M., Gibbs, Z.M., Wang, H. et al. (2016). Thermoelectric performance of co-doped SnTe with resonant levels. *Applied Physics Letters* 109: 042102.

**50** Korkosz, R.J., Chasapis, T.C., S-h, L. et al. (2014). High ZT in p-type (PbTe) $1\text{–}2_x$ $(PbSe)_x$ $(PbS)_x$ thermoelectric materials. *Journal of the American Chemical Society* 136: 3225–3237.

**51** Wood, C. (1988). Materials for thermoelectric energy conversion. *Reports on Progress in Physics* 51: 459.

**52** Snyder, G.J., Christensen, M., Nishibori, E. et al. (2004). Disordered zinc in $Zn_4Sb_3$ with phonon-glass and electron-crystal thermoelectric properties. *Nature Materials* 3 (7): 458–463.

**53** Androulakis, J., Hsu, K.F., Pcionek, R. et al. (2006). Nanostructuring and high thermoelectric efficiency in p-Type Ag $(Pb1\text{–}ySn_y)$ $mSbTe_{2+m}$. *Advanced Materials* 18: 1170–1173.

**54** Hsu, K.F., Loo, S., Guo, F. et al. (2004). Cubic $AgPb_mSbTe_{2+m}$: bulk thermoelectric materials with high figure of merit. *Science* 303: 818–821.

**55** Shakeel, M., Rehman, K., Ahmad, S. et al. (2021). A low-cost printed organic thermoelectric generator for low-temperature energy harvesting. *Renewable Energy* 167: 853–860.

**56** Wang, H. and Yu, C. (2019). Organic thermoelectrics: materials preparation, performance optimization, and device integration. *Joule* 3: 53–80.

**57** Choi, J., Jung, Y., Dun, C. et al. (2020). High-performance, wearable thermoelectric generator based on a highly aligned carbon nanotube sheet. *ACS Applied Energy Materials* 3: 1199–1206.

**58** Kojima, H., Abe, R., Fujiwara, F. et al. (2018). Universality of the giant Seebeck effect in organic small molecules. *Materials Chemistry Frontiers* 2: 1276–1283.

**59** Zhang, Q., Sun, Y., Xu, W., and Zhu, D. (2014). Organic thermoelectric materials: emerging green energy materials converting heat to electricity directly and efficiently. *Advanced Materials* 26: 6829–6851.

**60** Lindorf, M., Mazzio, K.A., Pflaum, J. et al. (2020). Organic-based thermoelectrics. *Journal of Materials Chemistry A* 8: 7495–7507.

**61** Park, T., Park, C., Kim, B. et al. (2013). Flexible PEDOT electrodes with large thermoelectric power factors to generate electricity by the touch of fingertips. *Energy & Environmental Science* 6: 788–792.

**62** Kim, G.H., Shao, L., Zhang, K., and Pipe, K.P. (2013). Engineered doping of organic semiconductors for enhanced thermoelectric efficiency. *Nature Materials* 12: 719–723.

**63** Zhang, Q., Sun, Y., Xu, W., and Zhu, D. (2012). Thermoelectric energy from flexible P3HT films doped with a ferric salt of triflimide anions. *Energy & Environmental Science* 5: 9639–9644.

**64** Zhang, Q., Sun, Y., Xu, W., and Zhu, D. (2014). What to expect from conducting polymers on the playground of thermoelectricity: lessons learned from four high-mobility polymeric semiconductors. *Macromolecules* 47: 609–615.

**65** Yan, H., Ohta, T., and Toshima, N. (2001). Stretched polyaniline films doped by (±)-10-camphorsulfonic acid: anisotropy and improvement of thermoelectric properties. *Macromolecular Materials and Engineering* 286: 139–142.

66 Patel, S.N., Glaudell, A.M., Kiefer, D., and Chabinyc, M.L. (2016). Increasing the thermoelectric power factor of a semiconducting polymer by doping from the vapor phase. *ACS Macro Letters* 5: 268–272.

67 Schlitz, R.A., Brunetti, F.G., Glaudell, A.M. et al. (2014). Solubility-limited extrinsic n-type doping of a high electron mobility polymer for thermoelectric applications. *Advanced Materials* 26: 2825–2830.

68 Shi, K., Zhang, F., Di, C.-A. et al. (2015). Toward high performance n-type thermoelectric materials by rational modification of BDPPV backbones. *Journal of the American Chemical Society* 137: 6979–6982.

69 Wang, S., Sun, H., Ail, U. et al. (2016). Thermoelectric properties of solution-processed n-doped ladder-type conducting polymers. *Advanced Materials* 28: 10764–10771.

70 Quinn, J.T.E., Zhu, J., Li, X. et al. (2017). Recent progress in the development of n-type organic semiconductors for organic field effect transistors. *Journal of Materials Chemistry C* 5: 8654–8681.

71 Liu, J., van der Zee, B., Alessandri, R. et al. (2020). N-type organic thermoelectrics: demonstration of $ZT > 0.3$. *Nature Communications* 11: 5694.

72 Jiang, Y., Geng, H., Li, W., and Shuai, Z. (2019). Understanding carrier transport in organic semiconductors: computation of charge mobility considering quantum nuclear tunneling and delocalization effects. *Journal of Chemical Theory and Computation* 15: 1477–1491.

73 Klues, M. and Witte, G. (2018). Crystalline packing in pentacene-like organic semiconductors. *CrystEngComm* 20: 63–74.

74 Liang, Z., Boland, M.J., Butrouna, K. et al. (2017). Increased power factors of organic–inorganic nanocomposite thermoelectric materials and the role of energy filtering. *Journal of Materials Chemistry A* 5: 15891–15900.

75 Kim, J. and Lim, J.-H. (2017). Organic–inorganic hybrid thermoelectric material synthesis and properties. *Journal of the Korean Ceramic Society* 54: 272–277.

76 Jin, H., Li, J., Iocozzia, J. et al. (2019). Hybrid organic–inorganic thermoelectric materials and devices. *Angewandte Chemie, International Edition* 58: 15206–15226.

77 Sun, Y., Sheng, P., Di, C. et al. (2012). Correction: organic thermoelectric materials and devices based on p- and n-type Poly(metal 1,1,2,2-ethenetetrathiolate)s. *Advanced Materials* 24: 1781.

78 Huewe, F., Steeger, A., Kostova, K. et al. (2017). Low-cost and sustainable organic thermoelectrics based on low-dimensional molecular metals. *Advanced Materials* 29: 1605682.

79 Erickson, K.J., Léonard, F., Stavila, V. et al. (2015). Thin film thermoelectric metal–organic framework with high Seebeck coefficient and low thermal conductivity. *Advanced Materials* 27: 3453–3459.

80 Hong, C.T., Kang, Y.H., Ryu, J. et al. (2015). Spray-printed CNT/P3HT organic thermoelectric films and power generators. *Journal of Materials Chemistry A* 3: 21428–21433.

# 7

# Carbon-Enhanced Piezoelectric Materials and Applications

*Mengying Xie[1], Chengcheng Tian[1], and Poh Choon Ooi[2]*

[1] *State Key Laboratory of Precision Measuring Technology and Instrument, College of Precision Instruments and Opto-electronics Engineering, Tianjin University, Tianjin, 300072, China*
[2] *Institute of Microengineering and Nanoelectronics (IMEN), Universiti Kebangsaan Malaysia, 43600 Bangi, Selangor, Malaysia*

## 7.1 Introduction

Piezoelectric materials are the most widely exploited and extensively used materials for industry, military, and daily life [1]. They have been widely used as energy harvesters [2, 3], sensors [1, 4], actuators [5], and capacitors [6]. Most high-performance piezoelectric materials are lead-based materials, and they are commonly toxic, rigid, and fragile, which could not be used in future portable electronics. Recently lead-free piezoelectric materials [7, 8] and composite materials that contain piezoelectric materials in a polymer matrix have been extensively explored. The piezoelectric materials can generate charges on the surface electrodes under deformation, and a mechanical strain will be induced as an electrical field that is applied to a pair of electrodes, as shown in Figure 7.1.

Some piezoelectric materials exhibit a tetragonal symmetry structure, and such a symmetric structure is associated with a dipole moment. The poling process is important as it leads to the alignment of ferroelectric domains in the direction of the applied electric field and provides a net remnant polarization. It is the remnant polarization that leads to piezoelectric effects. Suppose a strain is applied along with an axis of the poled piezoelectric material. The center of the cations and the center of anions are relatively displaced, resulting in a change in the piezoelectric polarization along with the direction of strain in the crystal. As a result, an electric current will be induced in an external circuit, and therefore, the piezoelectric materials can naturally be used for sensing and energy-harvesting applications.

The direct piezoelectric effect is a change of electric polarization and electric field due to applied mechanical stress, which can be expressed by Eq. (7.1).

$$D_i = d_{ij} T_j + \varepsilon_{ik} E_k \tag{7.1}$$

where $D$ is the dielectric displacement, $T$ is the stress tensor, $\varepsilon$ is the dielectric permittivity, $E$ is the electric field, and $d$ is the piezoelectric charge coefficient. Meanwhile,

*Enhanced Carbon-Based Materials and Their Applications*, First Edition.
Edited by Poh Choon Ooi, Mengying Xie, and Chang Fu Dee.

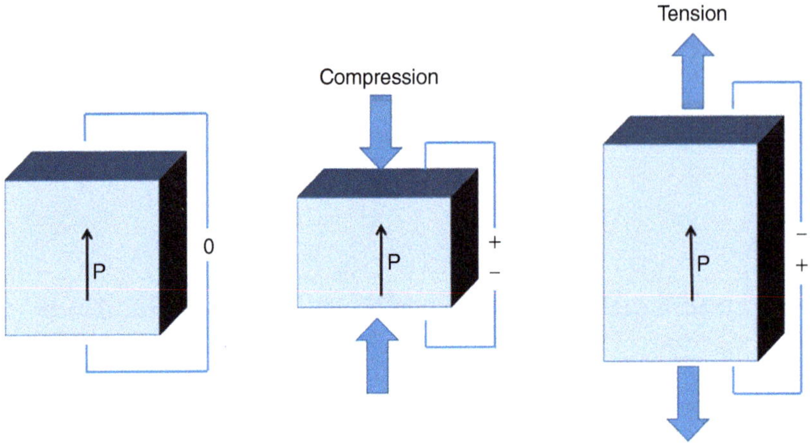


**Figure 7.1** Schematic diagram of piezoelectric effect.

*i*, *j*, and *k* refer to the different directions within the piezoelectric material. The piezoelectric charge coefficient $d_{ij}$ is the ratio of short-circuit charge per unit area flowing between the connected electrodes perpendicular to the *j* direction and the stress applied in the *i* direction. For example, when an applied strain, *F*, is applied along with the three-direction, the piezoelectric material can generate an electric charge *Q* between the connected electrodes perpendicular to the three-direction, flowing through a short circuit:

$$Q = d_{33} F_3 \tag{7.2}$$

Hence, the generated open-circuit voltage is given below:

$$V = \frac{c}{ab} g_{33} F_3 \tag{7.3}$$

where $g_{33}$ is the piezoelectric voltage constant, while *a, b,* and *c* are the dimensions of the piezoelectric materials. When the material is stressed along with the axis, an electric field will be developed along with the axis. The relationship between $d_{33}$ and $g_{33}$ is given below:

$$d_{33} = g_{33} \varepsilon_r \varepsilon_0 \tag{7.4}$$

It can be seen that the performance of a piezoelectric sensor is directly related to the piezoelectric voltage constant $g_{ij}$ and piezoelectric charge coefficient $d_{ij}$. Since piezoelectric materials respond to dynamic forces applied to their surfaces, they are commonly used as dynamic force sensors, touch sensors, vibration sensors, and accelerometers. In addition, for energy harvesting applications, a figure of merit $FoM_{ij}$ containing combinations of physical properties has been developed to describe the ability of materials to generate energy for practical applications.

$$FoM_{ij} = \frac{{d_{ij}}^2}{\varepsilon_r \varepsilon_0} \tag{7.5}$$

$FoM_{ij}$ assesses the electrical energy density under mechanical stress and indicates that the essential requirements for improved performance are low relative permittivity and high piezoelectric coefficient.

A piezoelectric material with high $d_{33}$ and low permittivity would benefit energy harvesting and sensing applications. Several approaches, including hierarchical porous structures [9, 10] and composite materials [11], have been investigated to improve performance. Studies have shown that the addition of carbon-based fillers can dramatically improve the piezoelectric properties. In addition, most piezoelectric materials are a class of dielectric materials that need to be polarized by applying external stimuli, i.e., mechanical, electrical, and thermal stimuli [12, 13]. On the other hand, some studies show that adding carbon fillers could eliminate the polarization process, which is called self-poling in most literature. In addition, carbon fillers can improve the mechanical properties of piezoelectric materials [14]. Therefore, this chapter focuses on the piezoelectric material with various carbon nanofillers and reviews the enhancement of the properties with different carbon nanofillers, as well as the applications of these piezoelectric composites.

## 7.2 Carbon-Enhanced Piezoelectric Materials

By incorporating carbon materials, particularly nanostructured carbon materials, the piezoelectric properties of a piezoelectric material can be improved significantly. Various 1D- and 2D-structured carbon materials, such as graphene, carbon nanotubes (CNTs), and quantum dots, have been mixed into the piezoelectric materials. The enhancement mechanism varies for different inorganic and organic piezoelectric materials. Therefore in the following content, the inorganic and organic piezoelectric materials are discussed separately. Table 7.1 summarizes the performance of piezoelectric with various carbon materials.

### 7.2.1 Inorganic Piezoelectric Materials

Inorganic piezoelectric materials are among the most important materials used in industry due to their high-stress sensitivity, high piezoelectric constant, good stability, and relatively high Curie temperature. Generally, inorganic piezoelectric ceramic materials can be classified into two groups, that is, lead-based and lead-free ceramics. Lead-based inorganic ceramics, such as lead zirconium titanate (PZT) and lead magnesium niobate–lead titanate (PMN–PT), generally possess superior piezoelectric properties and have been widely used in sensors, actuators, and energy harvesters. Commercial PZT ceramics have a high $d_{33}$ up to ~670 pC/N [10]. The $d_{33}$ of a PMN–PT single crystal can reach up to ~2000 pC/N [22]. Although bulk piezoelectric ceramics have high piezoelectric performance, they are too brittle to be integrated into flexible electronics. On the other hand, piezo-composite materials containing inorganic piezoelectric nano or microscale components distributed into a polymer matrix are candidates for flexible energy harvesters and self-powered sensing applications. Composite materials take advantage of the piezoelectricity of the fillers and flexibility from the polymer matrix.

**Table 7.1** Carbon-enhanced inorganic piezoelectric materials.

| Chemical compound | Ceramic/polymer piezoelectric materials | | Carbon materials | Fabrication method | Electrical output performance | | | References |
|---|---|---|---|---|---|---|---|---|
| | | | | | Open-circuit voltage (V) | Short-circuit current/current density | Power/power density | |
| Inorganic | Lead-based | PZT | MWCNT | Spin casting | 10 | 1.3 μA | — | [11] |
| | | Soft-PZT | GNP | Casted | 27 | 429 μA | 402 mW $m^{-2}$ | [15] |
| | Lead-free | $BaTiO_3$ | MWCNT | Spin casting | 3.2 | 250–350 nA | — | [16] |
| | | $BaTiO_3$ | Carbon | Molding | 31 | 1.8 μA | 45.4 μW$cm^{-2}$ | [17] |
| | | ZnO | MWCNT | Molding | 150 | 6.5 μA | — | [18] |
| Organic | Polymer | PVDF | WMCNT | Electrospinning | 6 | — | 81.8 nW | [19] |
| | | PVDF | rGO-Ag | Molding | 18 | 1.05 μA | 28 W$m^{-3}$ | [20] |
| | | PVDF-TrFE | GO | Drop casting | 4 | — | 4.41 μW $cm^{-2}$ | [21] |

Studies have proven that randomly dispersed ceramic fillers often lead to a lack of connectivity, giving rise to poor polarization and piezoelectric activity levels [10, 14]. In addition, the permittivity mismatch between the two phases results in an unfavorable electric-field distribution that restricts poling of the composite and further reduces piezoelectric properties. The connectivity among the piezoelectric phases is the key to improving the piezoelectric output performance [10]. Recently, researchers have been looking for methods at structural [10] and material levels [18, 23, 24] to increase connectivity and enhance the piezoelectric coefficient. Here in this chapter, we emphasize material-level enhancement. Studies have shown that the addition of carbon materials can enhance the piezoelectric coefficient of ceramics. These conductive fillers can form effective conduction paths to provide channels for collecting and transferring the charges across the surface of piezoelectric particles to the electrodes, owing to their high conductivity and large surface area. However, there is a threshold for the enhancement. The excess amount of conductive fillers would cause agglomeration, negatively affecting the output performance of the piezoelectric. Another advantage of adding carbon material is to improve the distribution of the piezoelectric phase in the polymer matrix.

The most common and easiest method is simply mixing the 1D/2D material into a piezoelectric composite. For example, 30 vol% PMN–PT powder was simply mixed with 1 vol% multi-walled carbon nanotubes (MWCNTs) and 69 vol% poly(vinylidene fluoride) (PVDF) to form the composite film. The output voltage and current of this mixture were more than two times higher than the PMN–PT/PVDF composite. This enhancement could be attributed to the strengthening interfacial interaction between PMN–PT and PVDF due to the reinforcement of the MWCNT particles that form a conductive network [18]. Park et al. developed a flexible, large area, and highly efficient nanocomposite energy generator by dispersing MWCNT and PZT particles in polydimethylsiloxane (PDMS), as shown in Figure 7.2a. The output voltage and current of the flexible generator were ~10 V and 1.3 μA, respectively, when subjected to regular bending/unbending deformation [23]. The remarkable flexibility and stretchability of this piezoelectric generator were due to the mechanical reinforcement of the PDMS matrix by the addition of CNTs. Moreover, the CNTs as a dispersing agent, stress reinforcement, and conduction path can improve the output performance of the nanocomposite-based piezoelectric device.

Recently, studies have proved that the direct growth of piezoelectric particles on carbon material reaches better performance than the mixed composite since it increases the piezoelectric substance. Figure 7.2b shows that Han et al. directly grew the PZT nanoparticles (PZTNP) on the CNTs by nucleating at the carboxylic groups in the defects of CNTs. The PZTNP–CNT (80 wt%) was then mixed with 20 wt% polymer binder polytetrafluoroethylene (PTFE) to form a composite [24]. This method increased the maximum amount of piezoelectric material and maintained flexibility simultaneously. The voltage output of the direct growth of PZTNP–CNT was five times higher than that of the simply mixed counterparts. Furthermore, Han et al. proved that the enhancement results from the flexoelectric effect generated in addition to the piezoelectric effect attributed to the PZT-atomically-interconnected CNT [24].

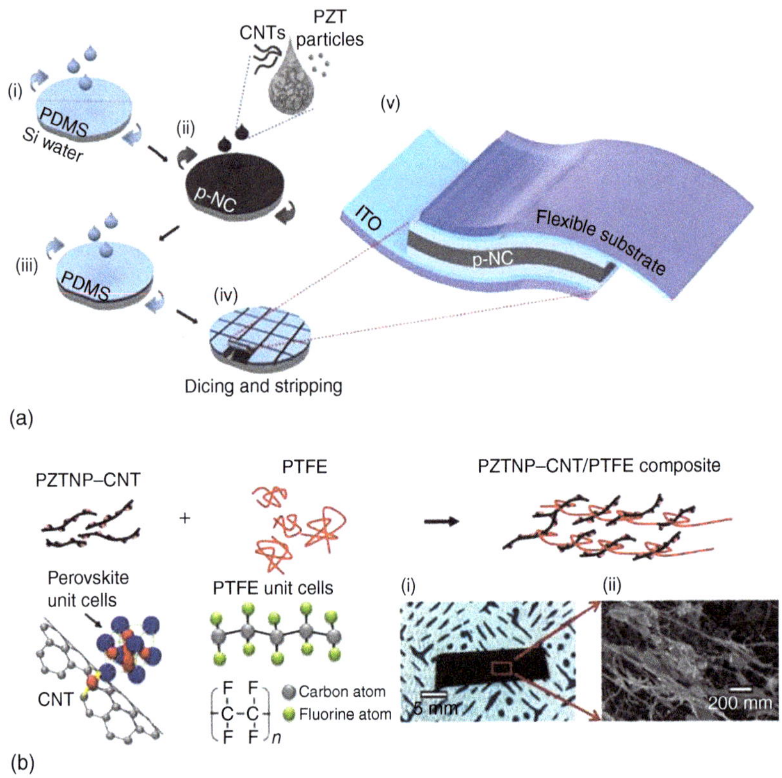


**Figure 7.2** Carbon-enhanced lead-based piezoelectric ceramics. (a) Schematic illustration of the process for fabricating flexible nanocomposite generator (NCG) devices. Source: Park et al. [23]/with permission of John Wiley and Sons. (b) Schematic diagram of the fabrication process for a PZTNP–CNT film using a PTFE binder. Source: Han et al. [24]. Reproduced from American Chemical Society/CC BY 4.0.

Due to the toxicity of lead, although lead-based piezoelectric ceramics have superior piezoelectric properties, they pose environmental threats when facing recycling and health threats in wearable applications. The wurtzite ZnO nanowire is a traditional lead-free piezoelectric material but possesses low $d_{33}$ of 9.9 pC/N [25]. Because of the low $d_{33}$, the primary way to improve the outputs is to connect unites in parallel to improve the output current or connect unites in series to improve the output voltage [26]. Studies have shown that carbon additives, especially nanofillers, could enhance the piezoelectric properties [16, 27]. For instance, Kim et al. mixed MWCNT into ZnO/PDMS to form a composite film for hybrid piezoelectric/triboelectric energy-harvesting applications [27]. The output performance was enhanced by increasing the amount of MWCNT. In Kim's work, the MWCNT played an important role in evenly distributing the ZnO nanoflowers in the composites and providing conduction paths to collect and transfer the charges across the surface of ZnO nanoflowers to the electrodes. The composite film with 0.015 g of MWCNT presented the highest voltage and current of 150 V and 6.5 μA, respectively. When the MWCNT content is further increased, the output will be dropped ascribed to the agglomeration.

Lead-free ceramics [28], such as $BaTiO_3$ (BT) and $KNaNbO_3$ (KNN), have merged to become a potential replacement for lead-based piezoelectric materials. $BaTiO_3$ has a $d_{33}$ of approximately ~350–500 pC/N [29]. KNN is an emerging lead-free ceramic, and pure KNN possesses a relatively high $d_{33}$ of 85 pC/N [30]. Park et al. mechanically mixed $BaTiO_3$ nanoparticles and MWCNT into PDMS to form a flexible and stretchable composite [31], generating an open-circuit voltage ($V_{oc}$) of 3.2 V and short-circuit current ($I_{sc}$) of 250–350 nA. As shown in Figure 7.3a, the CNT plays

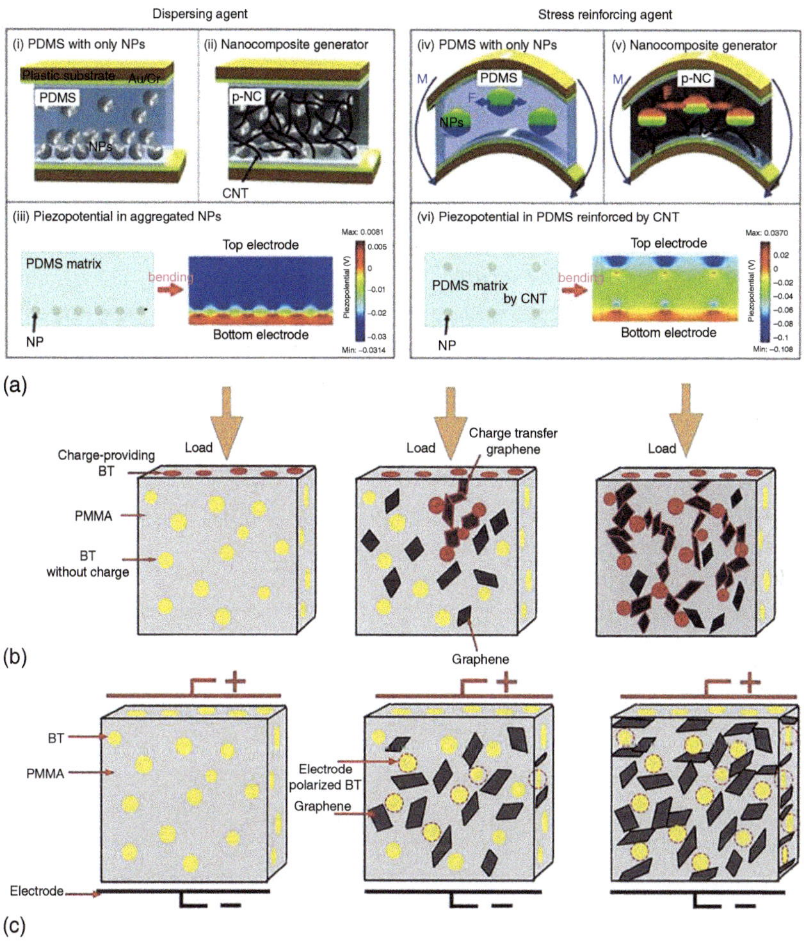


**Figure 7.3** Mechanism of the enhancement with carbon materials. (a) Schematics of the cross-sectional structure of NCG devices and the calculated piezopotential distributions for explaining the role of CNTs. The CNTs act as dispersing and stress-reinforcing agents. Source: Park et al. [31]/with permission of John Wiley and Sons. (b) Improving mechanism of the piezoelectric coefficient of the piezoelectric composites by graphene. (c) Improving the mechanism of the effective polarization voltage of the piezoelectric composites by graphene. Source: Tang et al. [32]/with permission of Elsevier.

an important role in improving the piezoelectric coefficient by (i) forming CNT networks to help the $BaTiO_3$ nanoparticles well distributed in the composite, (ii) entangling with nanoparticles in PDMS to reinforce the stress applied to the composite, (iii) forming conduction path to reduce the internal resistance. In addition, Park et al. have replaced MWCNT with single-walled carbon nanotubes (SWCNTs)/reduced graphene oxide (rGO) and found that the performance with rGO produced a lower output voltage than only CNT-based generator, which was likely deduced from the difference in the degree of mixing, which, in turn, was caused by the geometrical difference between the CNT networks and the laminated rGO structures [31].

In addition to CNTs, graphene is a carbon-based filler that possesses remarkable electrical, mechanical, and thermal properties. Graphene is a promising material in multifunctional wearable electronics and biomedical engineering owing to its biocompatibility and high-storage capability [33]. Several studies have shown that graphene can significantly increase the piezoelectric coefficient [15, 32]. Tang et al. added graphene into $BaTiO_3$ and polymethyl methacrylate (PMMA) mixture to form a bio-piezoelectric composite for biomedical applications [32]. With increasing graphene content, the piezoelectric coefficient increased up to ~0.8 pC/N. Here graphene increased the piezoelectric coefficient by increasing the conductivity, dielectric constant, and effective polarization voltage of the composite. The detailed improving mechanism can be found in Figure 7.3b,c. With an increase in the graphene content, the number of conductive networks among $BaTiO_3$ particles increases (Figure 7.3b). Figure 7.3c presents the mechanism of increasing effective polarization voltage due to the micro-capacitors. In the graphene conductive network, a pair of graphene forms the two poles of the micro-capacitor. The BT/PMMA-filled medium in the middle is directly polarized so that the effective polarization voltage of the whole composites increases. With the increase of graphene content, the number of micro-capacitors increased, the number of piezoelectric networks increased, and the number of $BaTiO_3$ particles polarized by graphene poles further increased. Accordingly, the effective polarization voltage of the composites increased.

### 7.2.2 Organic Piezoelectric Materials

Compared with the above inorganic piezoelectric materials, organic piezoelectric materials generally possess lower $d_{33}$. However, their natural flexibility makes them more suitable to be utilized in flexible and portable electronic devices [19, 34]. Among the organic piezoelectric materials, the PVDF and its copolymers are the most widely used for flexible and portable electronics. The piezoelectricity of PVDF as a semicrystalline polymer mainly depends on its five polymorphic crystal phases: $\alpha$, $\beta$, $\gamma$, $\delta$ and $\varepsilon$. The conformation of polymeric chains determines the five polymorphs. The $\alpha$-phase is the primary phase of PVDF and is the most thermally stable phase at room temperature and pressure [35]. In the $\alpha$-phase, the polymer chain has a TGTG′ (T-trans, G-gauche$^+$, and G′-gauche$^-$) dihedral conformation; hence, this antiparallel orientation of the dipoles results in a nonpolar PVDF, as shown in Figure 7.4a. The $\beta$-phase has all-trans planar zigzag conformation (TTTT).

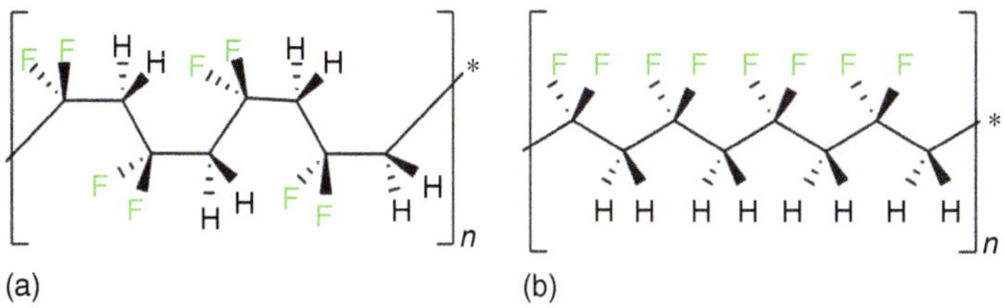


**Figure 7.4** Structure of (a) nonpolar $\alpha$-phase and (b) polar $\beta$-phase PVDF.

The parallel arrangement of dipoles exhibits the strongest electric dipole moment among all crystal phases, as shown in Figure 7.4b. The $\gamma$-phase is an intermediate conformation of PVDF $\alpha$ and $\beta$ phases (TTTGTTTG′). It is worth noting that the piezoelectric properties of PVDF are attributed to the existence of polar crystal phases ($\beta$-phase and $\gamma$-phase), in which the $\beta$-phase is the most desirable.

The main disadvantage of PVDF is its low electromechanical coupling, manifested in its small piezoelectric coefficient $d_{33}$, and it is difficulty in forming the $\beta$-phase [26]. Various polymer blends have been developed using alternative blending monomers. The type and composition ratio of the comonomer determine the crystal structure and crystallization of the polymer, and the crystallization directly affects the piezoelectric and thermoelectric properties. Common PVDF copolymers include trifluoroethane (TrFE), chlorotrifluoroethylene (CTFE), or hexafluoropropylene (HFP) comonomers [36–38]. The $d_{33}$ of commercial PVDF and PVDF–TrFE was reported as approximately 20~30 pC/N [39]. Nevertheless, the low net dipole moment was obtained using conventional preparation methods, specifically solvent casting, high-temperature annealing, and high-pressure crystallization. The poor $\beta$-phase could be associated with individual electric dipoles from different $\beta$-crystal units canceling out each other due to random alignment in bulk [40]. Therefore, efforts have been made to achieve effective $\beta$-phase nucleation in a PVDF matrix [41, 42], including the use of electrical poling [43] and uniaxial or biaxial mechanical drawing at elevated temperature [35], spin coating [44], and electrospinning [17]. The addition of nanofillers would disrupt the polymer molecular chain and enhance the $\beta$-phase crystallinity. The piezoelectric potential of the $\beta$-phase can be further enhanced by incorporating ferroelectric inorganic particles with a high dielectric constant. Despite the doping of the ceramic powders in the PVDF matrix, the dielectric constant of PVDF is not high enough because of the weak interfacial interaction between the ceramic powders and the polymeric matrix [45]. The incorporation of carbon-based nanomaterials can induce the PVDF electroactive crystalline phase through interfacial interaction and enhance the dielectric constant of the composites due to the percolation theory [46].

The information on $\beta$-phase nucleation is highly essential due to its polar electroactive conformation to utilize the piezoelectric properties for energy-harvesting purposes. In this context, X-ray diffraction (XRD) and Fourier-transform infrared spectroscopy (FTIR) are two common characterization methods that could be used to assess the formation and content of $\beta$-phase for the prepared piezoelectric film [47, 48]. For example, Kaspar et al. conducted the XRD measurement for the

PVDF and PVDF with graphite flakes samples; the $\alpha$- and $\beta$-phase can be observed at around 18° and 21°, respectively [48]. The pure PVDF $\beta$-phase concentration was recorded as 82.521% increased to 94.631% when 0.5 wt% carbon powder was incorporated.

The FTIR measurements were also conducted in the same study by Kaspar et al. The prominent peaks of the PVDF $\beta$-phase can be seen at 840, 1074, 1276, and 1431 $cm^{-1}$. A very small peak of $\alpha$-phase can be observed at 763 $cm^{-1}$ [48]. The relative fraction of the PVDF $\beta$-phase can be estimated with Eq. (7.6) [49].

$$F(\beta) = \frac{A_\beta}{(K_\beta + K_\alpha)A_\alpha + A_\beta} \tag{7.6}$$

where $F(\beta)$ is the fraction of $\beta$-phase, $A_\alpha$ and $A_\beta$ are the absorbances at respective wavenumbers of 763 and 840 $cm^{-1}$. Meanwhile, $K_\alpha$ and $K_\beta$ are the absorption coefficients at corresponding wavenumbers of $6.1 \times 10^4$ and $7.7 \times 10^4$ $cm^2$ $mol^{-1}$. Besides XRD and FTIR characterizations, the Raman spectra method can also be carried out to evaluate the presence of the desired $\beta$-phase. Kasper et al. further ascertained the existence of $\beta$-phase in their prepared samples by Raman measurement. They concluded that the prominent peaks were associated with $\alpha$- and $\beta$-phase at the focused location [48].

#### 7.2.2.1 Carbon Nanotubes

Similar to the additives for the inorganic piezoelectric material, nanostructured carbon-based materials, such as carbon nanoparticles, CNTs, and graphene, were commonly mixed into the organic piezoelectric material to improve the device output performance. Zhou et al. coated a thin layer of carbon shell around 0-D piezoceramic NPs (carbon-coated $BaTiO_3$ (BT@C)), carbon-coated PZT (PZT@C), and carbon-coated $KNaNbO_3$(KNN@C)) and studied the effect of the carbon shell thickness on the output performance [50], as shown in Figure 7.5a. An active interface was created for mobile charges between piezoceramic NPs and matrix, providing an effective area for charges to move and accumulate during the poling process and periodic pressing/release stages. The optimal $V_{oc}$, $I_{sc}$, and peak power density of the BT@C nanogenerator with approximately 15 nm carbon shell reach up to 31 V, 1.8 μA, and 45.4 μW $cm^{-2}$, which is almost 3.5, 4.2, and 20 times higher than those of the nanogenerators based on $BaTiO_3$, respectively. They also studied the PZT@C and KNN@C, and the same significant enhancement in the peak power density indicated that the core–shell strategy is universally applicable for fabricating 0-D piezoceramic NPs-based nanogenerators with high–power density [50]. Later, $BaTiO_3$ was directly grown on carbon nanoparticles to form a core–shell structure and then mixed with PVDF. This unique core–shell-structured composite increased the content of the polar $\beta$-phase in PVDF–TrFE and significantly enhanced the interfacial polarization between $BaTiO_3$ nanoparticles and the PVDF–TrFE matrix during the poling process. This nanogenerator obtained an optimal output voltage of 17 V and power of 14.3 μW under a bending-releasing mode [52]. CNT can provide a 3D conductive network within the polymer matrix

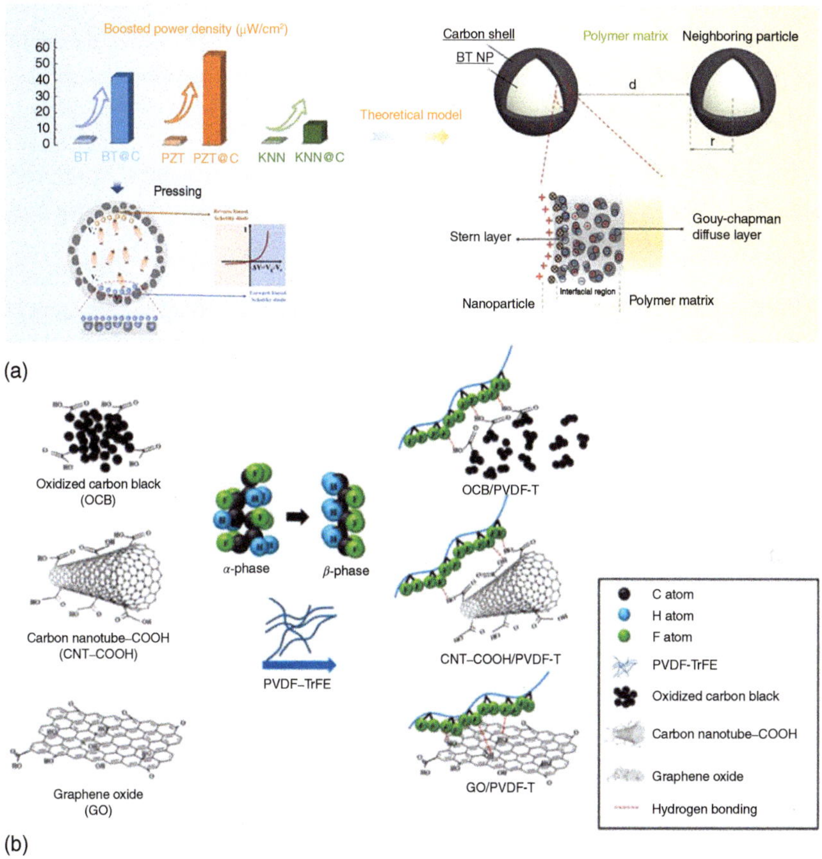


**Figure 7.5** (a) Schematic core–shell structure of interface modulated 0D piezoceramic nanoparticles/PDMS-based piezoelectric composites. Source: Zhou et al. [50]/Elsevier. (b) Schematic of carbon nanomaterials inducing the transformation of the $\alpha$-phase to $\beta$-phase of PVDF–TrFE. Source: Li et al. [51]/MDPI/Public Domain CC By 4.0.

due to the high surface-to-volume ratio and highly conductive properties. Various carbon nanomaterials, including oxidized carbon black with carboxylic acid or hydroxyl groups on the surface, modified carbon nanotubes (CNT-COOH), and GO, were mixed into PVDF–TrFE to induce a higher proportion of the $\beta$-phase structure and improve the overall piezoelectricity and mechanical strength of the film, as shown in Figure 7.5b [51]. Three phases (lanthanum-doped PZT/MWCNT/PVDF) composites were fabricated by tape casting followed by the hot press, and the measured $d_{33}$ was found to be increased with increasing CNT content [53].

In addition, porous materials possess unique advantages of exceptional repeated compression-recovery performance and charge distribution at the cell interface after polarization, contributing to superb piezoelectric performance. By combining supercritical carbon dioxide technology and bead-forming technology, controllable and homogeneously microporous PVDF/MWCNTs 3D foam has been fabricated [54].

Electrostatic interactions between the functional groups of MWCNT and the $CF_2$ dipoles of PVDF promoted the formation of more stretched PVDF chains, thus constructing more $\beta$-phase PVDF. The well-dispersed MWCNT particles could first play a heterogeneous nucleation role in the PVDF matrix, so effectively reducing the critical free-energy barrier of the cells and increasing the number of cell nucleation; simultaneously enhancing the strength of the cell walls, thus stabilizing the cell structure and protecting the cells from deterioration under the stretching force produced during the process of cell growth.

In addition, it has been reported that the flow field in melt processing can promote conformation ordering [55]. Flow-induced crystallization can trigger the preferred alignment of PVDF chains with TTTT conformation, allowing the dipoles of the polymer chains to align normally to the direction of the applied stress. Since microinjection molding can generate a strong flow field, it facilitated PVDF chains to align in extended TTTT confirmation. With the incorporation of dopamine-functionalized carbon nanotubes on the oriented molecules, the restricted molecular relaxation was restricted due to the strong interaction between the O–H and F–C groups [40]. Therefore, the composite presented extraordinary piezoelectric properties ($V_{oc} \sim 2.6$ V) without polarization treatment.

Furthermore, the surface modification of the fillers in the piezoelectric matrix can lead to an increase in piezoelectric properties without electrical poling as well [56]. Chowdhury et al. used polyethylene glycol (PEG) to modify the surface of 3D lithium-doped zinc oxide (Li–ZnO) and then mechanically mixed it with PVDF and MWCNT [57]. The surface-modified Li–ZnO significantly increased the *in situ* $\beta$-phase PVDF growth, while the MWCNT improved the electrical conductivity of the composite and reinforced the composite during mechanical stressing. The $V_{oc}$ and $I_{sc}$ can reach up to 10.1 V and 2740 μA when subjected to bending motions. Beyond the improvement in the dielectric constant and piezoelectric coefficient, studies have shown that the geometry of piezoelectric materials has a significant impact on the output performance of the piezoelectric nanogenerators [20]. Han et al. theoretically and experimentally investigated helical PVDF–TrFE nanobelt, graphene/PVDF–TrFE nanobelt, and MWCNT/PVDF–TrFE nanobelt using a direct-write technique (Figure 7.6) [58]. Due to the geometrical stress

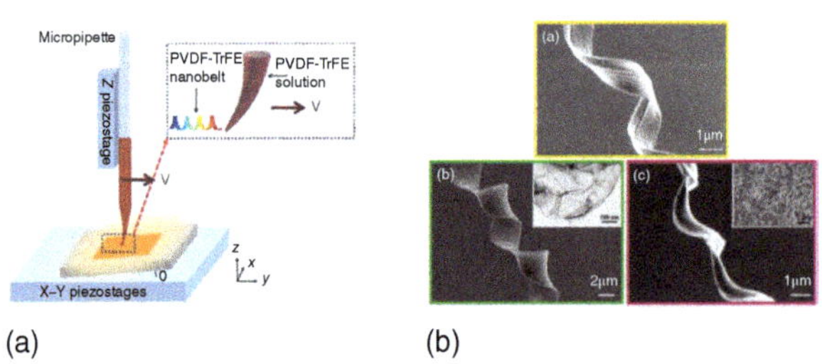


**Figure 7.6** (A) Schematic of the 3D-printing system for the PVDF–TrFE nanobelt. (B) SEM images of PVDF–TrFE nanobelt, graphene/PVDF–TrFE hybrid nanobelt, and MWCNT/PVDF–TrFE hybrid nanobelt. Source: Han et al. [58]. Reproduced with permission of John Wiley & Sons, Inc.

confinement effect, the helical nanobelt structure possesses a higher $d_{33}$, and the addition of carbon materials further promoted the formation of the $\beta$-phase of the PVDF–TrFE. The generated voltage of this graphene/PVDF–TrFE nanobelt can reach 35 mV [58].

#### 7.2.2.2 Graphene and Graphene-Based Materials

Graphene has been employed to nucleate the electroactive phase in PVDF that exhibits superior electrical and mechanical properties [21, 59, 60]. For instance, a hybrid composite was fabricated by mixing the PVDF–HFP powder into one to five layers of graphene and carbon black with *N*,*N*-dimethylformamide (DMF) [60]. A step-wise poling was employed to align the dipoles of the $\beta$-phase further. The filler effectively enhanced the phase transformation in the initial crystallization stage, and the high fraction of $\beta$-phase could be achieved at a low concentration of fillers. Consequently, the $V_{oc}$ of the composite was 81% higher than that of the pure PVDF–HFP films. Another 3D-structured composite with graphene aerogel (GA) was prepared using a vacuum-assisted impregnation process followed by freeze-drying. The GA/PVDF composite presented the highest $d_{33}$ of 16 pN/C with an average current of 0.51 μA and showed good stability under repeated pressing [21]. Furthermore, a plasmonic nanogenerator was developed by coupling plasmonic characteristics of graphene–silver (GAg) nanocomposite with the piezoelectricity of PVDF [61]. Figure 7.7 shows the GAg-induced self-polarization among the PVDF chains. The plasmonic behavior of GAg nanocomposite makes the nanogenerator sensitive to visible light illumination, and this plasmonic nanogenerator exhibited the highest conversion efficiency up to ~46.6% when illuminating green light. Another similar flexible self-poled piezoelectric nanogenerator based on an rGO–Ag/PVDF nanocomposite was developed [62]. The nanogenerator represented

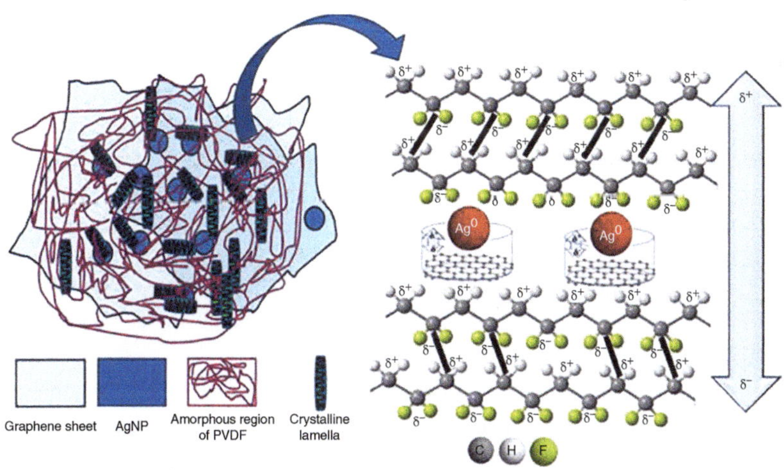


**Figure 7.7** Schematic of GAg-induced self-polarization among the PVDF chains within GAg-doped PVDF hybrid nanocomposite and the existence of GAg within the crystalline lamella of the nucleated electroactive phase through this self-polarization. Source: Sinha et al. [61]/with permission of American Chemical Society.

a peak $V_{oc}$ of 18 V, $I_{sc}$ of 1.05 μA, and power density of 28 $Wm^{-3}$ when connected to a 1 MΩ resistor.

The introduction of GO/rGO fillers into the PVDF matrix could result in strong interaction between the $-CF_2$ or $-CH_2$ groups of PVDF polymer and the –COOH or –OH groups contributed by GO and rGO. Consequently, the mixture of PVDF–GO/rGO can achieve almost pure $\beta$-phase orientation under the stimuli of uniaxially stretched at 80 °C. The presence of GO/rGO could enhance the device performance, including conductivity, permittivity, and dielectric constant, through the formation of conductive networks in the PVDF matrix [63]. In another study, the addition of GO lamellae to the PVDF created heterogeneous nucleating agents in crystallization during the electrospinning process, which reinforces the residual orientation of the $\beta$-phase. The distribution of GO lamellae on the outer shell and PVDF chains self-orientation at the inner fiber section resulted in the production of core–shell-structured GO/PVDF nanofibers with enhanced $\beta$-phase content. The piezoelectricity of composite material increased four-fold compared to pure PVDF [44].

On the other hand, studies have shown that rGO is more effective in improving the dielectric constant than GO [64–66]. By comparing the FTIR and XRD results of PVDF/GO and PVDF/rGO, researchers found that the inclusion of GO can obtain more $\beta$-phase than rGO because rGO contains less amount of carbonyl group to nucleate all PVDF chains into the required trans–trans conformation of $\beta$-phase [64]. When subjected to the same acceleration at the same frequency, the peak–peak voltages of PVDF/rGO, PVDF/GO, and PVDF were ~1.3, 0.9, and 0.7 V, respectively [64]. Another flexible PVDF/Bismuth aluminate ($Bi_2Al_4O_9$)/rGO composite was fabricated, and the fraction of $\beta$-phase enhanced from 53% to 76% with the incorporation of $Bi_2Al_4O_9$ and rGO nanofillers [67]. Later, Mishra et al. studied the combined effect of chemical addition and mechanical deformation by mechanically stretching hybrid PVDF/GO and PVDF/rGO composites at high temperatures. They achieved almost pure $\beta$-phase when stretched uniaxially at 80 °C under a small strain of ~200% [63]. Unlike mixing carbon material into piezoelectric material, Bhavanasi et al. developed bilayer films of poled PVDF–TrFE and GO layer. The charged GO with high Young's modulus and high dielectric constant enhanced the output performance of the PVDF–TrFE. The device possessed two times the $V_{oc}$ and 2.5 times the power output than that of the pure PVDF–TrFE film, which is attributed to the electrostatic contribution of GO film, internal stress in PVDF–TrFE film, and increased Young's modulus of the bilayer film when subjected to a force [68].

#### 7.2.2.3 Quantum Dots

Zero-dimensional nanostructured materials, such as quantum dots (QDs), have raised great attention [69] since their ultrasmall size produces quantum confinement effects and edge effects [70–72]. As a result of their spatial properties, they are subject to many extraordinary optical, electronic, and chemical properties that are not found in other materials. The size dependence shows that those properties can be easily controlled by tuning the size of QDs [73]. Since the metal-based QDs, such as cadmium selenide (CdSe) and cadmium telluride (CdTe), have drawbacks

**Figure 7.8** SEM images of the etched fracture surfaces of GQD/PVDF composite, with (a) 1.0 wt%, (b) 2.0 wt%, and (c) 3.0 wt% GQD loadings, respectively, crystallized at 300 MPa, 230 °C for 30 minutes. Source: Lu et al. [75]. Reproduced with permission from Royal Society of Chemistry.

of environmental toxicity, weak biocompatibility, and lack of stability, the carbon QDs (CQDs) [74] and graphene QDs (GQDs) [75, 76] have attracted much attention for applications of environmental engineering, bioimaging, and medical treatment. CQDs are quasi-spherical nanoparticles smaller than 10 nm in diameter, and the GQDs are single or few-layered graphene with lateral dimensions less than 100 nm. Ma et al. fabricated a novel PVDF/CQD-based poling-free piezoelectric photoluminescent composite using a solution-casting method followed by a high-pressure crystallization process [74]. The 3D quasi-spherical CQDs have *in situ* induced the self-assembly of polymorphic substructures in the PVDF crystallites at high pressure. Without any electrical polarization treatment, the maximum $V_{oc}$ density of the durable composite system reached 19.2 V cm$^{-2}$ and $I_{sc}$ density of 550 nA cm$^{-2}$, far exceeding that of pure PVDF. In addition, the composites showed excellent fluorescent effect with high stability, and its multicolor photoluminescence is well retained even after the endurance for a rigorous treatment at high pressure and high temperature. Similar to the PVDF/CQD composite, a facile approach developed the GQD self-assembled composite [75]. As shown in Figure 7.8, at high pressure, the size-distributed GQD aggregates *in situ* catalyzed the self-assembly of PVDF molecules into crystalline beta form 1D nanowires and 3D micro/nanowire architectures. The GQD/PVDF with 3 wt% GQD-generated $V_{oc}$ and $I_{sc}$ of 0.75 V and 25.7 nA is more than four times higher than the pure PVDF, respectively. In addition, the wettability of the surface is able to be modulated by controlling GQD concentration with chemical etching [75].

## 7.3 Fabrication Methods

Mixing carbon fillers into organic piezoelectric materials can enhance the output performance. On the other hand, fabrication methods, particularly electrospinning technology, can aid with the carbon fillers to further enhance the performance of the organic piezoelectric composite materials. Electrospinning is a simple, cost-effective, and versatile technology for producing ultra-long continuous micro/nanoscale fibers. A typical electrospinning device [77, 78] consists of a

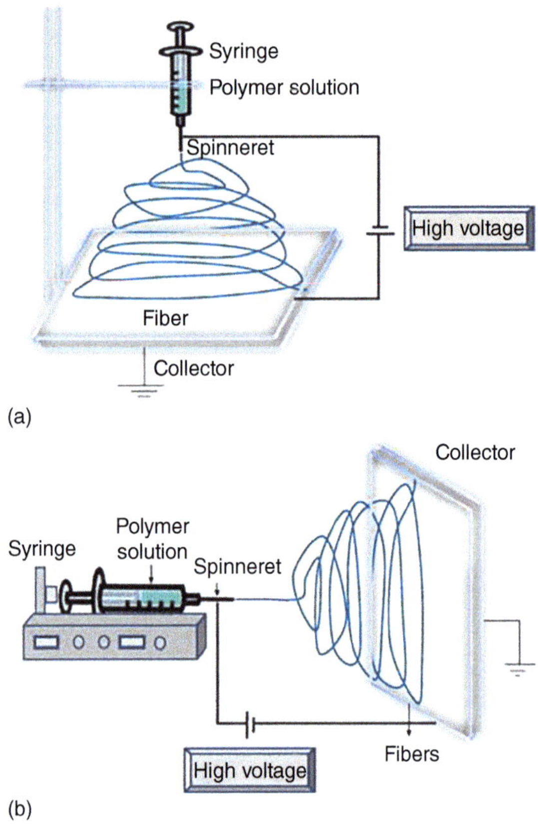


**Figure 7.9** Schematic for deposition of PVDF nanofibers around conductive core yarn by electrospinning (a) vertical and (b) horizontal electrospinning set up. Source: Bharadwaj and Kundu [78]/with permission of Elsevier.

syringe and a metal needle connected to a high-voltage DC power supply, as shown in Figure 7.9. When the polymer solution is ejected through the metal tip, under the action of the high-voltage electric field, the surface of the droplet at the tip of the syringe forms a tapered shape called the "Taylor cone." The formation of this shape is due to the combination of electrostatic repulsion between charges and Coulomb force [77]. When the polymer solution jet is ejected from the tip of the syringe and deposited on the grounded collector, mechanical stretching and electrical polarization occur simultaneously. Therefore, electrospinning can be used to induce the redirection of piezoelectric materials and form dipoles, so no subsequent polarization process is required [79–83]. The parameters of electrospinning, such as the applied voltage and the distance between the electrode and the collector, solvent materials, polymer molecular weight and concentration, and additives, have considerable influence on the morphology and properties of the fiber. Subsequent treatments, such as annealing, cooling, polarization, and compression, can be used to improve further the crystallinity of the polar phase of the piezoelectric fiber [82, 83].

Electrospinning, due to its unique electric polarization and mechanical stretching effect, is conducive to the formation of $\beta$-phase or induces the transformation of $\alpha$- to $\beta$-phase [84]. The $\beta$-phase content depends on the solvent type, additives, and electrospinning parameters [85–87]. It provides the alignment of molecular dipoles with a single-processing stage and the formation of a $\beta$-phase that has all dipolar moments pointing in the same direction.

For example, a novel sound-absorbing material was developed using electrospun PVDF membranes. Wu et al. showed that electrospinning effectively induced the $\beta$-phase formation and increased piezoelectricity of the PVDF membrane. Adding CNTs further improved the piezoelectricity through interfacial polarization, and the electrospun nanofiber membranes exhibited high-surface area, providing a large number of contact sites with the sound waves [88]. With the same method, Karumuthil et al. electrospun the hybrid nanocomposite consisting of PVDF, ZnO, and exfoliated graphene oxide (EGO) to form a piezoelectric film, as shown in Figure 7.10a. Consequently, the prepared film showed a better $\beta$-phase without thermal or electrical poling [89]. The addition of nanoparticles to the polymer solution increased the chances of enhancement of $\beta$-phase by acting as nucleating sites in the polymer. Electrospinning can again ease this enhancement due to the strain induced during fiber formation. EGO was added to get a uniform ZnO dispersion in the polymer to prevent agglomeration, hence inducing more nucleating sites in the matrix. In Figure 7.10a, when the polymer fiber is subjected to a force, the dipoles that are about to align become oriented, leading to charge separation in the bulk of the polymer. It leads to charge separation on the electrode, which is obtained as usable electrical energy at the external terminal of the device. When releasing the force, the aligned dipoles undergo a partial realignment, resulting in a decrease in potential. The cycle repeated every impact, thus generating alternating energy at the output of the device.

On the other hand, Roy et al. demonstrated electrospun PVDF/GO nanofibers for highly sensitive self-powered wearable pressure sensors and pyroelectric breathing sensors, as shown in Figure 7.10b [17]. In their work, GO was homogeneously dispersed in the PVDF matrix, and the $\beta$-phase nucleation occurred in the PVDF/GO composite due to the interaction between the negatively charged GO nanosheet and the PVDF fiber dipoles. In addition, the intensity of the electroactive phase gradually increased with increasing GO concentration since the electrostatic interaction between the $\pi$-electron cloud and the oxygen-containing functional groups in GO and the $-CH_2/-CF_2$ dipoles of PVDF. Resultantly, the device could generate a maximum output power density of $\sim$6.2 mW m$^{-2}$ when subjected to compressive stress, enhancing its range of applications. Moreover, it was demonstrated that doping with GO improves the pyroelectric energy harvesting and sensing performance of the device under repeated temperature fluctuations. The PVDF/GO-based nanogenerator possessed a maximum pyroelectric output power density of $\sim$1.2 nW m$^{-2}$ and could sense temperature changes during respiration.

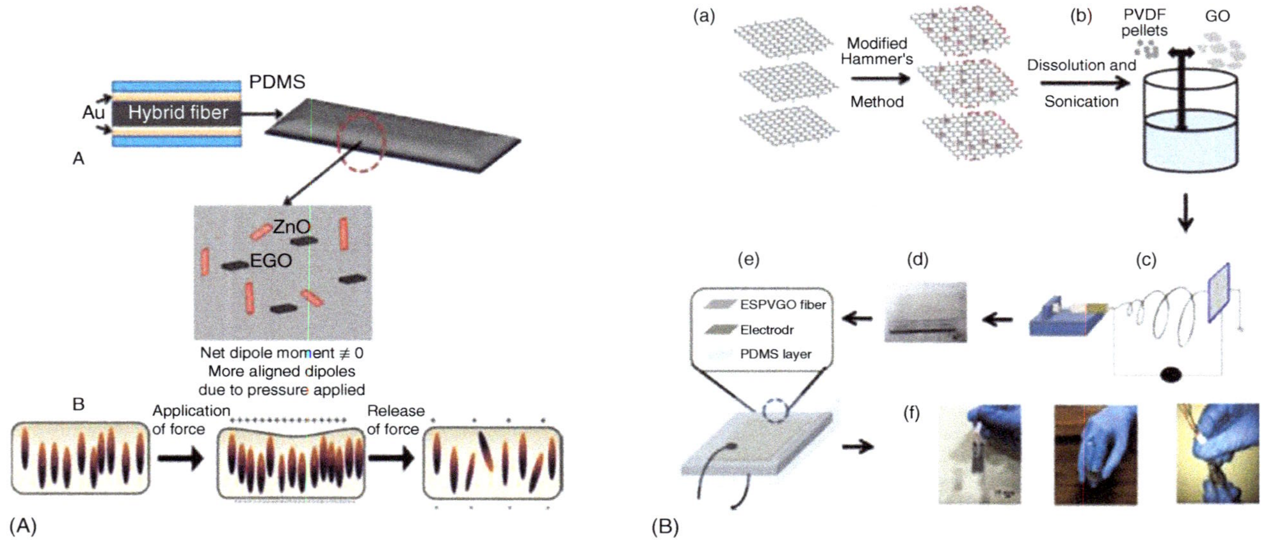


**Figure 7.10** (A) Schematic representation of electrospun fiber-based hybrid nanogenerator. Electrospun fiber with nanoparticles and feasible mechanism of charge generation. Source: Cherumannil Karumuthil et al. [89]/with permission of American Chemical Society. (B) Schematic route for fabricating electrospun PVDF/GO nanofibers for highly sensitive self-powered wearable pressure, and pyroelectric breathing sensors. Source: Roy et al. [17]. Reproduced with permission of American Chemical Society.

## 7.4 Applications

Since piezoelectric materials can directly convert mechanical energy into electrical energy due to the piezoelectric effect, piezoelectric materials with enhanced performance are significant in a wide range of applications, such as energy harvesters, wearable electronics, and biomedical devices [90].

### 7.4.1 Energy Harvesters

The most common application of piezoelectric composite is an energy harvester that converts mechanical waste energy into useful electrical energy and provides an alternative to conventional batteries with the potential to deliver sustainable energy to supply low-power electronic devices using ambient forms of energy [10, 91]. For example, as shown in Figure 7.11a, the core–shell BT@C/PVDF–TrFE composite is attached to the rear wheel of a bicycle to convert energy from wheel rotation into electrical energy, which can direct light up 12 white LEDs in a dark environment [52]. It can also charge a 1 μF capacitor to 3.25 V in 64 seconds. Another PVDF/$Bi_2Al_4O_9$/rGO-based nanocomposite generated a maximum output power density of 0.457 $\mu W\,cm^{-2}$ at 12 MΩ resistance, which is around 20 times more than that of bare PVDF film [52].

### 7.4.2 Biomechanical Sensor

Piezoelectric materials can be used as self-powered sensors by collecting mechanical signals, especially physiological signals. For physiological signal monitoring, highly sensitive, lightweight, low cost, flexible, compliant, biocompatible, and nontoxic are highly desirable for self-powered sensors [41]. Flexible lead-free piezoelectric nanocomposites (nontoxic) can be easily attached to the human body, thus having the potential to monitor breath, heart rate, and human motion. The monitoring of human kinematics, for example, the movement of limbs, joints, and feet, can reflect the force distribution during patients' daily activities and provides posture misalignment to guide rehabilitation training. It is becoming important for the future industrial and daily life but not limited to medical care. From the different strengths and frequencies, the output varies with the different forms of human movement; therefore, human kinematics can be monitored in real-time.

For example, a fully functioning shoe-insole nanogenerator was developed with soft-PZT/graphene nanopowder (GNP)/silicone. It successfully demonstrated its practical functionality by exhausting the real-life situation of human walking/running and generating peak $V_{oc}$, $I_{sc}$, and power density of about 27 V, 429 μA, and 402 $mW\,m^{-2}$, respectively [15]. In another study, MWCNT/ZnO/PDMS was attached to a slipper and could generate the voltage and current of ~75 V/3.2 μA, ~150 V/8 μA, and ~400 V/30 μA while walking, running, and jumping, respectively (Figure 7.11b) [27]. In addition, the movement of the joint can provide a bending force; thus, the self-powered sensor can generate electrical signals to detect

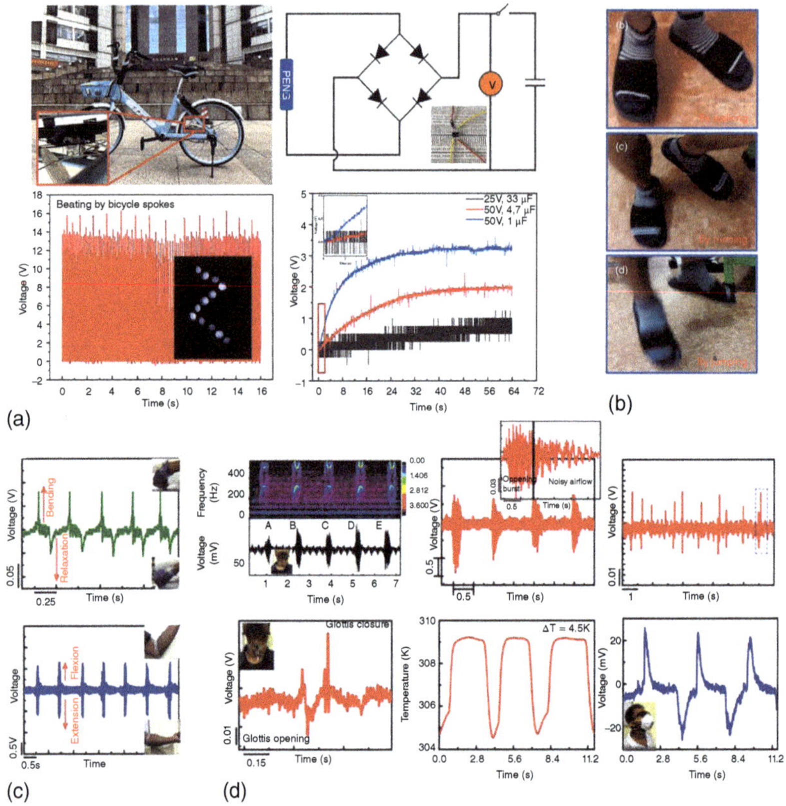


**Figure 7.11** (A) Piezoelectric nanogenerator is attached to the rear wheel of a bicycle to convert energy from wheel rotation into electrical energy. The generator lighted 12 white LEDs and charged different capacitors with a rectifier circuit. Source: Zhou et al. [52]. Reproduced with permission of American Chemical Society. (B) Demonstration of the practical or real-time applications of the MWCNT/ZnO/PDMS device to harvest human activities, such as walking, running, and jumping. Source: Kim et al. [27]. Reproduced with permission of American Chemical Society. (c) Photograph of wrist joint bending and stretching when the piezoelectric sensor was attached to the surface of the wrist and elbow joint. (d) Recorded output voltage response curves of the piezoelectric sensor attached to the throat of a male participant while speaking of "A," "B," "C," "D," and "E." The response of the sensor when monitoring a coughing action. Responsive curves of a swallowing exercise. The temperature fluctuation and recorded $V_{oc}$ of the sensor (photo of the GPPNG attached to an N95 breathing mask is shown in the inset) are driven by the human respiratory at room temperature. Source: Roy et al. [17]. Reproduced with permission of American Chemistry Society.

movement. For example, Roy et al. fixed the electrospun self-powered PVDF/GO multifunctional sensor to the wrist and elbow. The sensor was able to distinguish the flexion and extension of the elbow (Figure 7.11c) [17]. It can also be utilized to detect and monitor phonation (speech) while attaching the piezoelectric sensor to the throat (as shown in Figure 7.11d), which is helpful for speech rehabilitation training [92]. This is because, during the pronunciation of different alphabetic characters, the stresses originating from vocal vibrations due to the complex muscle movement all over the throat are different. Furthermore, respiration rate is one of the key parameters for diagnosing the illness. Unusual breathing patterns are also associated with different health conditions. For example, Cheyne–Stokes' breathing is an accurate sign of left ventricular dysfunction, especially for older patients. Since most piezoelectric materials are pyroelectric, the piezoelectric sensor can also monitor human breath [17]. Furthermore, weak physiological signals, such as pulse/heart rate, can be real-time detected with these sensors [51].

### 7.4.3 Other Applications

Piezoelectric polymer materials, especially PVDF, have become one of the most promising polymer matrices for dielectric nanocomposites due to their high dielectric constant, low dielectric loss, and thermal stability [93]. The addition of electrically conductive nanofillers (such as CNT [94] and graphene [95]) is an effective approach for obtaining polymeric dielectric nanocomposites. The good dispersion of MWCNT particles in the PVDF matrix was achieved with ball milling, which promoted the heterogeneous nucleation and enhancement effects of MWCNT on PVDF, resulting in the formation of more unique and denser cells, e.g., from $3.4 \times 10^5$ cells $cm^{-3}$ of pure foam to $1.7 \times 10^6$ cells $cm^{-3}$ of the composite foam when MWCNT content was 1.5%, and high compression strength, i.e., 3.1 MPa with 1.5% MWCNT at 10% compressive strain, about 1.1 MPa higher than that of pure foam. During the bead foaming process, the induced crystallization effect of MWCNT as well as the plasticizing and squeezing effect of $scCO_2$ on the PVDF matrix contributed to the increase of the $\beta$-phase crystal in the PVDF matrix, thus endowing the composite foaming part with good piezoelectric property, i.e., 6.5 V of output voltage under periodic compression [94]. In another study, a supercapacitor was developed using ZnO nanorods@carbon black/nickel foam as the positive electrode and activated carbon-coated nickel foam as the negative electrode in the $Na_2SO_4$ electrolyte [96]. Recently, piezoelectric energy harvesters can generate an electrical charge as a result of a change in a polarization of a material due to the application of stress. The electrical charge can be coupled with electrochemical systems for a wide range of potential applications, including water splitting, water treatment, and corrosion protection [97, 98]. Chen et al. studied the strong piezocatalysis in $BaTiO_3$/carbon hybrid nanocomposites for dye wastewater decomposition [99]. When the carbon content was 2 wt%, the $BaTiO_3$/carbon hybrid nanocomposites exhibited the optimum piezocatalytic performance, with the ~75.5% dye decomposition ratio and the ~0.04901 $min^{-1}$ reaction rate constant after 40 minutes of vibration stimulation. At the same time, pure $BaTiO_3$ is 48.4% and 0.01942 $min^{-1}$, respectively. The improvement of piezocatalytic performance in $BaTiO_3$/carbon hybrid nanocomposites can be ascribed to the action of carbon's charge transfer, which promotes the effective separation of the piezoelectrically-induced electric charges [99]. On the other hand, as mentioned in Section 7.2.2, the flexible plasmonic nanogenerator based on GAg/PVDF can also realize a new class of self-powered optical sensors [61].

## 7.5 Conclusion

This chapter focuses on the enhanced carbon materials, particularly nanomaterials, introduced into the piezoelectric materials (inorganic and organic) to improve their piezoelectricity. Carbon nanotubes, graphene-based, and various types of quantum dots among the nanomaterials are reported to enhance the $\beta$-phase of the PVDF and its copolymer. Here, the mechanism of the piezoelectricity improvement is reviewed accordingly. Fabrication methods, such as electrospinning, are promising techniques that can be used to induce the redirection of piezoelectric materials and hence do not require electrical poling. The significant electrical signals generated by piezoelectric materials are crucial for many applications, primarily energy harvesters, wearable electronics, and biomedical devices. However, the challenges, especially a charge-storage issue, must be solved before the broad implementation of the energy harvesters. The charges generated by the piezoelectric energy harvester need to be stored in an appropriate energy-storage system to ensure the undisrupted electrical energy supply that meets the Internet of Things technology.

## Acknowledgment

The authors would like to acknowledge the support of the National Natural Science Foundation of China (grant number 62001325).

## References

1 Tressler, J.F., Alkoy, S., and Newnham, R.E. (1998). Piezoelectric sensors and sensor materials. *Journal of Electroceramics* 2 (4): 257–272. https://doi.org/10.1023/A:1009926623551.

2 Hu, Y. and Wang, Z.L. (2014). Recent progress in piezoelectric nanogenerators as a sustainable power source in self-powered systems and active sensors. *Nano Energy* 14: 3–14. https://doi.org/10.1016/j.nanoen.2014.11.038.

3 Briscoe, J. and Dunn, S. (2015). Piezoelectric nanogenerators–a review of nanostructured piezoelectric energy harvesters. *Nano Energy* 14: 15–29.

4 Gautschi, G. (2002). Piezoelectric sensors. In: *Piezoelectric Sensorics*, 73–91. Springer.

5 Uchino, K. (1996). *Piezoelectric Actuators and Ultrasonic Motors*, vol. 1. Springer Science & Business Media.

6 Sharma, K., Arora, A., and Tripathi, S.K. (2019). Review of supercapacitors: materials and devices. *Journal of Energy Storage* 21: 801–825.

7 Maeder, M.D., Damjanovic, D., and Setter, N. (2004). Lead free piezoelectric materials. *Journal of Electroceramics* 13 (1): 385–392.

8 Panda, P.K. (2009). Environmental friendly lead-free piezoelectric materials. *Journal of Materials Science* 44 (19): 5049–5062.

**9** Zhang, Y. et al. (2017). Enhanced pyroelectric and piezoelectric properties of PZT with aligned porosity for energy harvesting applications. *Journal of Materials Chemistry A* 5 (14): 6569–6580. https://doi.org/10.1039/C7TA00967D.

**10** Xie, M., Zhang, Y., Kraśny, M.J. et al. (2018). Flexible and active self-powered pressure, shear sensors based on freeze casting ceramic–polymer composites. *Energy & Environmental Science* 11 (10): 2919–2927. https://doi.org/10.1039/C8EE01551A.

**11** Ramaratnam, A. and Jalili, N. (2006). Reinforcement of piezoelectric polymers with carbon nanotubes: pathway to next-generation sensors. *Journal of Intelligent Material Systems and Structures* 17 (3): 199–208. https://doi.org/10.1177/1045389X06055282.

**12** Guo, H., Ma, C., Liu, X., and Tan, X. (2013). Electrical poling below coercive field for large piezoelectricity. *Applied Physics Letters* 102 (9): 92902. https://doi.org/10.1063/1.4794866.

**13** Lee, C. and Tarbutton, J.A. (2014). Electric poling-assisted additive manufacturing process for PVDF polymer-based piezoelectric device applications. *Smart Materials and Structures* 23 (9): 95044. https://doi.org/10.1088/0964-1726/23/9/095044.

**14** Ray, M.C. and Batra, R.C. (2007). A single-walled carbon nanotube reinforced 1–3 piezoelectric composite for active control of smart structures. *Smart Materials and Structures* 16 (5): 1936.

**15** Choudhry, I., Khalid, H.R., and Lee, H.-K. (2020). Flexible piezoelectric transducers for energy harvesting and sensing from human kinematics. *ACS Applied Electronic Materials* 2 (10): 3346–3357. https://doi.org/10.1021/acsaelm.0c00636.

**16** Chen, S. et al. (2020). Fabrication and piezoresistive/piezoelectric sensing characteristics of carbon nanotube/PVA/nano-ZnO flexible composite. *Scientific Reports* 10 (1): 8895. https://doi.org/10.1038/s41598-020-65771-x.

**17** Roy, K. et al. (2019). A self-powered wearable pressure sensor and pyroelectric breathing sensor based on GO interfaced PVDF nanofibers. *ACS Application Nano Materials* 2 (4): 2013–2025. https://doi.org/10.1021/acsanm.9b00033.

**18** Das, S., Biswal, A.K., Parida, K. et al. (2018). Electrical and mechanical behavior of PMN-PT/CNT based polymer composite film for energy harvesting. *Applied Surface Science* 428: 356–363. https://doi.org/10.1016/j.apsusc.2017.09.077.

**19** Liu, F., Hashim, N.A., Liu, Y. et al. (2011). Progress in the production and modification of PVDF membranes. *Journal of Membrane Science* 375 (1–2): 1–27.

**20** Kim, S.M. et al. (2012). Radially dependent effective piezoelectric coefficient and enhanced piezoelectric potential due to geometrical stress confinement in ZnO nanowires/nanotubes. *Applied Physics Letters* 101 (1): 13104. https://doi.org/10.1063/1.4731779.

**21** Guo, Y. et al. (2020). Ultralight graphene aerogel/PVDF composites for flexible piezoelectric nanogenerators. *Compos. Commun.* 22: 100542.

**22** Park, S.-E. and Shrout, T.R. (1997). Characteristics of relaxor-based piezoelectric single crystals for ultrasonic transducers. *IEEE Transactions on Ultrasonics,*

*Ferroelectrics, and Frequency Control* 44 (5): 1140–1147. https://doi.org/10.1109/58.655639.

**23** Park, K.-I., Jeong, C.K., Ryu, J. et al. (2013). Flexible and large-area nanocomposite generators based on lead zirconate titanate particles and carbon nanotubes. *Advanced Energy Materials* 3 (12): 1539–1544. https://doi.org/10.1002/aenm.201300458.

**24** Han, J.K., Jeon, D.H., Cho, S.Y. et al. (2017). Flexible piezoelectric generators by using the bending motion method of direct-grown-PZT nanoparticles on carbon nanotubes. *Nanomaterials (Basel, Switzerland)* 7 (10). Department of Physics and Research Institute of Physics and Chemistry, Chonbuk National University, Jeonju 54896, Korea: https://doi.org/10.3390/nano7100308.

**25** Zhao, M.-H., Wang, Z.-L., and Mao, S.X. (2004). Piezoelectric characterization of individual zinc oxide nanobelt probed by piezoresponse force microscope. *Nano Letters* 4 (4): 587–590. https://doi.org/10.1021/nl035198a.

**26** Liu, H., Zhong, J., Lee, C. et al. (2018). A comprehensive review on piezoelectric energy harvesting technology: materials, mechanisms, and applications. *Applied Physics Reviews* 5 (4): https://doi.org/10.1063/1.5074184.

**27** Kim, D.H., Dudem, B., and Yu, J.S. (2018). High-performance flexible piezoelectric-assisted triboelectric hybrid nanogenerator via polydimethylsiloxane-encapsulated nanoflower-like ZnO composite films for scavenging energy from daily human activities. *ACS Sustainable Chemistry & Engineering* 6 (7): 8525–8535. https://doi.org/10.1021/acssuschemeng.8b00834.

**28** Banerjee, S., Bairagi, S., and Wazed Ali, S. (2021). A critical review on lead-free hybrid materials for next generation piezoelectric energy harvesting and conversion. *Ceramics International* 47 (12): 16402–16421. https://doi.org/10.1016/j.ceramint.2021.03.054.

**29** Acosta, M. et al. (2017). $BaTiO_3$-based piezoelectrics: fundamentals, current status, and perspectives. *Applied Physics Reviews* 4 (4): 41305. https://doi.org/10.1063/1.4990046.

**30** Pan, Z. et al. (2015). Enhanced piezoelectric properties and thermal stability in the $(K_{0.5}Na_{0.5})NbO_3$:ZnO lead-free piezoelectric composites. *Journal of the American Ceramic Society* 98 (12): 3935–3941. https://doi.org/10.1111/jace.13831.

**31** Park, K.-I. et al. (2012). Flexible nanocomposite generator made of $BaTiO_3$ nanoparticles and graphitic carbons. *Advanced Materials* 24 (22): 2999–3004. https://doi.org/10.1002/adma.201200105.

**32** Tang, Y., Chen, L., Duan, Z. et al. (2020). Graphene/barium titanate/polymethyl methacrylate bio-piezoelectric composites for biomedical application. *Ceramics International* 46 (5): 6567–6574. https://doi.org/10.1016/j.ceramint.2019.11.142.

**33** Geim, A.K. and Novoselov, K.S. (2007). The rise of graphene. *Nature Materials* 6 (3): 183–191. https://doi.org/10.1038/nmat1849.

**34** Kang, G. and Cao, Y. (2014). Application and modification of poly(vinylidene fluoride) (PVDF) membranes–a review. *Journal of Membrane Science* 463: 145–165.

**35** Li, L., Zhang, M., Rong, M., and Ruan, W. (2014). Studies on the transformation process of PVDF from α to β phase by stretching. *RSC Advances* 4 (8): 3938–3943. https://doi.org/10.1039/C3RA45134H.

**36** Hu, Z., Tian, M., Nysten, B., and Jonas, A.M. (2009). Regular arrays of highly ordered ferroelectric polymer nanostructures for non-volatile low-voltage memories. *Nature Materials* 8 (1): 62–67. https://doi.org/10.1038/nmat2339.

**37** Xu, H. (2001). Dielectric properties and ferroelectric behavior of poly(vinylidene fluoride-trifluoroethylene) 50/50 copolymer ultrathin films. *Journal of Applied Polymer Science* 80 (12): 2259–2266. https://doi.org/10.1002/app.1330.

**38** Ameduri, B. (2009). From vinylidene fluoride (VDF) to the applications of VDF-Containing polymers and copolymers: Recent developments and future trends. *Chemical Reviews* 109 (12): 6632–6686. https://doi.org/10.1021/cr800187m.

**39** Thongsanitgarn, P., Watcharapasorn, A., and Jiansirisomboon, S. (2010). Electrical and mechanical properties of PZT/PVDF 0–3 composites. *Surface Review and Letters* 17 (01): 1–7.

**40** Guo, J., Nie, M., and Wang, Q. (2020). Self-Poling polyvinylidene fluoride-based piezoelectric energy harvester featuring highly oriented β-phase structured at multiple scales. *ACS Sustainable Chemistry & Engineering* 9 (1): 499–509.

**41** Mokhtari, F., Azimi, B., Salehi, M. et al. (2021). Recent advances of polymer-based piezoelectric composites for biomedical applications. *Journal of the Mechanical Behavior of Biomedical Materials* 122: 104669.

**42** Singh, H.H., Singh, S., and Khare, N. (2018). Enhanced β-phase in PVDF polymer nanocomposite and its application for nanogenerator. *Polymers for Advanced Technologies* 29 (1): 143–150.

**43** Ting, Y., Gunawan, H., Sugondo, A., and Chiu, C.-W. (2013). A new approach of polyvinylidene fluoride (PVDF) poling method for higher electric response. *Ferroelectrics* 446 (1): 28–38. https://doi.org/10.1080/00150193.2013.820983.

**44** Cardoso, V.F., Minas, G., and Lanceros-Méndez, S. (2013). Multilayer spin-coating deposition of poly(vinylidene fluoride) films for controlling thickness and piezoelectric response. *Sensors and Actuators A: Physical* 192: 76–80. https://doi.org/10.1016/j.sna.2012.12.019.

**45** Chan, H.L.W., Cheung, M.C., and Choy, C.L. (1999). Study on $BaTiO_3$/P (VDF-TrFE) 0–3 composites. *Ferroelectrics* 224 (1): 113–120.

**46** Chen, J.-J., Li, Y., Zheng, X.-M. et al. (2018). Enhancement in electroactive crystalline phase and dielectric performance of novel PEG-graphene/PVDF composites. *Applied Surface Science* 448: 320–330. https://doi.org/10.1016/j.apsusc.2018.04.144.

**47** Salimi, A. and Yousefi, A.A. (2003). Analysis method: FTIR studies of β-phase crystal formation in stretched PVDF films. *Polymer Testing* 22 (6): 699–704.

**48** Kaspar, P. et al. (2020). Characterization of polyvinylidene fluoride (PVDF) electrospun fibers doped by carbon flakes. *Polymers* 12 (12): 2766. https://doi.org/10.3390/polym12122766.

**49** Yadav, P., Raju, T.D., and Badhulika, S. (2020). Self-poled hBN-PVDF nanofiber mat-based low-cost, ultrahigh-performance piezoelectric nanogenerator for biomechanical energy harvesting. *ACS Applied Electronic Materials* 2 (7): 1970–1980. https://doi.org/10.1021/acsaelm.0c00272.

**50** Zhou, Z. et al. (2021). Interface modulated 0-D piezoceramic nanoparticles/PDMS based piezoelectric composites for highly efficient energy harvesting application. *Nano Energy* 82: 105709. https://doi.org/10.1016/j.nanoen.2020.105709.

**51** Li, J.-W. et al. (2020). Enhanced piezoelectric properties of poly (vinylidenefluoride-co-trifluoroethylene)/carbon-based nanomaterial composite films for pressure sensing applications. *Polymers (Basel)* 12 (12): 2999.

**52** Zhou, Z. et al. (2020). Controllable core–shell $BaTiO_3$@carbon nanoparticle-enabled P(VDF-TrFE) composites: a cost-effective approach to high-performance piezoelectric nanogenerators. *ACS Applied Materials & Interfaces* 12 (1): 1567–1576. https://doi.org/10.1021/acsami.9b18780.

**53** Pal, A., Sasmal, A., Manoj, B. et al. (2020). Enhancement in energy storage and piezoelectric performance of three phase (PZT/MWCNT/PVDF) composite. *Materials Chemistry and Physics* 244: 122639.

**54** Xu, D., Zhang, H., Pu, L., and Li, L. (2020). Fabrication of poly (vinylidene fluoride)/multiwalled carbon nanotube nanocomposite foam via supercritical fluid carbon dioxide: synergistic enhancement of piezoelectric and mechanical properties. *Composites Science and Technology* 192: 108108.

**55** Cui, K., Ma, Z., Tian, N. et al. (2018). Multiscale and multistep ordering of flow-induced nucleation of polymers. *Chemical Reviews* 118 (4): 1840–1886.

**56** Cauda, V., Stassi, S., Bejtka, K., and Canavese, G. (2013). Nanoconfinement: an effective way to enhance PVDF piezoelectric properties. *ACS Applied Materials & Interfaces* 5 (13): 6430–6437. https://doi.org/10.1021/am4016878.

**57** Chowdhury, A.R., Jaksik, J., Hussain, I. et al. (2019). Surface-modified nanostructured piezoelectric device as a cost-effective transducer for energy and biomedicine. *Energy Technology* 7 (5): 1800767.

**58** Han, Y., Jiang, C., Lin, H. et al. (2020). Piezoelectric nanogenerators based on helical carbon materials and polyvinyledenedifluoride–trifluoroethylene hybrids with enhanced energy-harvesting performance. *Energy Technology* 8 (6): 1901249.

**59** Garain, S., Jana, S., Sinha, T.K., and Mandal, D. (2016). Design of *in situ* poled $Ce^{3+}$-doped electrospun PVDF/graphene composite nanofibers for fabrication of nanopressure sensor and ultrasensitive acoustic nanogenerator. *ACS Applied Materials & Interfaces* 8 (7): 4532–4540. https://doi.org/10.1021/acsami.5b11356.

**60** Cai, J. et al. (2019). Preparing carbon black/graphene/PVDF-HFP hybrid composite films of high piezoelectricity for energy harvesting technology. *Composites. Part A, Applied Science and Manufacturing* 121: 223–231.

**61** Sinha, T.K. et al. (2016). Graphene-silver-induced self-polarized PVDF-based flexible plasmonic nanogenerator toward the realization for new class of self powered optical sensor. *ACS Applied Materials & Interfaces* 8 (24): 14986–14993. https://doi.org/10.1021/acsami.6b01547.

**62** Pusty, M., Sinha, L., and Shirage, P.M. (2019). A flexible self-poled piezoelectric nanogenerator based on a rGO–Ag/PVDF nanocomposite. *New Journal of Chemistry* 43 (1): 284–294.

**63** Mishra, S., Sahoo, R., Unnikrishnan, L. et al. (2020). Investigation of the electroactive phase content and dielectric behaviour of mechanically stretched PVDF-GO and PVDF-rGO composites. *Materials Research Bulletin* 124: 110732.

**64** Rahman, M.A., Lee, B.-C., Phan, D.-T., and Chung, G.-S. (2013). Fabrication and characterization of highly efficient flexible energy harvesters using PVDF–graphene nanocomposites. *Smart Materials and Structures* 22 (8): 85017.

**65** Shang, J. et al. (2013). Fabrication and enhanced dielectric properties of graphene–polyvinylidene fluoride functional hybrid films with a polyaniline interlayer. *Journal of Materials Chemistry A* 1 (3): 884–890. https://doi.org/10.1039/C2TA00602B.

**66** Ongun, M.Z., Oguzlar, S., Doluel, E.C. et al. (2020). Enhancement of piezoelectric energy-harvesting capacity of electrospun β-PVDF nanogenerators by adding GO and rGO. *Journal of Materials Science: Materials in Electronics* 31 (3): 1960–1968.

**67** Anand, A., Meena, D., and Bhatnagar, M.C. (2020). Synthesis and characterization of flexible PVDF/$Bi_2Al_4O_9$/RGO based piezoelectric materials for nanogenerator application. *Journal of Alloys and Compounds* 843: 156019. https://doi.org/10.1016/j.jallcom.2020.156019.

**68** Bhavanasi, V., Kumar, V., Parida, K. et al. (2016). Enhanced piezoelectric energy harvesting performance of flexible PVDF-TrFE bilayer films with graphene oxide. *ACS Applied Materials & Interfaces* 8 (1): 521–529.

**69** Zhang, Z., Zhang, J., Chen, N., and Qu, L. (2012). Graphene quantum dots: an emerging material for energy-related applications and beyond. *Energy & Environmental Science* 5 (10): 8869–8890. https://doi.org/10.1039/C2EE22982J.

**70** Lim, S.Y., Shen, W., and Gao, Z. (2015). Carbon quantum dots and their applications. *Chemical Society Reviews* 44 (1): 362–381.

**71** LeCroy, G.E. et al. (May 2014). Toward structurally defined carbon dots as ultracompact fluorescent probes. *ACS Nano* 8 (5): 4522–4529. https://doi.org/10.1021/nn406628s.

**72** Fernando, K.A.S. et al. (Apr. 2015). Carbon quantum dots and applications in photocatalytic energy conversion. *ACS Applied Materials & Interfaces* 7 (16): 8363–8376. https://doi.org/10.1021/acsami.5b00448.

**73** Hong, N.H. (ed.) (2019). Chapter 1 - Introduction to nanomaterials: basic properties, synthesis, and characterization. In: *Micro and Nano Technologies*, N. H. B. T.-N.-S. M. M. Hong, Ed., 1–19. Elsevier.

**74** Ma, S. (2018). *et al.*, Nanogenerators begin to light up: a novel poling-free piezoelectric system with multicolor photoluminescence as an efficient mechatronics development platform. *Advanced Materials Interfaces* 5 (19): 1800587.

**75** Lu, C. et al. (2016). Self-powered graphene quantum dot/poly (vinylidene fluoride) composites with remarkably enahanced mechanical-to-electrical conversion. *RSC Advances* 6 (71): 67400–67408.

**76** Badatya, S., Kumar, A., Sharma, C. et al. (2021). Transparent flexible graphene quantum dot-(PVDF-HFP) piezoelectric nanogenerator. *Materials Letters* 290: 129493.

**77** Gao, H., Chen, Z., Zeng, R. et al. (2018). A self-adaptive restoration algorithm for image corrupted with strong noise based on group sparsity residual constraint | 基于组稀疏残差约束的自适应强噪声图像复原算法. *Huanan Ligong Daxue Xuebao/Journal South China Univ. Technol. (Natural Sci.)* 46 (8): 11–18. https://doi.org/10.3969/j.issn.1000-565X.2018.08.002.

**78** Bhardwaj, N. and Kundu, S.C. (2010). Electrospinning: a fascinating fiber fabrication technique. *Biotechnology Advances* 28 (3): 325–347. https://doi.org/10.1016/j.biotechadv.2010.01.004.

**79** Ribeiro, C. et al. (2018). Electroactive poly(vinylidene fluoride)-based structures for advanced applications. *Nature Protocols* 13 (4): 681–704. https://doi.org/10.1038/nprot.2017.157.

**80** Park, S.-H., Lee, H.B., Yeon, S.M. et al. (2016). Flexible and stretchable piezoelectric sensor with thickness-tunable configuration of electrospun nanofiber mat and elastomeric substrates. *ACS Applied Materials & Interfaces* 8 (37): 24773–24781. https://doi.org/10.1021/acsami.6b07833.

**81** Li, T. et al. (2018). Pure OPM nanofibers with high piezoelectricity designed for energy harvesting *in vitro* and *in vivo*. *Journal of Materials Chemistry B* 6 (33): 5343–5352. https://doi.org/10.1039/c8tb01702f.

**82** Lin, M.-F., Xiong, J., Wang, J. et al. (2018). Core-shell nanofiber mats for tactile pressure sensor and nanogenerator applications. *Nano Energy* 44: 248–255. https://doi.org/10.1016/j.nanoen.2017.12.004.

**83** Pan, X. and Sagan, H. (2016). Digital image clustering algorithm based on multi-agent center optimization. *Journal of Digital Information Management* 14 (1): 8–14.

**84** He, Z., Rault, F., Lewandowski, M. et al. (2021). Electrospun PVDF nanofibers for piezoelectric applications: a review of the influence of electrospinning parameters on the β phase and crystallinity enhancement. *Polymers* 13 (2): 174. https://doi.org/10.3390/polym13020174.

**85** Andrew, J.S. and Clarke, D.R. (2008). Effect of electrospinning on the ferroelectric phase content of polyvinylidene difluoride fibers. *Langmuir* 24 (3): 670–672. https://doi.org/10.1021/la7035407.

**86** Shao, H., Fang, J., Wang, H., and Lin, T. (2015). Effect of electrospinning parameters and polymer concentrations on mechanical-to-electrical energy conversion of randomly-oriented electrospun poly(vinylidene fluoride) nanofiber mats. *RSC Advances* 5 (19): 14345–14350. https://doi.org/10.1039/c4ra16360e.

**87** Huang, S. et al. (2008). Electrospinning of polyvinylidene difluoride with carbon nanotubes: synergistic effects of extensional force and interfacial interaction on crystalline structures. *Langmuir* 24 (23): 13621–13626. https://doi.org/10.1021/la8024183.

**88** Wu, C.M. and Chou, M.H. (2016). Polymorphism, piezoelectricity and sound absorption of electrospun PVDF membranes with and without carbon nanotubes. *Composites Science and Technology* 127: 127–133. https://doi.org/10.1016/j.compscitech.2016.03.001.

**89** Cherumannil Karumuthil, S., Prabha Rajeev, S., Valiyaneerilakkal, U. et al. (2019). Electrospun poly (vinylidene fluoride-trifluoroethylene)-based polymer nanocomposite fibers for piezoelectric nanogenerators. *ACS Applied Materials & Interfaces* 11 (43): 40180–40188.

**90** Galpayage Dona, D.G., Wang, M., Liu, M. et al. (2012). Recent advances in fabrication and characterization of graphene-polymer nanocomposites. *Graphene* 1 (2): 30–49.

**91** Bowen, C.R., Kim, H.A., Weaver, P.M., and Dunn, S. (2014). Piezoelectric and ferroelectric materials and structures for energy harvesting applications. *Energy & Environmental Science* 7 (1): 25. https://doi.org/10.1039/C3EE42454E.

**92** Chen, X. et al. (2017). A stretchable and transparent nanocomposite nanogenerator for self-powered physiological monitoring. *ACS Applied Materials & Interfaces* 9 (48): 42200–42209. https://doi.org/10.1021/acsami.7b13767.

**93** Xia, W. and Zhang, Z. (2018). PVDF-based dielectric polymers and their applications in electronic materials. *IET Nanodielectrics* 1 (1): 17–31.

**94** Chen, L. et al. (2020). Chemically functionalized multi-walled CNTs induced phase behaviors of poly (vinylidene fluoride) nanocomposites and its dielectric properties. *Synthetic Metals* 269: 116555.

**95** Yang, C., Hao, S.-J., Dai, S.-L., and Zhang, X.-Y. (2017). Nanocomposites of poly(vinylidene fluoride) - Controllable hydroxylated/carboxylated graphene with enhanced dielectric performance for large energy density capacitor. *Carbon N. Y.* 117: 301–312. https://doi.org/10.1016/j.carbon.2017.03.004.

**96** Jayababu, N. and Kim, D. (2021). ZnO nanorods@conductive carbon black nanocomposite based flexible integrated system for energy conversion and storage through triboelectric nanogenerator and supercapacitor. *Nano Energy* 82: 105726. https://doi.org/10.1016/j.nanoen.2020.105726.

**97** Zhang, Y., Xie, M., Adamaki, V. et al. (2017). Control of electro-chemical processes using energy harvesting materials and devices. *Chemical Society Reviews* 46 (24): https://doi.org/10.1039/c7cs00387k.

**98** Su, Y., Zhang, L., Wang, W. et al. (2018). Enhanced $H_2$ evolution based on ultrasound-assisted piezo-catalysis of modified $MoS_2$. *Journal of Materials Chemistry A* 6 (25): 11909–11915. https://doi.org/10.1039/c8ta03208d.

**99** Chen, L. et al. (2021). Strong piezocatalysis in barium titanate/carbon hybrid nanocomposites for dye wastewater decomposition. *Journal of Colloid and Interface Science* 586: 758–765.

# 8

# Actuators Based On the Carbon-Enhanced Materials

*Mauro Giorcelli*[1,2] *and Mattia Bartoli*[1,2]

[1] *Center for Sustainable Future Technologies – CSFT@POLITO, Via Livorno 60, 10144 Torino, Italy*
[2] *Consorzio Interuniversitario Nazionale per la Scienza e Tecnologia dei Materiali (INSTM), Via G. Giusti 9, 50121 Florence, Italy*

## 8.1 Introduction

Actuators are inspired by natural autonomous systems. They could be defined as "movers", components of a machine able to change their shape under external stimuli (e.g. light, heat, electric, magnetism, and solvent), producing mechanical work to control a mechanism or system [1–3].

Carbon is well recognized as a versatile material to produce actuators [4]. The well-known carbon properties, for example, their electrical and thermal conductivity, make carbon an ideal candidate to produce actuators. Carbon weak reactivity contributes also to the actuators' long duration. At standard temperature and pressure, it resists oxidation and does not react with acids or any alkali metals.

Based on sp bonds, carbon materials can be generally classified into different allotropes, such as (i) graphite, carbon nanotube, and graphene with entire $sp^2$ hybrid bonds, (ii) amorphous carbon with $sp^2$ and $sp^3$ hybrid bonds, (iii) diamond with pure $sp^3$ hybrid bonds, and (iv) carbon chain with sp hybrid bonds. These carbon allotropes have unique merits, including remarkable optical, thermal, and mechanical properties, which are promising for actuator design [5, 6].

These different carbon families are used to produce actuators taking advantage of their different peculiarities.

Here in this chapter, we will introduce the principles for actuation on a molecular scale. Then, we will move to a dedicated section for carbon nanomaterials used to produce actuators. This section will introduce graphene and related materials, carbon nanotubes, and fullerenes, exploring their properties useful to actuators production. A section dedicated to carbon-based actuation will present actuators based on main carbon allotropes, such as carbon nanotube, graphene, graphene oxide, and fullerene. At the end of this chapter, challenges and prospects of actuators based on carbon nanostructures will be presented.

*Enhanced Carbon-Based Materials and Their Applications*, First Edition.
Edited by Poh Choon Ooi, Mengying Xie, and Chang Fu Dee.

## 8.2 Actuation on the Molecular Scale

Molecular actuation stands at the ground level of all biological locomotion systems, from bacteria [7] to human cells [8]. The locomotion of living beings is due to the actuation realized by protein complexes activated through phosphorylation cascade reactions [9]. Biological actuators are significantly elegant and totally unreproducible on a laboratory scale by using common synthetic ways. Nonetheless, actuation on a molecular scale has been deeply studied, and some species have been produced to mimic the complex movements mediated by proteins. Compounds, such as rotaxanes [10] and catenanes [11], represent reliable actuators at a molecular level. These systems showed the great advantage of being light stimuli responsive [12]. Light molecular actuation was accepted worldwide as one of the most astonishing achievements of supramolecular chemistry. In 2016, Savauge, Stoddart, and Feringa were awarded for the first realization of molecular machines.

Furthermore, molecular actuation systems could be based on the isomerization of carbon–carbon double bonds in highly hindered aromatic molecules [13]. Despite their remarkable features, lab-produced molecular actuators lack scalability due to the highly complex multistep production process. A different approach could be based on polymer and polymers composites [14]. This approach is far simpler than the others and can use carbon materials to mediate the stimuli response. Accordingly, in the following sections, an overview of nanostructured carbon-based polymeric actuators is reported.

## 8.3 Carbon Nanomaterials

Among all nanostructured carbon materials, the allotropic shapes of carbon are the fittest for molecular actuation. In this section, their main characteristics are briefly overviewed.

### 8.3.1 Graphene and Related Materials

Graphene is defined as a single plane of graphite. On the graphene plane, carbon atoms composed the elementary cell by three $\sigma$-type bonds with the p orbitals perpendicular to the graphene plane itself. This atomic arrangement leads to the full delocalization of the $\pi$ bond between carbon atoms on the overall graphene plane. As a consequence, electrons of the $\pi$ bond are free to move all along the graphene plane [15–17]. This is the motivation for graphene's high electrical conductivity and carrier rate [18].

The unreached properties of neat graphene are counterbalanced by poor workability and the scarce availability of high-quality materials [19]. Plenty of scientific literature claims the use of graphene by defining nanographite and few-layered materials as if they are real graphene [20, 21]. Overcoming the issues is not easy, several derived species have tentatively replaced real neat graphene. Graphene oxide (GO) represents a reasonable alternative. GO is a highly oxidized graphene form rich in oxygens-containing residual groups [22]. The functionalities of GO basal lattice are represented by epoxide and hydroxyl groups, while carbonyl and carboxylic residues are more detectable on the edges. GO structure is affected by the production way adopted.

GO productive way was established in the middle of the nineteenth century with the studies of the chemical reactivity of graphite run by Brodie in 1859 aiming to determine the atomic weight of carbon atoms [23]. Brodie studied the oxidation of graphite by using a mixture of nitric acid and potassium chlorate for a long period, up to several days. He produced a water-soluble species with a light-yellow powder consistency made by shining transparent flakes. He named it upon graphitic acid, but actually, the all-world known it as GO. A half-century later, Staudenmaier described a different methodology for producing "graphitic acid" by using a concentrated sulfuric acid solution of $KClO_3$ [24]. Later, in the middle of the twenty-first century, Hummers and Offeman described another approach for the oxidation of graphite [25]. They treated graphite fuming sulfuric acid solution of potassium permanganate and sodium nitrate. Coming back to the present, Hummers method is the most widely used to produce GO in all laboratories and industries. Anyhow, all of these three methodologies are still used to produce GO with unique features. A comprehensive study was conducted by Poh et al. [26] describing the different properties of reduced GO (rGO) produced from different GO precursors. The authors showed the effect of the three methodologies on the residual functionalities of rGO, correlating them with the electronic properties. This study shows the essential differences between GO materials in composition and their nonstoichiometric structure [27]. Accordingly, several studies proposed different structures of GO, named after their authors (Hofmann, Ruess, Scholz–Boehm, Nakajima–Matsuo, Lerf–Klinowski, and Szabo) [28], as summarized in Figure 8.1.

Scientists agreed that the Lerf–Klinowski is the more realistic structure to describe the GO, as proved by several analytic studies [30].

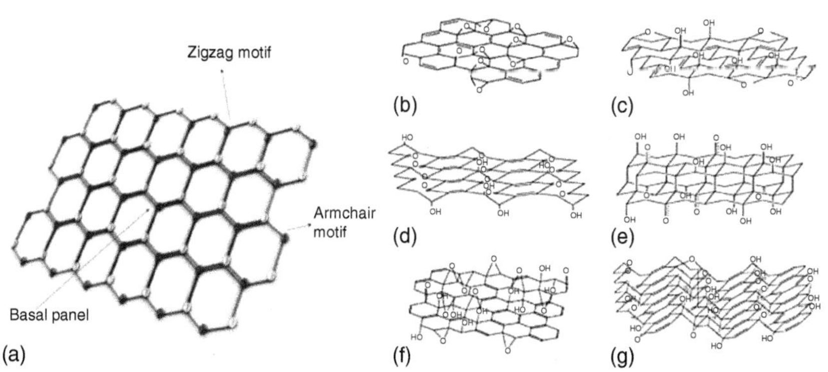


**Figure 8.1** Structures of neat graphene (a) and GO different structures as proposed by Hofmann (b), Ruess (c), Scholz–Boehm (d), Nakajima–Matsuo (e), Lerf–Klinowski (f), and Szabo (g). Source: Lavagna et al. [29]/MDPI/Public Domain CC by 4.0.

The previously mentioned rGO is the other graphene-related material that has been the rising star of graphene research [31]. rGO is obtained through a reduction of GO by using different reductive species [32, 33]. The reduction process has removed the oxygenated functional groups of GO leading to carbon to oxygen ratio down to 0.4 wt% [34].

### 8.3.2 Carbon Nanotubes

Carbon nanotubes (CNTs) are an allotropic state of carbon that could be represented by a cylindrical graphite sheet rolled up in a tubular structure according to chiral vector $\vec{C}$ reported in Eq. (1).

$$\vec{C} = n\vec{a}_1 + m\vec{a}_2 \tag{8.1}$$

where $n$ and $m$ are positive integers named chiral indices, and $\vec{a}_1, \vec{a}_2$ are vectors on a graphene plane, as reported in Figure 8.2.

Chiral indices $(n,m)$ could be used to define the three structures of a CNT according to the orientation of its longitudinal axis with respect to the hexagonal lattice. Accordingly, three different CNT types could be identified named zigzag

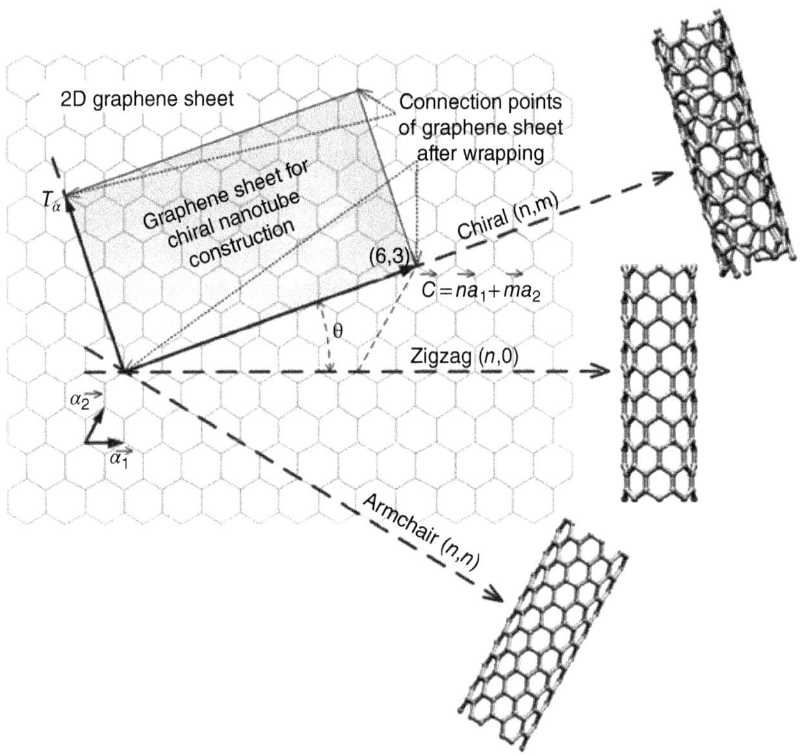


**Figure 8.2** Folding a graphene sheet accordingly with a chiral vector to lead to the three main CNTs structures named armchair, zigzag, and chiral. Source: Jan Prasek et al., 2011 [35]/The Royal Society of Chemistry.

($m$ = 0), armchair ($n$ = m), and chiral ($n \neq m \neq 0$). Chiral CNTs showed the geometrical properties of enantiomeric molecules. Furthermore, Belin and Epron [36] reported predictability of electronic properties based on chiral indices, which were summarized in two different cases as follows:

(i) $|n - m| = 3q, q \in \mathbb{N}, q \neq 0$
(ii) $|n - m| = 3q \pm 1, q \in \mathbb{N}, q \neq 0$

In the first case, CNTs showed a metallic behavior, while in the second, they showed semiconductor properties. Interestingly, armchair CNTs always satisfy the first condition showing a metallic behavior.

CNTs could be composed of single or by multiple graphene sheets taking the name of single-walled CNTs (SWCNTs) and multi-walled CNTs (MWCNTs), as sketched in Figure 8.3, respectively.

SWCNTs show a diameter ranging from 0.3 nm [37] to over 1 nm [38], while MWCNTs can reach up to over 100 nm [39, 40]. CNTs length is highly variable and depends on the production process but can range from a few nanometers [41] up to half a meter [42]. Furthermore, CNTs can be topped with fullerene-type molecules [43].

Due to their unique $sp^2$ carbonaceous hexagonal structure, CNTs showed some desired (electrical, mechanical, and thermal) properties, as reported in Table 8.1, making them standard for material excellency.

CNTs are characterized by a high-aspect ratio of up to $10^6$ [51] and exhibit superior mechanical properties, especially high toughness and elastic moduli [48]. Additionally, CNTs have very attractive electrical conductivity ranging from semiconducting to metallic behavior [52, 53]. Many procedures have been used to produce CNTs on a medium and large scale ranging from arc discharge [54], laser ablation [55], flame synthesis approach [56], and chemical vapor deposition (CVD) [57].

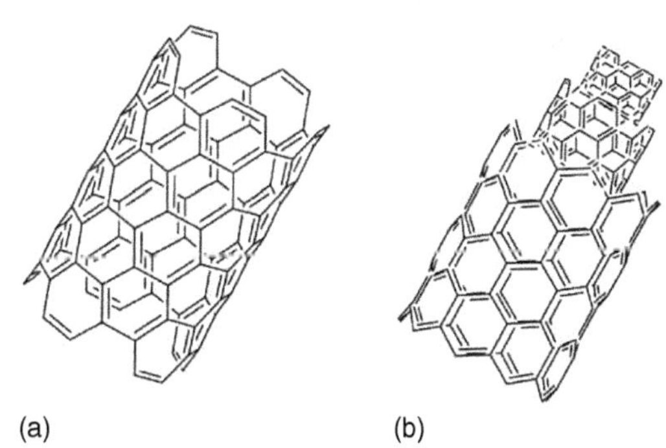


**Figure 8.3** Longitudinal section of (a) SWCNT and (b) MWCNT. Source: Giorcelli et al. [4]/ MDPI/Public Domain CC by 4.0.

**Table 8.1** Electrical, mechanical, and thermal properties of SWCNT and MWCNT.

| | **Young modulus (GPa)** | **Tensile strength (GPa)** | **Resistivity (Ωm)** | **Thermal conductivity (W mK$^{-1}$)** |
|---|---|---|---|---|
| SWCNT | 900–1700 [44] | 75 [45] | $10^{-6}$ [46] | 1750–5800 [47] |
| MWCNT | 690–1800 [48] | 150 [45] | $10^{-5}$ [49] | 3000 [50] |

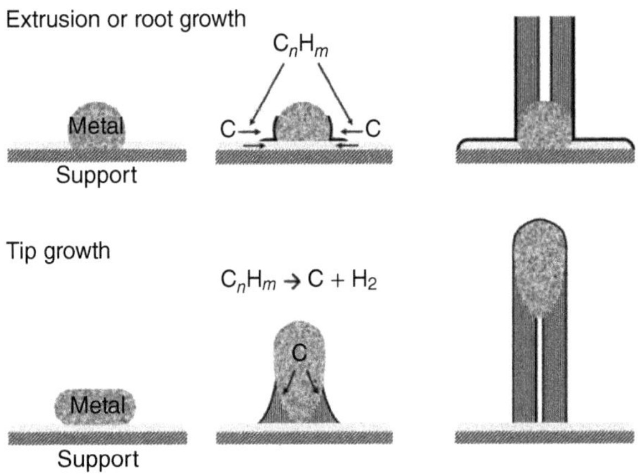


**Figure 8.4** Mechanism of CNTs growth during CVD process: root and tip growth. Source: Sinnot et al. [65]/with permission of Elsevier.

CVD is considered the low-cost process for the production of CNTs, and it is mostly used for the deposition of CNTs thin-film deposition. This approach has its strong point in the high yield of CNTs produced and the temperature requirement [43]. Furthermore, the CVD route allows great control over both morphology and structure of the CNTs leading to the growth of aligned CNTs [58, 59] with reasonably low-residence times [60].

The mechanism of CNTs growth during the CVD process is the key to this approach's success for the CNTs production [58, 61]. This procedure requires the use of metal catalysts, such as iron [62], cobalt [63], or nickel [64], based on one. As reported in Figure 8.4, the CNTs growth could proceed through two different mechanisms named root and tip growth, respectively [66].

Both mechanisms involve the molecular decomposition of the carbon precursor, generally hydrocarbons, on the side of catalytic particles leading to the formation of metal carbides that become supersaturated [67].

Root growth mechanism takes place when the particle adheres to the surface is strong, and carbon precipitates from the particle top with the growth of the filament and the metal particle attached to the substrate [68].

Tip growth occurs when the particle adherence to the surface is weak, and carbon precipitation occurs at the bottom surface with the growing filament lifting the particle as it grows, capping the top of CNTs [69].

### 8.3.3 Fullerenes

Among all allotropic carbon forms, fullerenes are the smallest and have the highest bonding strain [70]. Fullerene shape drives the chemical reactivity close to one of polyolefinic systems allowing cycloaddition reactions on their surface [71]. Fullerene was firstly identified using mass spectroscopy by Kroto et al. in 1985[72].

As illustrated in Figure 8.5, several fullerene types are available, but $C_{60}$ has gained the greatest attention due to its properties.

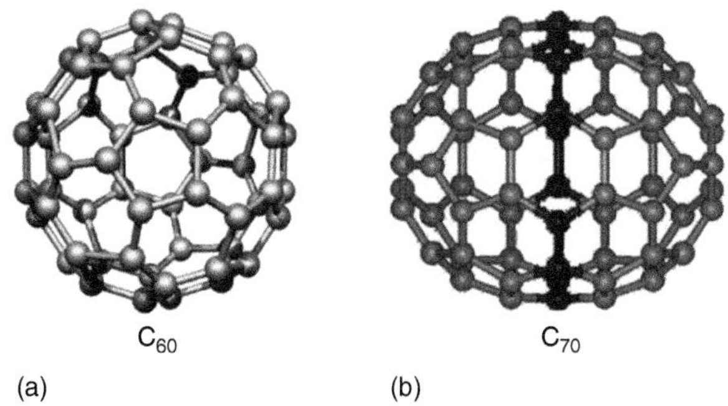


**Figure 8.5** 3D representation of (a) $C_{60}$, also known as buckyball, and (b) fullerene $C_{70}$. Source: Astefanei et al. [73]/with permission of Elsevier.

$C_{60}$ is counterbalanced by a high-cost price that slows down its use. Nonetheless, fullerenes have been used in several actuator applications.

Takehara et al. [74] developed an industrial-scale procedure for the production of fullerene using a combustion system based on pre-mixed laminar sooty flames at reduced pressure. The authors claimed to achieve a production of up to tons for the year in a continuous pilot system.

This technology was very promising, and Frontier Carbon Corporation established mass production of fullerenes with a 0.4 ton/yr capacity by using a pilot machine developed by TDA Research Inc. [75].

## 8.4 Carbon-Based Actuation

### 8.4.1 Carbon Nanotube-Based Actuators

CNTs are used to produce actuators exploiting their features as pristine materials or dispersed into composites.

The first example reported of actuation by using CNTs was the study of Baughman et al. [76] dated 1999. The authors used an aggregated sheet of SWCNTs known as buckypaper for use as electronic actuation. Accordingly, they reported a displacement of up to 0.12% under a square-wave potential of 0.5 V. In 2002, Roth et al. [55] described and actuated using a single MWCNT and atomic force microscopy (AFM) probe, as shown in Figure 8.6.

Authors claimed a displacement of up to 12 nm by applying a maximum potential of up to 1 V. Several studies investigated the relationship between electrical input and actuations of CNTs [78], proving how the aspect ratio is a key point together with the network CNTs properties [79, 80]. Senga et al. [81] proved by using a transmission electron microscope, the transition between the flattened state and the tubular state in thick CNTs, enlightening the relationship with thermal energy, as shown in Figure 8.7.

The combination of ionic liquids with CNTs and polymers is a very efficient way to a well-dispersed CNTs gels, as reported by Sugino et al. [82]. The authors described

**Figure 8.6** Atomic force microscopy caption of a single FM image of an MWCNT. Source: Roth and Baughman [77]. Reproduced with permission of Elsevier.

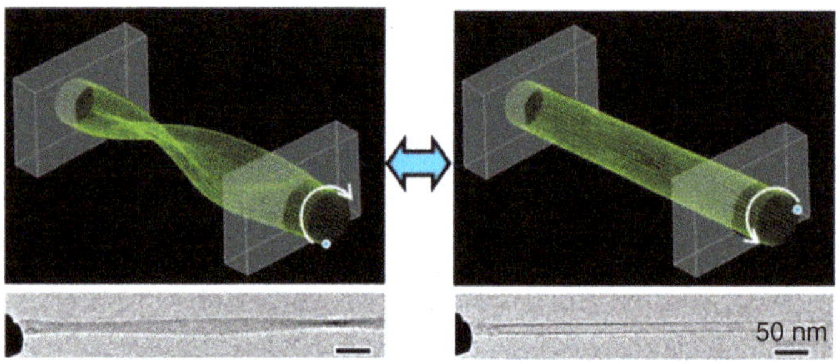


**Figure 8.7** Schematic nanotorsional actuator based on CNT. Source: Senga et al. [81]. Reproduced from permission of Elsevier.

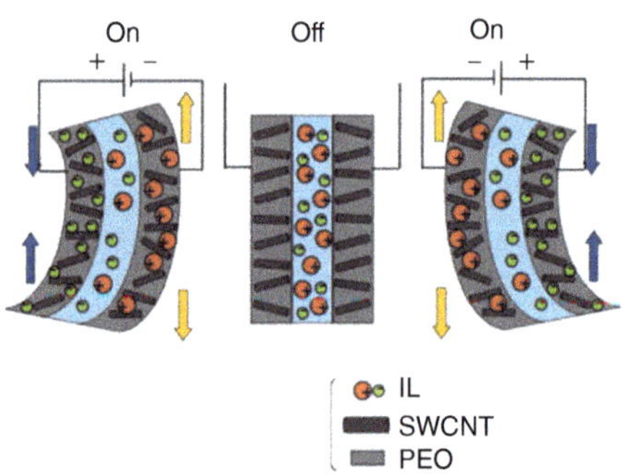


**Figure 8.8** Scheme of the actuators based on the ion-transfer mechanism. Source: Terasawa et al. [84]/with permission of Elsevier.

the advantages of the simultaneous use of SWCNTs and ionic liquids combined with a conductive polymer electroactuator that reached a strain value three times higher than previous results [83]. Terasawa et al. [84] used an ionic surfactant to produce a CNT-containing gel, as shown in Figure 8.8

The authors performed the actuation by using poly(vinylidene fluoride-*co*-hexafluoropropylene) or poly(ethylenoxide) (PEO) materials showing better performance than the last one. The authors proposed a model based on ionic transport number, and ionic volume was proposed to explain the actuation proving that the self-diffusion coefficient, rate of ion transport in the electrodes, and electrolyte types played a vital role in the system.

Among CNT polymer composites, elastomers are the most studied composite [85]. Fang et al. [86] developed a dielectric elastomer mixed with aligned SWCNTs sheets modified by laser ablation to achieve a mechanical anisotropy. This treatment enlightens an appreciable beneficial effect during the electrochemical actuation by reaching a strain of up to 33%. Anisotropic actuation was performed by Cakmak et al. [87] by developing soft linear actuators able to operate using a low electrical field of up to 100 mV. The authors assembled their actuators by creating an SWCNT electrode supported onto an acrylic layer and observing a strain of up to 40% in dielectric electroactive polymers during the actuation. Sulfonated fluoropolymers (i.e. Nafion) have gained tremendous interest due to their facile synthesis and filming ability [88].

Lian et al. [89] studied the actuation using functionalized SWCNTs dispersed into Nafion, doubling the storage modulus and the strain compared to the neat matrix.

Accordingly, Baughman et al. [76] used carboxylic surface-functionalized SWCNTs supported onto carbon fibers for the production of Nafion–silica-based shape memory actuator. The authors reported both electrical and thermally induced actuation magnified by the presence of CNTs. A common procedure in CNTs actuation is the layering technique. Mukai et al. [90] described the production of a millimetric layer of CNTs supporting the growth of poly(pyrrole) through electropolymerization. The authors were performing the actuation in a watery medium using a potential of 0.5 mV and measuring a displacement of up to 0.1%. Sachyani et al. [91] produced a three-layered structure based on Kapton and SWCNTs. The authors achieved an angular strain of up to 300° by applying 1 V. This remarkable achievement was due to the unaligned SWCNTs that improved the bending modulus of composites. Zhang et al. [92] used a nonconductive polymer composite containing CNTs for the realization of a performing actuator. Specifically, the authors dispersed super-aligned SWCNTs sheet into poly(dimethylsiloxane), producing a transparency-switchable actuator. By applying a voltage of up 1 V, the transmittance composites changed from 0.7% to 67% at the wavelength of 550 nm, and a displacement up to 8.4 mm was observed. The actuator is also highly cyclable and stable up to 100 cycles. Polyimides and other technical polymers could also be used for the production of actuators containing CNTs. Ning et al. [93] produced aligned CNT matrices supported onto a polyimide film for performing thermomechanical actuation. Zhou et al. [94] produced a multi-responsive actuator based on super-aligned SWCNTs. The authors realized a U-shaped device that shows twisting deformation under both thermal and electrical stimuli. Moreover, SWCNTs could be combined with other carbon sources, such as carbides [95], or inorganic species [96], improving the responses of the actuators by up to 60% [97].

Due to their low cost and larger availability, MWCNTs have been more used than SWCNTs for the production of actuators. Capeluto et al. [98] used MWCNTs dispersed into azopolymeric matrix to produce a light-responsive transparent film actuator. Cui et al. [99] realized a memory shape light-responsive actuator able to lift 550 times its weight under infrared stimulation.

Ji et al. [100] dispersed MWCNTs into poly(dimethylsiloxane), producing a thick film able to generate a linear strain actuation of up to 4% under high voltage of up to 100 V. Conductive poly(pyrrole) has been used in combination with a high load of MWCNTs of up to 25 wt% for producing an electroactuator able to achieve a displacement of up to 18% [101]. Shirasu et al. [102] described an actuator by using a thermoset polymer. Authors assembled a U-shaped epoxy-based device by dispersing aligned MWCNTs and achieving a bending displacement of up to 10% under a voltage of 6 V. Furthermore, MWCNTs were obtained with ruthenium oxides, and the inorganic tailored carbon materials were used as actuators increasing the performances of non-functionalized CNTs up to twice [103].

Aouraghe et al. [104] assembled a low-voltage and fast-responding electrothermal actuator by combining a flexible and tough CNT film with remarkable thermal and electrical conductivity with poly(dimethylsiloxane). Accordingly, the authors produced an asymmetric bilayer-structured actuator by coating polymer with a thin layer of 11 μm of CNTs achieving interesting values of thermal expansion coefficient up to $9.3 \times 10^{-4}\,°C^{-1}$, Young's modulus of up to 2.07 MPa. The as-produced actuator showed a large angular deformation reaching a bending angle of up around 324° together with a high electro-heating performance of up to 351 °C by using a low driving voltage of 8 V. Furthermore, the authors claimed that the heat produced and actuation observed could be controlled by adjusting the applied voltages and showing the same values in 20 cycles in the same conditions. Finally, the authors realized a real soft robotic hand able to lift up to five times its weight.

Improved performances of MWCNT-based actuators were reported by using 1-ethyl-3-methylimidazolium tetrafluoroborate ionic liquid-based systems by Terasawa et al. [105]. The authors compared the CNT-based actuators with carbon-black-based devices, showing a strain improvement of up to 0.8%.

Randomly distributed CNT systems were used to combine actuation ability avoiding the time and resource-consuming alignment procedure. Chen's research group described a couple of interesting examples using buckypaper/Nafion actuators [106] or without a polymer matrix under a very high voltage of up to 1100 V [107]. Both studies achieved an actuation of up to 1%.

### 8.4.2 Graphene and Graphene Oxide Actuators

Graphene-based materials actuation was evaluated by Saane and Onck [108] using density functional theory calculations based on an adaptive intermolecular reactive empirical bond order potential methodology.

The authors proved the better performances of graphene compared to the inorganic material, such as nanoporous gold film achieving a smaller actuation stroke under an electrical field but generating more mechanical work. Graphene-containing composites used as electroactive actuators have been reported by Tungkavet et al. [109]. The authors described a highly sensitive to electric stimuli actuator with a monotonic response to electrical solicitations.

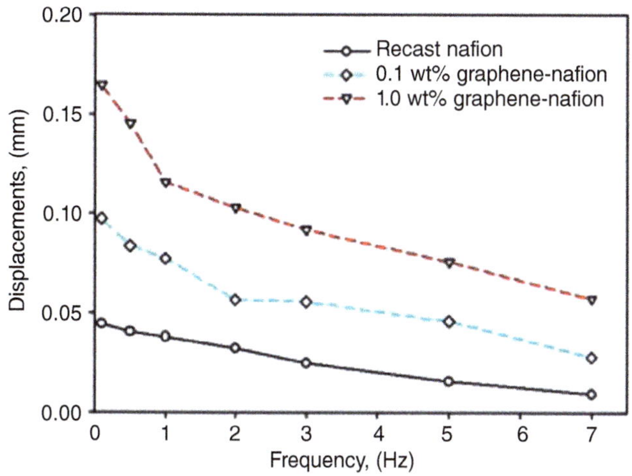


**Figure 8.9** Actuation performances of graphene-based Nafion actuator. Source: Jung et al. [110]/with permission of Elsevier.

As seen in Figure 8.9, graphene-containing Nafion membrane represents a good example of actuator benchmark and for properties comparison, such as mechanical strength between devices, with a displacement of up to 60% equivalent to 17 mm [110].

Consequently, graphene has been used to develop wearable poly(pyrrole)-based composite fibers. This material was used as net actuators [111]. Sen et al. [112] combined cellulose and graphene nanoplatelets of up to 0.5 wt% for the production of electroactuator that under DC excitation voltages of up to 7 V can reach a strain of up to 1%. Titanium-doped graphene was used as actuator material mixed into an elastomeric matrix, reaching a strain of up to 72% using 39 V [113]. Interestingly, Liu et al. [114] have run a photochemical actuation using bilayered graphene nanoplatelets dispersed into poly(dimethylsiloxane) under near-infrared light.

The last great unsolved issue in the use of graphene was the poor dispersion in the polymeric matrix but GO could represent a good compromise between dispersibility and net graphene properties. The preparation of related hydrogels could easily realize actuators based on GO, and they are able to actuate under both receiving thermal and UV stimuli [115]. Actuation using near-infrared light was studied by Chen et al. [116] using GO dispersed into poly(*N*-isopropylacrylamide), achieving remarkable results in the retention of shape memory. Similarly, Surana et al. [117] obtained very close results by using Nafion-based materials.

The interactions between residual functional groups of GO and polar functionalities of the Nafion matrix were investigated by Lee et al. [118], while other authors run similar studies by using sulfonated poly(styrene) [107] [119] or poly(pyrrole) [108].

Zhang et al. [120] developed a soft bilayer actuator through a simple fabrication process using GO. Firstly, the authors produced a mixture of ionic liquid and GO and injected it into a mold by capillarity under the molten state. Afterward, the molded

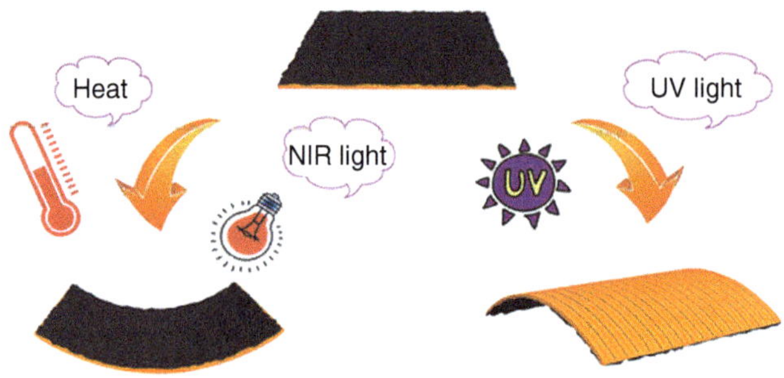


**Figure 8.10** GO-based membrane actuator. Source: Zhang et al. [120]/with permission from American Chemical Society.

materials were cured by using UV light, and a bilayer membrane was recovered, as shown in Figure 8.10.

This material was able to display various molecular alignment, and multi-stimulus responses due to the mismatch in coefficient of thermal expansion GO and the azobenzene-doped liquid crystal network used for the production of the membrane. The actuator showed fast and reversible complex deformations under the heat, UV, and near IR light stimuli.

Vural et al. [121] assembled a tandem bimorph actuator based on proteins and GO reaching a maximum bend three times higher than conventional GO.

Terasawa and Asaka [122] used GO in combination with vapor-grown carbon fibers running the actuation mediated by an ionic liquid medium, reaching an improvement of up to 56% compared with the same system containing CNTs instead. Wang et al. [123] reported the production of multi-responsive graphene-based poly(dimethylsiloxane) (PDMS) composite able to actuate under different chemical stimuli with a recovery of up to 98% after three cycles.

Moving from GO to its reduced form, the rGO buckypaper analog was used by Selvakumar et al. [124] as the biomorphic electrical actuator. Similarly, Wang et al. [125] studied an rGO-based epoxy composite with shape memory properties that showed a high recoverability after hundreds of actuation cycles. Kaur and Engles [126] described the use of visible light actuation by using rGO dispersed into chitosan, reaching a displacement of up to 4%.

However, the lack of nanotechnologies that enable flexible integration of advanced photothermal materials with actuators of complex configurations significantly restricts their practical applications. In this study, laser-induced graphene (LIG) tape is reported as stick-on photothermal labels for developing light-driven actuators based on the Marangoni effect. With the help of direct laser writing technology, graphene patterns with superior photothermal properties are prepared on the PI tape. The patterned LIG tape can be stuck on any desired objects and generates an asymmetric photothermal field under light irradiation, forming a photothermal Marangoni actuator. Additionally, the PI tape with LIG patterns can be folded into ·D origami actuators that permit photothermal Marangoni actuation, including both translation and rotation. The graphene-based photothermal

Marangoni actuators feature biocompatibility, which is confirmed by MDA-MB-cells proliferation experiments. Owing to the excellent photothermal property of LIG patterns, the as-produced photothermal actuators can be manipulated by a variety of light sources, holding great promise for developing light-driven soft robots However, the lack of nanotechnologies that enable flexible integration of advanced photothermal materials with actuators of complex configurations significantly restricts their practical applications. In this study, LIG tape is reported as stick-on photothermal labels for developing light-driven actuators based on the Marangoni effect. With the help of direct laser writing technology, graphene patterns with superior photothermal properties are prepared on the PI tape. The patterned LIG tape can be stuck on any desired objects and generates an asymmetric photothermal field under light irradiation, forming a photothermal Marangoni actuator. Additionally, the PI tape with LIG patterns can be folded into ·D origami actuators that permit photothermal Marangoni actuation including both translation and rotation. The graphene-based photothermal Marangoni actuators feature biocompatibility, which is confirmed by MDA-MB- cells proliferation experiments. Owing to the excellent photothermal property of LIG patterns, the as-produced photothermal actuators can be manipulated by a variety of light sources, holding great promise for developing light-driven soft robots However, the lack of nanotechnologies that enable flexible integration of advanced photothermal materials with actuators of complex configurations significantly restricts their practical applications. In this study, LIG tape is reported as stick-on photothermal labels for developing light-driven actuators based on the Marangoni effect. With the help of direct laser writing technology, graphene patterns with superior photothermal properties are prepared on the PI tape. The patterned LIG tape can be stuck on any desired objects and generates an asymmetric photothermal field under light irradiation, forming a photothermal Marangoni actuator. Additionally, the PI tape with LIG patterns can be folded into ·D origami actuators that permit photothermal Marangoni actuation including both translation and rotation. The graphene-based photothermal Marangoni actuators feature biocompatibility, which is confirmed by MDA-MB-cells proliferation experiments. Owing to the excellent photothermal property of LIG patterns, the as-produced photothermal actuators can be manipulated by a variety of light sources, holding great promise for developing light driven soft robots.

Wand et al. [127] produced a light-driven actuator based on a LIG tape stuck on a photothermal label. This actuator was based on the mass transfer along with an interface between two fluids due to a gradient of the surface tension, a phenomenon known as the Marangoni effect. The authors produced a new actuator by using direct laser writing technology for the realization of graphene patterns with very high photothermal properties on polymeric tape. The patterned graphene-based tape displayed the intriguing properties of being stuck on several surfaces exploiting the actuation under the asymmetric photothermal field and light irradiation. Furthermore, the authors folded the laser patterned tape actuator into an origami-shaped structure achieving photothermal Marangoni actuation with translation and rotation displacement.

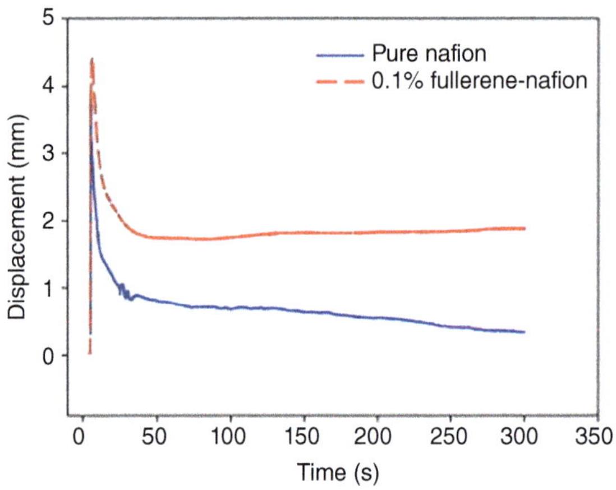


**Figure 8.11** Straightening-back phenomena of neat Nafion (blue line) and fullerene-based actuator (red line) under a DC solicitation of up to 2.0 V. Source: Jung et al. [129]/with permission of Elsevier.

### 8.4.3 Fullerene-Based Actuators

As mentioned above in the chapter, fullerene's high-surface area, porosity, thermal stability, nontoxicity, and biocompatibility properties are solid platforms for producing actuators.

Huang et al. [128] encapsulated scandium nitride into $C_{80}$ for the production of a single-molecule memory and logic device used for parallel molecular computing architectures assembling.

Jung et al. [129] realized an actuator based on $C_{60}$ dispersed into a Nafion membrane showing a harmonic response to an oscillating electrical field. Compared to the neat Nafion actuator, authors reported a more extensive bending deformation in the case of fullerene-based actuator together with a decrement in shortening back phenomenon, as reported in Figure 8.11.

Panwar et al. [130] created a very complex polymeric host able to react with oxygenated $C_{60}$. The related fullerene-containing material was used as an electroactuator in combination with a polymer blend of poly(vinylidene fluoride), poly(vinylpyrrolidone), and sulfonated poly(styrene).

## 8.5 Challenges and Prospectives of Actuators Based on Carbon Nanostructures

Actuators based on carbonaceous materials are a field evolving day by day. Nanostructured and nanocarbon represent the platforms that allow to realize polymer actuators on a molecular scale. As discussed in this chapter, allotropic carbon materials show a combination of size and properties that perfectly match electro, thermal, and light-driven actuation.

The vast class of CNTs, graphene, and fullerenes actuators comprised so much different materials that a real classification is far beyond our intent. Nonetheless, some common characteristics, such as toughness, reversibility, and tunability, can be four in all of them.

We strongly believe that the future will be bright and full of new enhancements and applications ranging from biological systems to nanoelectronic devices for these materials.

## References


1 Chaikof, E.L. (2010). Muscle mimic. *Nature* 465: 44–45.

2 Wong, W.S., Li, M., Nisbet, D.R. et al. (2016). Mimosa origami: a nanostructure-enabled directional self-organization regime of materials. *Science Advances* 2: e1600417.

3 Kempaiah, R. and Nie, Z. (2014). From nature to synthetic systems: shape transformation in soft materials. *Journal of Materials Chemistry B* 2: 2357–2368.

4 Giorcelli, M. and Bartoli, M. (2019). Carbon nanostructures for actuators: an overview of recent developments. *Actuators*, MDPI 8 (2): 46–59.

5 Medhekar, N.V., Ramasubramaniam, A., Ruoff, R.S., and Shenoy, V.B. (2010). Hydrogen bond networks in graphene oxide composite paper: structure and mechanical properties. *ACS Nano* 4: 2300–2306.

6 Rafiee, M.A., Rafiee, J., Wang, Z. et al. (2009). Enhanced mechanical properties of nanocomposites at low graphene content. *ACS Nano* 3: 3884–3890.

7 Berg, H.C. (1975). How bacteria swim. *Scientific American* 233: 36–45.

8 Adelstein, R.S., Conti, M.A., and Barylko, B. (1978). The role of myosin phosphorylation in regulating actin-myosin interaction in human blood platelets. *Thrombosis and Haemostasis* 40: 241–244.

9 Krebs, E.G., Kent, A.B., and Fischer, E.H. (1958). The muscle phosphorylase B kinase reaction. *The Journal of Biological Chemistry* 231: 73–83.

10 Tian, H. and Wang, Q.-C. (2006). Recent progress on switchable rotaxanes. *Chemical Society Reviews* 35: 361–374.

11 Gil-Ramírez, G., Leigh, D.A., and Stephens, A.J. (2015). Catenanes: fifty years of molecular links. *Angewandte Chemie International Edition* 54: 6110–6150.

12 Balzani, V., Credi, A., and Venturi, M. (2009). Light powered molecular machines. *Chemical Society Reviews* 38: 1542–1550.

13 Zhang, Y., Ma, Y., and Sun, J. (2013). Reversible actuation of polyelectrolyte films: expansion-induced mechanical force enables cis–trans isomerization of azobenzenes. *Langmuir* 29: 14919–14925.

14 Noh, T.-G., Tak, Y., Nam, J.-D., and Choi, H. (2002). Electrochemical characterization of polymer actuator with large interfacial area. *Electrochimica Acta* 47: 2341–2346.

15 Mintmire, J.W., Dunlap, B.I., and White, C.T. (1992). Are fullerene tubules metallic? *Physical Review Letters* 68: 631.

16 Yan, J.-A., Ruan, W., and Chou, M. (2009). Electron-phonon interactions for optical-phonon modes in few-layer graphene: first-principles calculations. *Physical Review B* 79: 115443.
17 Dresselhaus, M., Jorio, A., and Saito, R. (2010). Characterizing graphene, graphite, and carbon nanotubes by Raman spectroscopy. *Annual Review of Condensed Matter Physics* 1: 89–108.
18 Rhee, K.Y. (2020). *Electronic and Thermal Properties of Graphene*. Multidisciplinary Digital Publishing Institute.
19 Lee, H.C., Liu, W.-W., Chai, S.-P. et al. (2016). Synthesis of single-layer graphene: a review of recent development. *Procedia Chemistry* 19: 916–921.
20 Narita, A., Wang, X.-Y., Feng, X., and Müllen, K. (2015). New advances in nanographene chemistry. *Chemical Society Reviews* 44: 6616–6643.
21 Sun, Z., Fang, S., and Hu, Y.H. (2020). 3D graphene materials: from understanding to design and synthesis control. *Chemical Reviews* 128 (8): 10336–10453.
22 Brisebois, P. and Siaj, M. (2020). Harvesting graphene oxide–years 1859 to 2019: a review of its structure, synthesis, properties and exfoliation. *Journal of Materials Chemistry C* 8: 1517–1547.
23 Brodie, B.C. (1859). On the atomic weight of graphite. *Philosophical Transactions of the Royal Society of London* 149: 249–259.
24 Staudenmaier, L. (1898). Verfahren zur darstellung der graphitsäure. *Berichte der Deutschen Chemischen Gesellschaft* 31: 1481–1487.
25 William, S., Hummers, J., and Offeman, R.E. (1958). Preparation of graphitic oxide. *Journal of the American Chemical Society* 80: 1339–1339.
26 Poh, H.L., Šaněk, F., Ambrosi, A. et al. (2012). Graphenes prepared by Staudenmaier, Hofmann and Hummers methods with consequent thermal exfoliation exhibit very different electrochemical properties. *Nanoscale* 4: 3515–3522.
27 Araújo, M.P., Soares, O., Fernandes, A. et al. (2017). Tuning the surface chemistry of graphene flakes: new strategies for selective oxidation. *RSC Advances* 7: 14290–14301.
28 Szabó, T., Berkesi, O., Forgó, P. et al. (2006). Evolution of surface functional groups in a series of progressively oxidized graphite oxides. *Chemistry of Materials* 18: 2740–2749.
29 Lavagna, L., Meligrana, G., Gerbaldi, C. et al. (2020). Graphene and lithium-based battery electrodes: a review of recent literature. *Energies* 13: 4867.
30 Lerf, A., He, H., Riedl, T. et al. (1997). 13C and 1H MAS NMR studies of graphite oxide and its chemically modified derivatives. *Solid State Ionics* 101: 857–862.
31 Thakur, S. and Karak, N. (2015). Alternative methods and nature-based reagents for the reduction of graphene oxide: a review. *Carbon* 94: 224–242.
32 Guex, L.G., Sacchi, B., Peuvot, K.F. et al. (2017). Experimental review: chemical reduction of graphene oxide (GO) to reduced graphene oxide (rGO) by aqueous chemistry. *Nanoscale* 9: 9562–9571.

**33** Liu, S., Tian, J., Wang, L., and Sun, X. (2011). A method for the production of reduced graphene oxide using benzylamine as a reducing and stabilizing agent and its subsequent decoration with Ag nanoparticles for enzymeless hydrogen peroxide detection. *Carbon* 49: 3158–3164.

**34** Lee, X.J., Hiew, B.Y.Z., Lai, K.C. et al. (2019). Review on graphene and its derivatives: synthesis methods and potential industrial implementation. *Journal of the Taiwan Institute of Chemical Engineers* 98: 163–180.

**35** Sanginario, A., Miccoli, B., and Demarchi, D. (2017). Carbon nanotubes as an effective opportunity for cancer diagnosis and treatment. *Biosensors* 7: 9.

**36** Belin, T. and Epron, F. (2005). Characterization methods of carbon nanotubes: a review. *Materials Science and Engineering B* 119: 105–118.

**37** Zhao, X., Liu, Y., Inoue, S. et al. (2004). Smallest carbon nanotube is 3 Å in diameter. *Physical Review Letters* 92: 125502.

**38** Fagan, J.A., Hároz, E.H., Ihly, R. et al. (2015). Isolation of >1 nm diameter single-wall carbon nanotube species using aqueous two-phase extraction. *ACS Nano* 9: 5377–5390.

**39** Liu, W.-W., Chai, S.-P., Mohamed, A.R., and Hashim, U. (2014). Synthesis and characterization of graphene and carbon nanotubes: a review on the past and recent developments. *Journal of Industrial and Engineering Chemistry* 20: 1171–1185.

**40** Navas, H., Picher, M., Andrieux-Ledier, A. et al. (2017). Unveiling the evolutions of nanotube diameter distribution during the growth of single-walled carbon nanotubes. *ACS Nano* 11: 3081–3088.

**41** Inam, F., Reece, M.J., and Peijs, T. (2011). Shortened carbon nanotubes and their influence on the electrical properties of polymer nanocomposites. *Journal of Composite Materials* 46: 1313–1322.

**42** Zhang, R., Zhang, Y., Zhang, Q. et al. (2013). Growth of half-meter long carbon nanotubes based on Schulz–Flory distribution. *ACS Nano* 7: 6156–6161.

**43** Manawi, Y.M., Samara, A., Al-Ansari, T., and Atieh, M.A. (2018). A review of carbon nanomaterials' synthesis via the chemical vapor deposition (CVD) method. *Materials* 11: 822.

**44** Krishnan, A., Dujardin, E., Ebbesen, T. et al. (1998). Young's modulus of single-walled nanotubes. *Physical Review B* 58: 14013.

**45** Han, J. (2004). Structures and properties of carbon nanotubes. In: *Carbon Nanotubes*, 16–45. Boca Rathon: CRC Press.

**46** Thess, A., Lee, R., Nikolaev, P. et al. (1996). Crystalline ropes of metallic carbon nanotubes. *Science* 273: 483–487.

**47** Cao, J., Yan, X., Xiao, Y., and Ding, J. (2004). Thermal conductivity of zigzag single-walled carbon nanotubes: Role of the umklapp process. *Physical Review B* 69: 073407.

**48** Treacy, M.J., Ebbesen, T.W., and Gibson, J.M. (1996). Exceptionally high Young's modulus observed for individual carbon nanotubes. *Nature* 381: 678–680.

49 Mansfield, E., Feldman, A., Chiaramonti, A.N. et al. (2015). Morphological and electrical characterization of MWCNT papers and pellets. *Journal of Research of the National Institute of Standards and Technology* 120: 304–315.
50 Choi, T.Y., Poulikakos, D., Tharian, J., and Sennhauser, U. (2005). Measurement of thermal conductivity of individual multiwalled carbon nanotubes by the 3-ω method. *Applied Physics Letters* 87: 013108.
51 Li, J., Ma, P.C., Chow, W.S. et al. (2007). Correlations between percolation threshold, dispersion state, and aspect ratio of carbon nanotubes. *Advanced Functional Materials* 17: 3207–3215.
52 Lekawa-Raus, A., Patmore, J., Kurzepa, L. et al. (2014). Electrical properties of carbon nanotube based fibers and their future use in electrical wiring. *Advanced Functional Materials* 24: 3661–3682.
53 Bandaru, P.R. (2007). Electrical properties and applications of carbon nanotube structures. *Journal of Nanoscience and Nanotechnology* 7: 1239–1267.
54 Arora, N. and Sharma, N. (2014). Arc discharge synthesis of carbon nanotubes: comprehensive review. *Diamond and Related Materials* 50: 135–150.
55 Scott, C.D., Arepalli, S., Nikolaev, P., and Smalley, R.E. (2001). Growth mechanisms for single-wall carbon nanotubes in a laser-ablation process. *Applied Physics A* 72: 573–580.
56 Hamzah, N., Yasin, M.M., Yusop, M.M. et al. (2017). Rapid production of carbon nanotubes: a review on advancement in growth control and morphology manipulations of flame synthesis. *Journal of Materials Chemistry A* 5: 25144–25170.
57 Kumar, M. and Ando, Y. (2010). Chemical vapor deposition of carbon nanotubes: a review on growth mechanism and mass production. *Journal of Nanoscience and Nanotechnology* 10: 3739–3758.
58 Musso, S., Fanchini, G., and Tagliaferro, A. (2005). Growth of vertically aligned carbon nanotubes by CVD by evaporation of carbon precursors. *Diamond and Related Materials* 14: 784–789.
59 Porro, S., Musso, S., Giorcelli, M. et al. (2007). Optimization of a thermal-CVD system for carbon nanotube growth. *Physica E: Low-dimensional Systems and Nanostructures* 37: 16–20.
60 Brukh, R. and Mitra, S. (2006). Mechanism of carbon nanotube growth by CVD. *Chemical Physics Letters* 424: 126–132.
61 Deck, C.P. and Vecchio, K. (2005). Growth mechanism of vapor phase CVD-grown multi-walled carbon nanotubes. *Carbon* 43: 2608–2617.
62 Durrer, L., Helbling, T., Zenger, C. et al. (2008). SWNT growth by CVD on Ferritin-based iron catalyst nanoparticles towards CNT sensors. *Sensors and Actuators B: Chemical* 132: 485–490.
63 Terrado, E., Redrado, M., Munoz, E. et al. (2006). Carbon nanotube growth on cobalt-sprayed substrates by thermal CVD. *Materials Science and Engineering: C* 26: 1185–1188.
64 Makris, T.D., Giorgi, L., Giorgi, R. et al. (2005). CNT growth on alumina supported nickel catalyst by thermal CVD. *Diamond and Related Materials* 14: 815–819.

**65** Sinnott, S., Andrews, R., Qian, D. et al. (1999). Model of carbon nanotube growth through chemical vapor deposition. *Chemical Physics Letters* 315: 25–30.
**66** Purohit, R., Purohit, K., Rana, S. et al. (2014). Carbon nanotubes and their growth methods. *Procedia Materials Science* 6: 716–728.
**67** Krasnikov, D.V., Kuznetsov, V.L., Romanenko, A.I., and Shmakov, A.N. (2018). Side reaction in catalytic CVD growth of carbon nanotubes: surface pyrolysis of a hydrocarbon precursor with the formation of lateral carbon deposits. *Carbon* 139: 105–117.
**68** Gavillet, J., Loiseau, A., Journet, C. et al. (2001). Root-growth mechanism for single-wall carbon nanotubes. *Physical Review Letters* 87: 275504.
**69** Liu, B., Lee, T., Jung, S. et al. (2005). Synthesis of well-aligned carbon nanotubes with open tips. *Carbon* 43: 1341–1346.
**70** Dresselhaus, M.S., Dresselhaus, G., and Eklund, P.C. (1996). *Science of Fullerenes and Carbon Nanotubes: Their Properties and Applications*. Elsevier.
**71** Nakamura, E. and Isobe, H. (2003). Functionalized fullerenes in water. The first 10 years of their chemistry, biology, and nanoscience. *Accounts of Chemical Research* 36: 807–815.
**72** Kroto, H.W., Heath, J.R., O'Brien, S.C. et al. (1985). C60: Buckminsterfullerene. *Nature* 318: 162–163.
**73** Astefanei, A., Núñez, O., and Galceran, M.T. (2015). Characterisation and determination of fullerenes: a critical review. *Analytica Chimica Acta* 882: 1–21.
**74** Takehara, H., Fujiwara, M., Arikawa, M. et al. (2005). Experimental study of industrial scale fullerene production by combustion synthesis. *Carbon* 43: 311–319.
**75** Murayama, H., Tomonoh, S., Alford, J.M., and Karpuk, M.E. (2005). Fullerene production in tons and more: from science to industry. *Fullerenes, Nanotubes, and Carbon Nanostructures* 12: 1–9.
**76** Baughman, R.H., Cui, C., Zakhidov, A.A. et al. (1999). Carbon nanotube actuators. *Science* 284: 1340.
**77** Roth, S. and Baughman, R.H. (2002). Actuators of individual carbon nanotubes. *Current Applied Physics* 2: 311–314.
**78** Seyyed Fakhrabadi, M.M., Rastgoo, A., and Taghi Ahmadian, M. (2014). Size-dependent instability of carbon nanotubes under electrostatic actuation using nonlocal elasticity. *International Journal of Mechanical Sciences* 80: 144–152.
**79** Suppiger, D., Busato, S., and Ermanni, P. (2008). Characterization of single-walled carbon nanotube mats and their performance as electromechanical actuators. *Carbon* 46: 1085–1090.
**80** Geier, S.M., Mahrholz, T., Wierach, P., and Sinapius, M. (2018). Morphology- and ion size-induced actuation of carbon nanotube architectures. *International Journal of Smart and Nano Materials* 9: 111–134.
**81** Senga, R., Hirahara, K., Yamaguchi, Y., and Nakayama, Y. (2012). Carbon nanotube torsional actuator based on transition between flattened and tubular states. *Journal of Non-Crystalline Solids* 358: 2541–2544.

**82** Sugino, T., Shibata, Y., Kiyohara, K., and Asaka, K. (2018). Actuation mechanism of dry-type polymer actuators composed by carbon nanotubes and ionic liquids. *Sensors and Actuators B: Chemical* 273: 955–965.

**83** Sugino, T., Kiyohara, K., Takeuchi, I. et al. (2009). Actuator properties of the complexes composed by carbon nanotube and ionic liquid: the effects of additives. *Sensors and Actuators B: Chemical* 141: 179–186.

**84** Terasawa, N., Hayashi, Y., Koga, T. et al. (2014). High-performance polymer actuators based on poly(ethylene oxide) and single-walled carbon nanotube–ionic liquid-based gels. *Sensors and Actuators B: Chemical* 202: 382–387.

**85** Pelrine, R., Kornbluh, R., Joseph, J. et al. (2000). High-field deformation of elastomeric dielectrics for actuators. *Materials Science and Engineering: C* 11: 89–100.

**86** Fang, X., Li, A., Yildiz, O. et al. (2017). Enhanced anisotropic response of dielectric elastomer actuators with microcombed and etched carbon nanotube sheet electrodes. *Carbon* 120: 366–373.

**87** Cakmak, E., Fang, X., Yildiz, O. et al. (2015). Carbon nanotube sheet electrodes for anisotropic actuation of dielectric elastomers. *Carbon* 89: 113–120.

**88** Yeo, S.C. and Eisenberg, A. (1977). Physical properties and supermolecular structure of perfluorinated ion-containing (nafion) polymers. *Journal of Applied Polymer Science* 21: 875–898.

**89** Lian, H., Qian, W., Estevez, L. et al. (2011). Enhanced actuation in functionalized carbon nanotube–Nafion composites. *Sensors and Actuators B: Chemical* 156: 187–193.

**90** Mukai, K., Yamato, K., Asaka, K. et al. (2012). Actuator of double layer film composed of carbon nanotubes and polypyrroles. *Sensors and Actuators B: Chemical* 161: 1010–1017.

**91** Sachyani, E., Layani, M., Tibi, G. et al. (2017). Enhanced movement of CNT-based actuators by a three-layered structure with controlled resistivity. *Sensors and Actuators B: Chemical* 252: 1071–1077.

**92** Zhang, W., Weng, M., Zhou, P. et al. (2017). Transparency-switchable actuator based on aligned carbon nanotube and paraffin-polydimethylsiloxane composite. *Carbon* 116: 625–632.

**93** Ning, W., Wang, Z., Liu, P. et al. (2018). Multifunctional super-aligned carbon nanotube/polyimide composite film heaters and actuators. *Carbon* 139: 1136–1143.

**94** Zhou, Z.-w., Yan, Q.-h., Liu, C.-h., and Fan, S.-S. (2017). An arm-like electrothermal actuator based on superaligned carbon nanotube/polymer composites. *New Carbon Materials* 32: 411–418.

**95** Wang, X., Zhang, H., and Liu, X. (2019). A free-standing carbon nitride actuator is driven by the ambient humidity. *Diamond and Related Materials* 97: 107434.

**96** Gendron, D., Bubak, G., Ceseracciu, L. et al. (2016). Significant strain and force improvements of single-walled carbon nanotube actuator: a metal chalcogenides approach. *Sensors and Actuators B: Chemical* 230: 673–683.

**97** Torop, J., Arulepp, M., Leis, J. et al. (2010). Nanoporous carbide-derived carbon material-based linear actuators. *Materials* 3: 9–25.

**98** Capeluto, M.G., Fernández Salvador, R., Eceiza, A. et al. (2017). Azopolymer film as an actuator for organizing multiwall carbon nanotubes. *Optical Materials* 66: 247–252.

**99** Cui, X., Chen, J., Zhu, Y., and Jiang, W. (2020). Natural sunlight-actuated shape memory materials with reversible shape change and self-healing abilities based on carbon nanotubes filled conductive polymer composites. *Chemical Engineering Journal* 382: 122823.

**100** Ji, X., El Haitami, A., Sorba, F. et al. (2018). Stretchable composite monolayer electrodes for low voltage dielectric elastomer actuators. *Sensors and Actuators B: Chemical* 261: 135–143.

**101** Rasouli, H., Naji, L., and Hosseini, M.G. (2017). The effect of MWCNT content on electropolymerization of PPy film and electromechanical behavior of PPy electrode-based soft actuators. *Journal of Electroanalytical Chemistry* 806: 136–149.

**102** Shirasu, K., Yamamoto, G., Inoue, Y. et al. (2017). Development of large-movements and high-force electrothermal bimorph actuators based on aligned carbon nanotube reinforced epoxy composites. *Sensors and Actuators A: Physical* 267: 455–463.

**103** Galantini, F., Bianchi, S., Castelvetro, V., and Gallone, G. (2013). Functionalized carbon nanotubes as a filler for dielectric elastomer composites with improved actuation performance. *Smart Materials and Structures* 22: 055025.

**104** Aouraghe, M.A., Mengjie, Z., Qiu, Y., and Fujun, X. (2021). Low-voltage activating, fast responding electro-thermal actuator based on carbon nanotube film/PDMS composites. *Advanced Fiber Materials* 3: 38–46.

**105** Terasawa, N., Ono, N., Mukai, K. et al. (2012). High performance polymer actuators based on multi-walled carbon nanotubes that surpass the performance of those containing single-walled carbon nanotubes: effects of ionic liquid and composition. *Sensors and Actuators B: Chemical* 163: 20–28.

**106** Chen, I.W.P., Cottinet, P.-J., Tsai, S.-Y. et al. (2012). Improved performance of carbon nanotube buckypaper and ionic-liquid-in-Nafion actuators for rapid response and high durability in the open air. *Sensors and Actuators B: Chemical* 171–172: 515–521.

**107** Chen, I.W.P., Liang, Z., Wang, B., and Zhang, C. (2010). Charge-induced asymmetrical displacement of an aligned carbon nanotube buckypaper actuator. *Carbon* 48: 1064–1069.

**108** Saane, S.S.R. and Onck, P.R. (2015). Atomistic modeling of the stiffness, strength and charge-induced actuation of graphene nanofoams. *Extreme Mechanics Letters* 5: 54–61.

**109** Tungkavet, T., Seetapan, N., Pattavarakorn, D., and Sirivat, A. (2015). Graphene/gelatin hydrogel composites with high storage modulus sensitivity for using as electroactive actuator: effects of surface area and electric field strength. *Polymer* 70: 242–251.

**110** Jung, J.-H., Jeon, J.-H., Sridhar, V., and Oh, I.-K. (2011). Electro-active graphene–Nafion actuators. *Carbon* 49: 1279–1289.

**111** Wang, Y., Bian, K., Hu, C. et al. (2013). Flexible and wearable graphene/polypyrrole fibers towards multifunctional actuator applications. *Electrochemistry Communications* 35: 49–52.

**112** Sen, I., Seki, Y., Sarikanat, M. et al. (2015). Electroactive behavior of graphene nanoplatelets loaded cellulose composite actuators. *Composites Part B: Engineering* 69: 369–377.

**113** Chen, T., Qiu, J., Zhu, K., and Li, J. (2016). Electro-mechanical performance of polyurethane dielectric elastomer flexible micro-actuator composite modified with titanium dioxide-graphene hybrid fillers. *Materials & Design* 90: 1069–1076.

**114** Liu, H., Niu, D., Jiang, W. et al. (2016). Illumination-oriented and thickness-dependent photomechanical bilayer actuators realized by graphene-nanoplatelets. *Sensors and Actuators A: Physical* 239: 45–53.

**115** Guo, L., Hao, Y.-W., Yang, P. et al. (2019). Fast fabrication of graphene oxide/reduced graphene oxide hybrid hydrogels for thermosensitive smart actuator utilizing laser irradiation. *Materials Letters* 237: 245–248.

**116** Chen, Z., Cao, R., Ye, S. et al. (2018). Graphene oxide/poly (*N*-isopropylacrylamide) hybrid film-based near-infrared light-driven bilayer actuators with shape memory effect. *Sensors and Actuators B: Chemical* 255: 2971–2978.

**117** Surana, K., Singh, P.K., Bhattacharya, B. et al. (2015). Synthesis of graphene oxide coated Nafion membrane for actuator application. *Ceramics International* 41: 5093–5099.

**118** Lee, J.-W., Kwon, T., Kang, Y. et al. (2018). Styrenic block copolymer/sulfonated graphene oxide composite membranes for highly bendable ionic polymer actuators with large ion concentration gradient. *Composites Science and Technology* 163: 63–70.

**119** Liu, A., Yuan, W., and Shi, G. (2012). Electrochemical actuator based on polypyrrole/sulfonated graphene/graphene tri-layer film. *Thin Solid Films* 520: 6307–6312.

**120** Zhang, L., Pan, J., Liu, Y. et al. (2020). NIR–UV responsive actuator with graphene oxide/microchannel-induced liquid crystal bilayer structure for biomimetic devices. *ACS Applied Materials & Interfaces* 12: 6727–6735.

**121** Vural, M., Lei, Y., Pena-Francesch, A. et al. (2017). Programmable molecular composites of tandem proteins with graphene oxide for efficient bimorph actuators. *Carbon* 118: 404–412.

**122** Terasawa, N. and Asaka, K. (2018). High-performance graphene oxide/vapor-grown carbon fiber composite polymer actuator. *Sensors and Actuators B: Chemical* 255: 2829–2837.

**123** Wang, W., Zhang, Y.-L., Han, B. et al. (2019). A complementary strategy for producing moisture and alkane dual-responsive actuators based on graphene oxide and PDMS bimorph. *Sensors and Actuators B: Chemical* 290: 133–139.

**124** Selvakumar, D., Alsalme, A., Alghamdi, A., and Jayavel, R. (2017). Reduced graphene oxide paper as bimorphic electrical actuators. *Materials Letters* 191: 182–185.

**125** Wang, W., Liu, D., Liu, Y. et al. (2015). Electrical actuation properties of reduced graphene oxide paper/epoxy-based shape memory composites. *Composites Science and Technology* 106: 20–24.

**126** Kaur, R.P. and Engles, D. (2018). Transport in a fullerene terminated aromatic molecular device. *Journal of Science: Advanced Materials and Devices* 3: 206–212.

**127** Wang, W., Han, B., Zhang, Y. et al. (2021). Laser-induced graphene tapes as origami and stick-on labels for photothermal manipulation via marangoni effect. *Advanced Functional Materials* 31: 2006179.

**128** Huang, T., Zhao, J., Feng, M. et al. (2012). A multi-state single-molecule switch actuated by rotation of an encapsulated cluster within a fullerene cage. *Chemical Physics Letters* 552: 1–12.

**129** Jung, J.-H., Vadahanambi, S., and Oh, I.-K. (2010). Electro-active nano-composite actuator based on fullerene-reinforced Nafion. *Composites Science and Technology* 70: 584–592.

**130** Panwar, V., Ko, S.Y., Park, J.-O., and Park, S. (2013). Enhanced and fast actuation of fullerenol/PVDF/PVP/PSSA based ionic polymer metal composite actuators. *Sensors and Actuators B: Chemical* 183: 504–517.

# 9

# Display Based on Carbon-Enhanced Materials

*Feng Lin[1,2], Hong Zhong[2], and Zhiming Wang[2]*

[1] *Yunan University, School of Materials and Energy, South Section, East Outer Ring Road, Chenggong District, Kunming 650500, China*
[2] *University of Electronic Science and Technology of China, Institute of Fundamental and Frontier Sciences, No. 4, Sec. 2, North Jianshe Rd, Chengdu 610054, China*

## 9.1 Introduction

In the modern information technology (IT) age, display technology is one of the most important research and industrial fields. Almost all the information transport and access are closely related to displays. The daily life of human beings is deeply influenced by display technologies, from electronic books, smartphones, and television to even wearable displays [1]. The global market scale of the display has already been over 100 billion dollars and will be over 200 billion in the near future, thus playing a critically important position in the world economy [2].

Based on the display mode, the display devices can be divided into transmission modes, such as liquid-crystal display (LCD); reflective modes, such as electronic book and e-paper; and light-emission mode, including organic light-emitting diode (OLED) displayand field-emission display (FED) [3]. In the last decades, LCD has been the mainstream display market, and OLED is labeled as the next-generation display. The traditional cathode ray tube display is only applied in particular areas, and the reflective electronic books display takes a very small part in the market. With the development of display technologies, high luminance, high contrast, high resolution, low cost, flexible, and even wearable displays are the trend of the future display. Carbon-based nanomaterials, including zero-dimensional carbon nanodots (CDs), one-dimensional carbon nanotubes (CNTs), and two-dimensional graphene and graphene oxide (GO), play important roles in different kinds of displays because of their excellent optical, electrical, and mechanical properties [3–6]. These unique properties greatly improved the emission, contrast, power consumption, efficiency, and flexibility performance of the display, thus promoting the development of current and even next generation of displays.

This part will first summarize the properties of zero-dimensional CDs and CDs-based displays, including photoluminescent and electroluminescent CDs-based LED. In the second part, we will summarize the properties of CNTs and the performance enhancement of CNTs on FED, LCD, and OLED as electron emission,

*Enhanced Carbon-Based Materials and Their Applications*, First Edition.
Edited by Poh Choon Ooi, Mengying Xie, and Chang Fu Dee.

alignment, and electrode materials. Finally, in the third part, we will discuss the properties of graphene and GO and their applications as liquid-crystal materials in LCD and transparent electrodes of OLED.

## 9.2 Display Based on CDs

CDs are typically defined as a fascinating class of zero-dimensional quasi-spherical carbon nanoparticles with ultrafine sizes below 10 nm, and the electronic bandgap structures of which are remarkably influenced by the quantum confinement effect [2, 7–9]. CDs were first discovered during the purification of single-walled carbon nanotubes (SWCNTs) derived from arc-discharge soot through electrophoresis preparation in 2004 [7]. Characteristically, the surface of CDs contains lots of carboxylic acid groups, which impart their great water solubility and suitability for functionalization (Figure 9.1a,b) [10]. CDs consist of three types: graphene quantum dots (GQDs), carbon nanodots (CNDs), and carbonized polymer dots (CPDs). Benefiting from the distinctive structure and unique optoelectronic characteristics, including long hot-electron lifetime, high electron mobility, tunable bandgaps, biocompatibility, water solubility, low toxicity, and strong stable fluorescence, CDs have attracted enormous attention from scientific researchers [15–22].

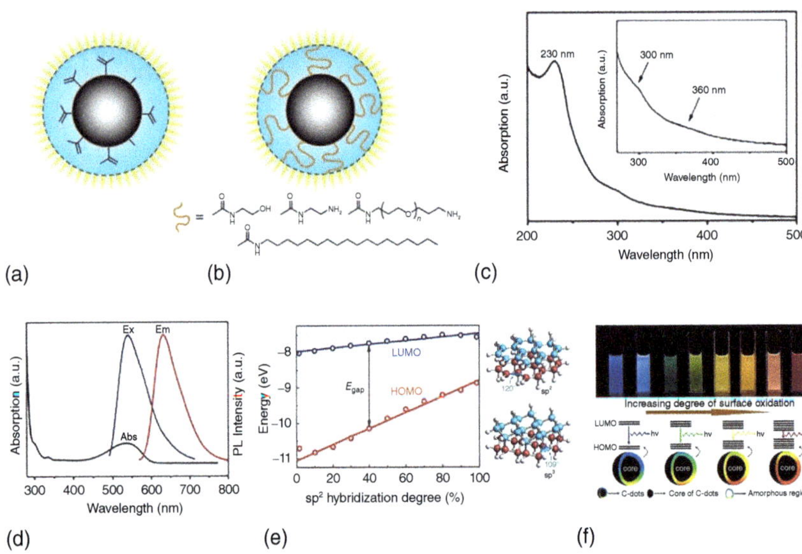


**Figure 9.1** Depiction of CDs (a) after surface oxidative treatment and (b) after functionalization with surface passivation reagents. Source: Baker et al. [10]/with permission of John Wiley and Sons. (c) The UV–Vis absorption spectra of CDs. Source: Fan et al. [11]/with permission of Elsevier. (d) UV–Vis absorption (black), PL excitation (blue) and emission (red) spectra of R-CDs. Source: Wang et al. [12]/with permission of John Wiley and Sons. (e) The CDs' energy gap of HOMO and LUMO states by the sp$^2$ hybridization factor of carbon domains. Source: Tepliakov et al. [13]/with permission of American Chemical Society. (f) The tunable PL of CDs by the degree of surface oxidation. Source: Ding et al. [14]/with permission of American Chemical Society.

As a consequence of the health concerns and the environmental conservation, CDs were regarded as potential replacements for toxic metal-based quantum dots (QDs) to develop well-fluorescent, eco-friendly, and low-toxicity alternatives in the display area. In particular, the optical properties of CDs can be tuned by individual shape, size, surface functional groups, and heteroatom, such as tunable PL efficiently stimulated by multiphoton up-conversion ranging from ultraviolet to near-infrared (NIR) [10, 23–26]. Such similar function as QDs shown in light-emitting diodes (LEDs) applications, CDs have already demonstrated the PL behaviors in LEDs as an active emission layer [18, 27–29].

### 9.2.1 Synthesis of CDs

The reported approaches of synthesizing CDs with tunable size can be generally classified into two main categories: up-down and bottom-up [30]. The former involves breaking down large carbon structures, such as nanodiamonds [31], graphite [2], graphite oxide [32], CNTs [33], carbon soot [34], and activated carbon [35]. The fabrication methods include arc discharge [7], laser ablation [36–38], electrochemical synthesis [39–42], nanometer etching [43, 44], and hydrothermal/solvothermal/special oxidation cleavage [45–47]. Xu et al. firstly discovered the example of CDs when they purified SWCNTs from arc-discharged soot [7]. As a follow-up, a variety of synthetic methods were developed. These synthetic methods inevitably have disadvantages, such as complex operation, uncontrollable reaction results, and the CDs products obtained, generally show low fluorescence quantum yield. Therefore, the top-down synthesis method is not suitable for large-scale production.

On the other hand, the "bottom-up" approaches synthesis of CDs from molecular or polymeric precursors, such as citrate [48], carbohydrates [49, 50], and polymer–silica nanocomposites, [51] through combustion [52], pyrolyzation [53], hydrothermal [54, 55], solvothermal [21, 56], and microwave-assisted pyrolysis [57, 58]. Compared with the top-down method, the bottom-up method can accurately design the reaction precursors and reaction conditions to regulate the size, shape, and heteroatom doping conditions of the product. As a result, it is easy to obtain CDs with high fluorescence quantum yield and other photoelectric properties, which provides the possibility for the large-scale production and application of CDs. It is well worth noting that cost is one of the important determinants of whether quantum dots can become competitors of semiconductor QDs. However, the above methods require expensive precursors, complex instruments, or postprocessing to synthesize fluorescent CDs. Thus, more economic-friendly methods need to be developed.

### 9.2.2 Optical Properties of CDs in Display

CDs possess similar optical properties despite the different reaction pathways and precursor systems in the various synthesized processes. Therefore, optical absorption and luminescence are the most important properties of CDs for display applications.

CDs typically show obvious absorption in the UV region (230–320 nm) with a tail extending into the visible region, due to the π-conjugated electrons in the $sp^2$ atomic framework and the chemical groups connected at the edge. The broad peak around 230 nm is ascribed to the $\pi$–$\pi^*$ transition of aromatic C=C bonds, whereas the shoulder at 300 nm is attributed to the $n$–$\pi^*$ transition of C=O bonds or other connected groups (Figure 9.1c) [11]. Furthermore, the size of π-domains, shape, composition, or structure of hybridization derivatives, such as functional groups and surface passivation, can also contribute to the absorption characteristics of CDs. The CDs generally show strong excitonic absorption bands in the UV-vis absorption spectra due to the quantum confinement effect (Figure 9.1d) [12]. In addition, the absorption peaks and range of CDs could be modulated by doping and surface passivation. For example, the absorption range of graphene quantum dots can be broadened after sulfur doping, and nitrogen doping could result in redshift absorption of carbon polymer dots [59, 60].

To date, CDs with various sizes and different PL emission wavelengths ranging from ultraviolet (UV) to NIR have been reported [26, 54, 61–63]. The quantum confinement effect is the major feature of CDs, which occurs when CDs are smaller than their exciton Bohr radius. For CDs with an intact carbon core and fewer surface chemical groups, the bandgap transition of conjugated π-domains is thought to be the major intrinsic PL center [64]. These π-domains are isolated by creating $sp^2$-hybridized islands rich in π-electrons through the reduction of graphene oxides [65]. They are created in a way that there are no π-connections between the $sp^2$ islands because any π-connections between the $sp^2$ islands would lead to interisland quenching of desired fluorescence emissions [66]. The localized $sp^2$ carbon clusters in the $sp^3$ carbon matrix serve as individual radiative emission centers instead of CD's whole carbonic core. A few aromatic rings of finite $sp^2$ clusters create quantized energy levels, inducing fluorescence through electron–hole recombination. Fu et al. modeled $sp^2$ carbon nanodomains of polycyclic aromatic hydrocarbon and postulated that size, distribution, and stacking of $sp^2$ nanodomains in $sp^3$ matrix induce CD's excitation-dependent PL with a large Stoke shift [67]. Because the different sizes of $sp^2$ nanodomains affect their gaps between the highest occupied molecular orbital (HOMO) and the lowest unoccupied molecular orbital (LUMO), the nanodomains work as selective emission centers at different wavelengths. In addition, the stacking of $sp^2$ nanodomains can trigger exciton self-trapping, inducing large Stoke shifts of CDs. In a recent study, Tepliakov et al. have proposed that carbon core cannot form freely isolated $sp^2$ nanodomains during the carbonization process but rather constitutes intermediate $sp^2$–$sp^3$ configurations [13]. In this model, the $sp^2$ hybridization degree of nanodomains affects CD's emission characteristics, and a higher degree of $sp^2$ hybridization induces higher delocalization of valence electrons over the domain area, increasing HOMO levels and reducing the bandgap (Figure 9.1e). These studies suggest that the growth, distribution, and $sp^2$ hybridization degree of nanodomains govern CD's absorption and emission features.

The other type of PL mechanism is based on the series of transitions in CDs' surface-defect states. In particular, oxygen-containing functional groups dominantly modulate the PL characteristics of CDs by working as emissive capture centers for excitons [14, 59, 64]. Ding et al. reported that CDs of similar size showed redshifted PL as the degree of surface oxidation increased (Figure 9.1f) [14]. In addition, the Cho group clearly demonstrated the PL variance of graphene oxide quantum dots from blue to green fluorescence by controlling the amount of oxygen content, validating the high dominance of surface-defect state (e.g. C=O and C–O–C groups) rather than carbon-core state [68]. The findings imply that a high degree of surface oxidation increases the population of emissive traps. Other types of surface functionalities (e.g. C=N) also modulate the emission of CDs by inducing distinctive energy levels [69]. As a result, the ensemble of multiple, broadly distributed emissive states derived from various functionalities accounts for the excitation-dependent PL of CDs.

### 9.2.3 CDs in LEDs Display Applications

Because of the multicolor, high-specific-surface areas, high electron mobility, tunable bandgaps, and high fluorescence quantum yield covering the visible light region, CDs can be used as high-efficient phosphors of LEDs [27, 54]. As the fourth generation of lighting source, LEDs have attracted wide attention in recent years due to their advantages of high luminous efficiency, long life, small size, energy saving, environmental protection, and potential applications in LCD and panchromatic display [70, 71]. As a new type of fluorescent carbon material, CDs are expected to replace expensive rare-earth phosphors and toxic semiconductor quantum dots in the field of LEDs. The above-mentioned properties, especially the wide emission range, make CDs more advantageous in white-light-emitting diodes (WLEDs).

Generally, LEDs based on CDs are usually divided into two categories. The first category is photoluminescent LEDs with different colors by using CDs as phosphors under the excitation of an external light source; the second category is using CDs as an active luminescent layer. Different colors of electroluminescent LEDs are realized by combining the injected electrons and holes in the luminescent layer.

#### 9.2.3.1 Photoluminescent LEDs

CDs-based PL LEDs are optoelectronic devices that emit light by exciting CDs phosphors with ultraviolet or blue light. However, due to the aggregation quenching phenomenon of CDs, the fluorescence of CDs solid powder is obviously weakened or even completely disappeared compared with the dispersion state, which greatly hinders the development of CDs in solid-state optoelectronic devices. In order to overcome the adverse effect of aggregation fluorescence quenching, many research groups prepared phosphors by dispersing CDs on substrates and applying them to LEDs. The solid substrates include borax, silica gel, starch, resin, and other polymers. Figure 9.2a shows the RGB fluorescence CDs in the solid matrix of borax and polymer [12]. Due to the low fluorescence quantum yield of solid CDs powders, the WLEDs of multicolor CDs phosphors with high performance have been gradually realized. Tang et al. [72] demonstrated first that GQDs can be used as a good light conversion material in WLEDs by covering GQDs on commercial

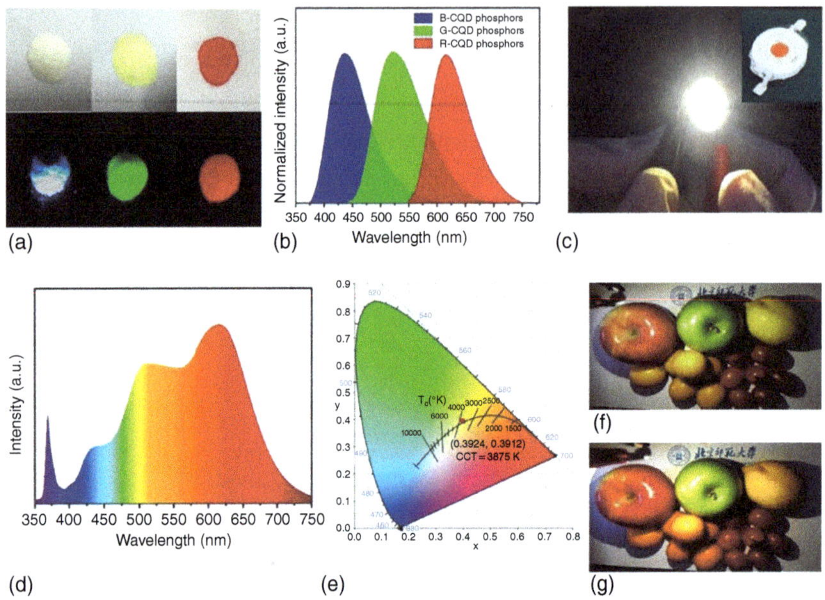


**Figure 9.2** Exciting CDs phosphors with ultraviolet and sunlight. (a) Images of B-, G-, and R-CDs phosphors under sunlight (above) and UV light (below) and (b) their corresponding emission spectra. (c) Photographs of the warm WLED lamp (inset) and the operating warm WLED lamp. (d) EL spectrum and (e) CIE color coordinate of the warm WLED lamp. Fruit color under (f) the commercial WLED lamp and (g) the CD warm WLED lamp. Source: Wang et al. [12]/with permission of John Wiley and Sons.

blue (410 nm) LED chips. In order to improve the color-rendering index (CRI) and regulate the color temperature, Sun et al. prepared WLEDs by mixing blue CDs phosphors with high fluorescence quantum yield, yellow and red copper indium zinc sulfur phosphors [73]. The CRI reached 93, which far exceeds commercial WLEDs (CRI < 75) based on yellow rare-earth phosphors. After that, they prepared WLEDs with adjustable color temperatures from 2805 to 7786 K by mixing blue CDs phosphors with yellow and red fluorescence polymer [74]. By using 1,3-dihydroxynaphthalene as the precursor and potassium periodate as the catalyst, Wang et al. prepared red fluorescence CDs with an edge-emitting fluorescence quantum yield of up to 53% [12]. By coating the tricolor CDs phosphors on the UV LED chips, full carbon-based warm WLEDs with a CRI of 97 were achieved, which is the highest value at present. A highly efficient trichromatic fluorescent film can be prepared by dispersing trichromatic CDs with high quantum yield into poly(methyl methacrylate). The warm WLED fabricated with tricolor CD fluorescent film exhibited an emission efficiency of 31.3 lm $W^{-1}$ (Figure 9.2b–g), which is comparable to the well-developed WLEDs based on rare-earth phosphors and semiconductor quantum dot phosphors.

Although solid-state fluorescence can be achieved by dispersing CDs in a solid matrix to avoid the aggregation quenching effect, introducing a solid matrix layer will reduce device efficiency and increase production costs. Therefore, the optimal method is to prepare CDs with high fluorescence quantum yield and solid-state fluorescence. Li et al. synthesized yellow solid fluorescent carbon nanorings (CNRs) with a pure hydroxyl functional group. It was demonstrated that this structure exhibited an active effect on overcoming the aggregation and quenching effect caused by π–π stacking [75]. Based on such solid-state fluorescence properties, they prepared WLEDs with chromaticity coordinates of 0.28, 0.27 by combining yellow solid-state fluorescent CDs and blue LED chips, as shown in Figure 9.3a–d. After that, Chen et al. reduced the aggregation quenching effect by modifying the surface of CDs and regulating the molecular spacing of CDs. Thus, solid-state fluorescent CDs with tunable fluorescence were also successfully prepared (Figure 9.3e,f) [76]. With this method, the fabrication process of LEDs was simplified effectively, and the cost of CDs-based optoelectronic devices could be reduced. He et al. prepared red solid fluorescent carbon dots with fluorescence emission at 638 nm by surface modification and blue–red double-emission WLEDs by combining them with a solid matrix [77]. In addition, solid-state fluorescent CDs that overcome aggregation and quenching through surface hydrogen bonding, heteroatom doping, and self-assembly of CDs

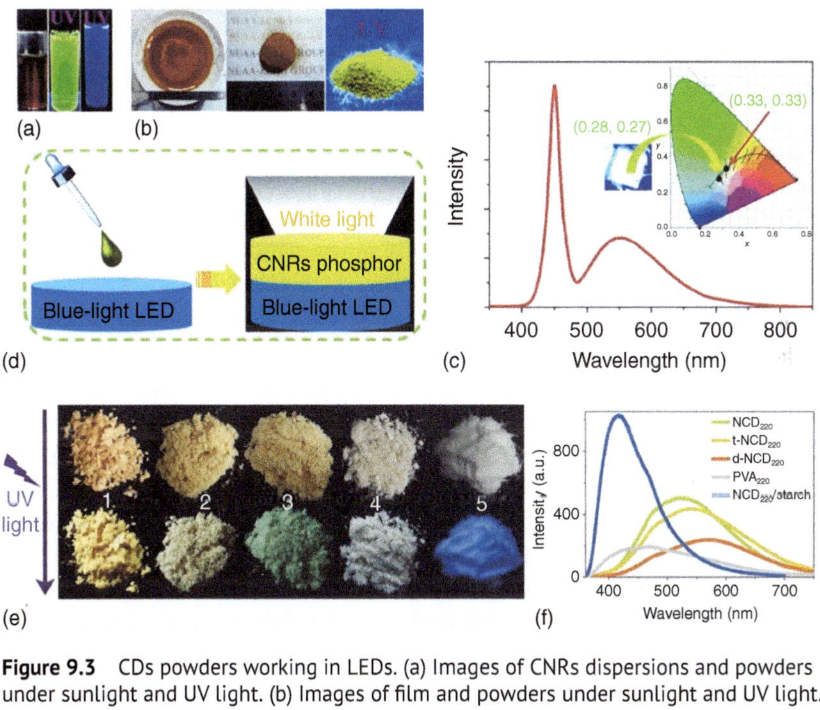


**Figure 9.3** CDs powders working in LEDs. (a) Images of CNRs dispersions and powders under sunlight and UV light. (b) Images of film and powders under sunlight and UV light. (c) Fabrication strategy of WLEDs based on blue-light-emitting diodes. (d) EL spectra of assembled LEDs. The insets are the digital image of the device and the corresponding CIE (Commission internationale de l'éclairage) chromaticity diagram. Source: Li et al. [75]. Reproduced with permission of John Wiley and Sons, Inc. (e) Photographs of d-NCD220 (No. 1), t-NCD220 (No. 2), NCD220 (No. 3), PVA220 (No. 4), and NCD220/starch composite (No. 5) powder under daylight (top) and UV light of 365 nm (bottom) and (f) SSF spectra of these powder. Source: Chen et al. [76]/with permission of John Wiley and Sons.

have also been prepared. However, the emission wavelengths of such solid-state fluorescent carbon dots are generally in the blue–green light region [78, 79].

PL LEDs based on CDs have achieved rapid development in recent years, however, it is challenging to produce full-carbon LEDs with high luminescence efficiency and high CRI due to the low fluorescence quantum yield of CDs in the long-wave region. Therefore, CDs-based PL LEDs are still in the early stage of development compared with the LEDs based on semiconductor quantum dots. Therefore, it is of great significance to develop a method for large-scale production of CDs with high fluorescence quantum yield (especially in the long-wave region) for the development and application of CDs in the field of LEDs.

#### 9.2.3.2 Electroluminescent LEDs

Due to the fluorescence excitation dependence caused by the defect states and surface molecular states on CDs, the carrier injection efficiency of CDs is low when they are applied to electroluminescent (EL) LEDs. Therefore, it is very important to prepare CDs with eigenstate band-edge emission. In 2017, Yuan et al. synthesized high-efficient eigenstate band-edge-emission fluorescent CDs from blue to red. They proved that CDs could be directly applied as active light-emitting layers of EL LEDs, as shown in Figure 9.4a–d [27]. The maximum brightness of the WLED is 2050 cd $m^{-2}$, and the current efficiency is 1.1 cd $A^{-1}$. The luminescence mechanism of LEDs based on CDs is that holes and electrons are injected into the CDs' active luminescence layer through the hole-injection layer of PEDOT: PSS and the electron transport layer of TPBI, respectively. Their work shows that CDs have great potential applications in electroluminescent LEDs and lays a solid foundation for the development of high-performance electroluminescent diodes based on high-efficiency eigenstate band-edge-emission fluorescence CDs.

Meaningfully, Gupta et al. reported EL LEDs prepared based on GQDs [81]. They doped blue fluorescent GQDs in poly(2-methoxy-5-(2-ethylhexoxy)-1,4-phenylacetylene) (MEH-PPV) and manipulated the performance of LED by changing the doping concentration of GQDs. The luminous intensity of the device increases with the increase of doping concentration, which directly reflects the increase of quantum efficiency in the device. GQDs dispersed in MEH-PPV undergo some additional electron-transfer processes during the luminescence process, which improves the charge-injection capability of the device, increases the density of the carrier, and finally leads to the reduction of the illuminating voltage of the device. Rhee et al. prepared a series of GQDs with a specific size distribution, as shown in Figure 9.4e–g [80]. The bandgap of this series of GQDs gradually decreases as the size increases, leading to a fluorescence color change of GQDs from blue to yellow–brown. They prepared EL LEDs by doping a series of GQDs in the host material of 4,4′-bis (9-carbazole) biphenyl (CBP). In this system, electrons and holes are injected into CBP from TPBI and GraHIL through electric bias. Since the energy levels of GQDS are contained in the energy levels of CBP, electrons, and holes can be further transferred to GQDS through CBP for luminescence. Zhang et al. prepared EL CDs-based LEDs with tunable luminous color [16]. By adjusting the injection current density, the luminescence color can be changed from blue to

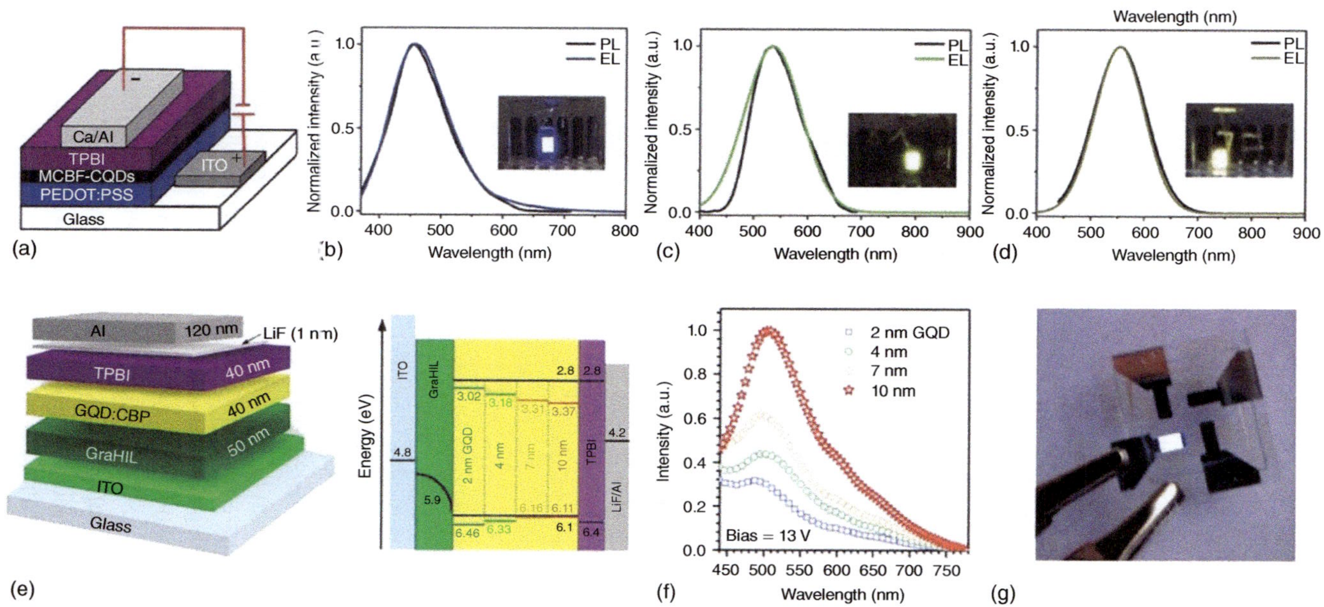


**Figure 9.4** Structures and EL spectra of LEDs employing CDs. (a) The device structure comprising ITO/PEDOT: PSS (anode), MCBF-CDs (active emission layer), TPBi (ETL), and Ca/Al (cathode). The normalized PL spectra and the corresponding output EL spectra of (b) B-BF-CDs, (c) G-BF-CDs, and (d) Y-BF-CDs. The photographs in the insets of (b–d) display the close-up view of the surface emission from blue, green, and yellow emissions of monochrome LEDs. Source: Yuan et al. [27]. Reproduced with permission of John Wiley and Sons, Inc. (e) Physical and electronic structures of LEDs employing the GQDs. GraHIL and TPBI (1,3,5-tri(phenyl-2-benzimidazolyl)-benzene) are hole and electron transporting layers, respectively. (f) Normalized EL intensity of a set of LEDs at a fixed bias (13 V). (g) Photo of white-light emission from devices employing the 10 nm GQDs. Source: Kwon et al. [80]/with permission of American Chemical Society.

white. They attributed the phenomenon to the existence of three recombination mechanisms in the CDs' luminescent layer of the device.

CDs-based EL LEDs have made great breakthroughs and progress in recent years. However, compared with $Cd^{2+}$ semiconductor quantum dots and perovskite materials with high fluorescence quantum yield (75–90%) and narrow emission wavelength (half-peak width less than 35 nm), CDs-based EL LEDs still have a performance gap. In other words, the application research of CDs in EL LEDs is still in the early stage. Nevertheless, through continuous in-depth research, the preparation of CDs with high fluorescence quantum yield and carrier mobility can effectively improve device performance. Furthermore, compared with the above two materials, CDs have good stability and environment-friendly characteristics; so, it is believed that CDs have the potential in replacing semiconductor quantum dots and perovskite quantum dot materials in the future and be widely used in EL LEDs.

## 9.3 Display Based on Carbon Nanotubes

CNTs have attracted enormous research interest in fundamental science and practical applications since their discovery in 1991 [82]. Composed of hexagonal carbon network roll, CNTs have a one-dimensional cylindrical structure with a high-aspect ratio, thus exhibiting excellent chemical stability, anisotropic optical properties, flexibility, transparency, and electric and thermal conductivity. The electron mobility of CNTs reaches over 100 000 $cm^2\ (V\,s)^{-1}$ [83]. These properties make CNT a promising candidate in display applications, especially in flexible and transparent displays. This section will summarize the display applications of CNT as emission material of FED, alignment and polarized material of LCD, channel materials of thin-film transistor (TFT) of LCD and OLED, and the transparent electrode.

### 9.3.1 CNTs Emission Material in Display

With the fast development of information technology, it is desired to fabricate light and thin display devices. FED is one of the promising technologies as it possesses the high-resolution advantages of traditional cathode-ray tube display and overcomes the limited view angle of LCD. The images of FED are produced by phosphor striking with electrons generated by field-emission electrodes. The complex fabrication and weak stability of conventional Spindt-type emitters are the main drawbacks that hinder the development of FED. Because of the rod structure, excellent current conductivity, and chemical stability, CNTs can be used as the field emitter of FED to accelerate the electrons and subsequently illuminate phosphor with a high voltage [84, 85]. The typical structure of CNT FED includes diode type and triode type, in which CNTs are vertically assembled as the cathode. The schematic of triode-type CNT FED is shown in Figure 9.5a [86]. Chemical vapor deposition (CVD), plasma-enhanced CVD, and screen printing are the main fabrication methods of CNTs cathode in the FED [91–93]. In order to fabricate the CNT emitters on a large scale, super-aligned CNTs (SACNTs) yarn and film

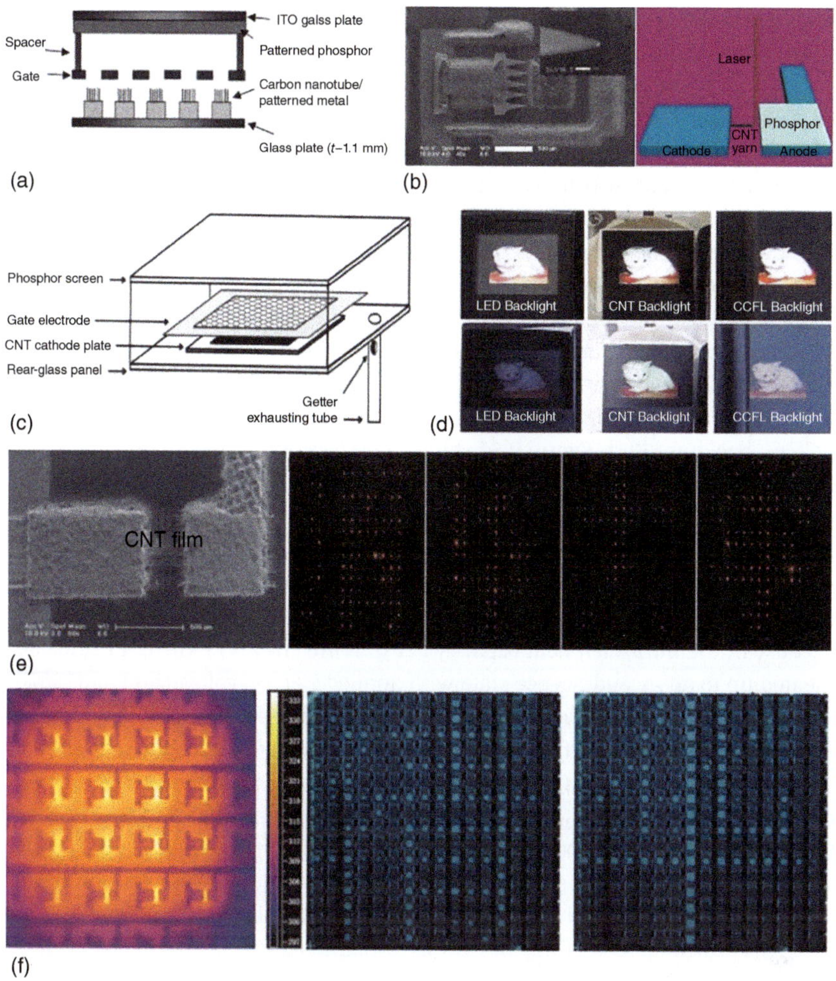


**Figure 9.5** CNT works as an emission material on display. (a) Schematic of triode-type CNT emitter of FED. Source: Choi et al. [86]/with permission of AIP Publishing. (b) SEM image of SACNT emitter and schematic of planar-type emission structure. Source: Adapted from Peng et al. [47]. (c) Schematic of CNT backlight of LCD. Source: Zhang et al. [87]/with permission of AIP Publishing. (d) LCD performance of LED, CNT, and CCFL backlight in indoor and outdoor conditions. Source: Zhang et al. [87]. Reproduced with permission of AIP Publishing. (e) SEM image of CNT incandescent display unit and character images shown on a display panel. Source: Liu et al. [88]. Reproduced with permission of John Wiley and Sons, Inc. (f) Infrared image of CNT film microheater array and patterns shown on the thermochromic display. Source: Refs. [89, 90]. Reproduced with permission of American Chemical Society, Reproduced with permission of John Wiley and Sons, Inc.

provide promising choices. Figure 9.5b shows the SEM image of the SACNT yarn emitter with a planar-type emission structure [94]. CNTs were also widely used as the electron emission layer in a plasma display panel as a high-performance field-emission material. The sustaining voltage was decreased by over 30 V more than the conventional one [95].

With the posttreatment of the CNT emitter by laser, the emission efficiency and brightness can be further improved. The high brightness of the CNT emitter makes it a potential candidate as the backlight of LCD [96–98]. The CNT backlight has a similar structure to the emitter of FED, and the structure can be even much simpler than a diode, and lateral structure can achieve the goal [99, 100]. Compared with the cold-cathode fluorescent lamp (CCFL) and LED backlight of LCD, the CNT backlight with triode structure can produce patterned light emission, thus reducing the power consumption by five times that of the conventional LCD backlight [1, 101]. Figure 9.5c shows the schematic of the CNT triode-structure lateral backlight. The CNT backlight realized high luminance of 12 000 cd $m^{-2}$, and the light transmission ratio was proved to be about 8%, which is two times that compared to the commercial devices [87]. The high brightness, local luminance controllability, and high transmission made the contrast ratio as high as 300 000 : 1, about 200 times compared to the CCFL backlight [96]. Figure 9.5d shows the LCDs with LED, CCFL, and CNT backlights in indoor and outdoor light conditions [87]. It can be observed that the CNT backlight of LCD shows higher contrast and brightness in both conditions. In addition, the low-temperature growth of the CNT backlight with plasma-enhanced CVD on common soda–lime glass reduced the fabrication cost and promoted its commercial application [101].

Because of the ultrasmall heat capacity per unit area of CNTs, the CNT film can be heated up to an incandescence state with applied electric voltage and fabricated as CNT incandescent display [102]. After screen printing of CNT film on electrode and laser cutting, the CNT film display unit and array were fabricated, as shown in Figure 9.5e [88]. The CNT display can be heated to an incandescence state and cooled down in about 1 ms, and the individual display pixel brightness reached 6400 cd $m^{-2}$ with a heating power of 0.08 W [88]. By applying small-mass CNT yarn and cool electrode end, the ramp-up time and cooling-down time were reduced to 0.79 and 0.36 ms, respectively [9]. When thermochromic pigments were loaded on the CNT incandescent film, the color of pigments changed with increasing temperature. Therefore, a CNT film array mixed with pigments was fabricated as the prototype of a thermochromic display [89]. The CNT film was firstly fabricated as a controllable heater array with screen printing and laser-cutting technique [90]. Then, the microheater array with pigments was heated independently with driving circuits to display different patterns, as shown in Figure 9.5f [89, 90]. The response time of the thermochromic display can be tuned by controlling the loading of pigments. It should be mentioned that both the CNT incandescent and thermochromic displays have relatively slow response time and small display contrast. Therefore, these displays cannot meet the requirement of dynamic image display at present but are suitable for applications, such as electronic books and advertisements.

### 9.3.2 CNTs as Alignment and Polarized Material in LCDs

The commercial LCDs always use rod-like liquid-crystal molecules to control the transmission of the backlight through the applied electric field. As one-dimensional

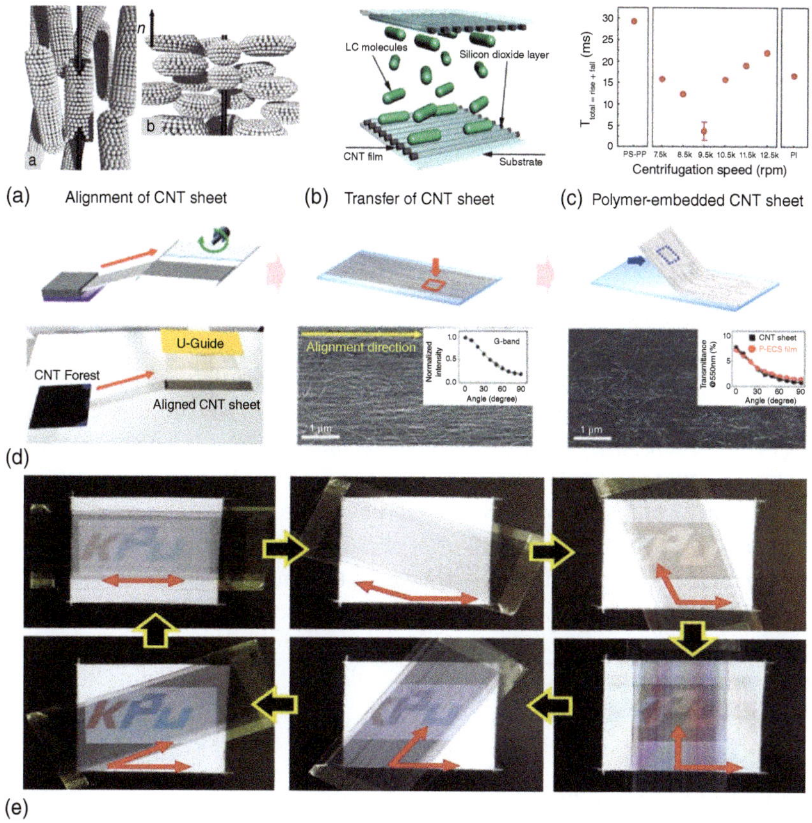


**Figure 9.6** CNTs as alignment material and polarizer in LCD. (a) Schematic of CNT alignment assistance in rod-like and disk-type LCs. Source: Lagerwall et al. [106]/with permission of John Wiley and Sons. (b) Schematic of CNT film as the alignment layer of LCs. Source: Fu et al. [107]/with permission of Elsevier. (c) The switching time of CNT-aligned LC cells as a function of centrifugation speed. Source: Lee et al. [108]/with permission of John Wiley and Sons. (d) Schematic of CNT film polarizer fabrication. Source: Adapted from Shin et al. [109]. CNTs were drawn from the CNT forest and transferred to a glass substrate to be embedded in the polymer as the alignment layer and polarizer. (e) Cleaning of letters on LCD of CNT film polarizer by rotating the polarizer angle. Source: Jung et al. [110]. Reproduced with permission of John Wiley and Sons, Inc.

cylinder molecules, CNTs also exhibit a liquid-crystalline phase and can be used as LC-like material in different parts of LCDs [103]. When mixed with other LC materials, the mutual alignment assistance between CNTs and LC molecules was widely demonstrated [104–107]. Figure 9.6a shows the schematic of CNT alignment in rod-like and disk-type LCs [106]. For vertically aligned LCDs, the addition of CNTs in rod-like liquid crystals increased the bend elastic constant of LCs, thus reducing the rising and decaying response time [105]. By decorating CNTs with copper oxide to reduce the ionic impurities and rotational viscosity in LCs, the response time of LCD can be further improved [104].

Besides that, CNTs can also be used as an alignment layer in the pre-orientation of LCs in the display. During the fabrication of LCD, a polyimide film is coated on the substrate and then rubbed to form a pre-tilt layer for LC molecules. Figure 9.6b is the schematic of CNTs film as the alignment layer of LCD [107]. By mixing CNTs with the alignment polymer layer, it formed a more effective alignment layer after rubbing [108]. The LCs formed a stable pre-tilt angle and showed a much improved switching speed of 3.8 ms, four times faster than the commercial polyimide layer, as shown in Figure 9.6c [108]. Because of the one-dimensional movement of electrons in CNT, the single CNT or CNTs film also exhibit polarized properties at a broad-spectrum range [111–114]. If the alignment order of CNTs is further improved, the CNTs film can be used as both alignment layer of LC molecules and optical polarizer of LCD [109, 110]. As shown in Figure 9.6d, the CNT forest was vertically grown on the substrate, and then transferred horizontally to glass or other polymer substrates. Finally, the CNTs film was partially embedded in polymer and applied as an alignment layer and polarizer [109]. The CNT film polarizer efficiency reached 91% with flexible, bendable, transparent, and conductive properties [109]. Figure 9.6e shows the display performance of the CNT film polarizer on a commercial LCD [110]. Although the polarization efficiency and transparency need to be further improved, the CNTs film polarizer shows promising applications on flexible, wearable, and transparent displays. Future research should focus more on improving the alignment degree and uniformity of CNTs to optimize the polarization and transparency of CNTs film polarizers.

### 9.3.3 CNT–TFT in LCD and OLED

LCD and OLED are the main part of the current display industry, and the TFT is the critical device for pixel controlling in the display. With the rapid development of display techniques, it is difficult for the amorphous silicon-based TFT to meet the requirements of large size, high-frame rate, flexibility, and low-power consumption. Because of the high carrier's mobility, transparency, and flexibility, CNTs have been widely used as semiconducting material for the fabrication of TFT in voltage-driven and current-driven displays, such as LCD, electrochromic display, and OLED [115]. In the TFT of the display, the source is connected with the data signal to drive the display units, and the gate is connected with the scan signal to control the on/off state [116, 117]. The CNTs network is used as the channel material of TFT.

For the voltage-driven display, such as LCD and electrochromic displays, the drain of TFT is connected with the pixel's bottom ITO electrode to control the orientation of LCs and transmission of backlight, as shown in Figure 9.7a [116, 123]. Figure 9.7b shows the structure of back-gated TFT with CNTs network as the channel materials and the corresponding display pattern on the LCD. For the current-driven display OLED, one addressing TFT is required to turn on the pixel, and the other driving TFT is connected with the OLED to provide current for the active molecular [117, 118, 124, 125]. The typical structure schematic of CNT–TFT-based OLED is shown in Figure 9.7c [118]. The CNT film can be directly grown on the substrate with the CVD method to fabricate TFT, but high temperature and high cost are required.

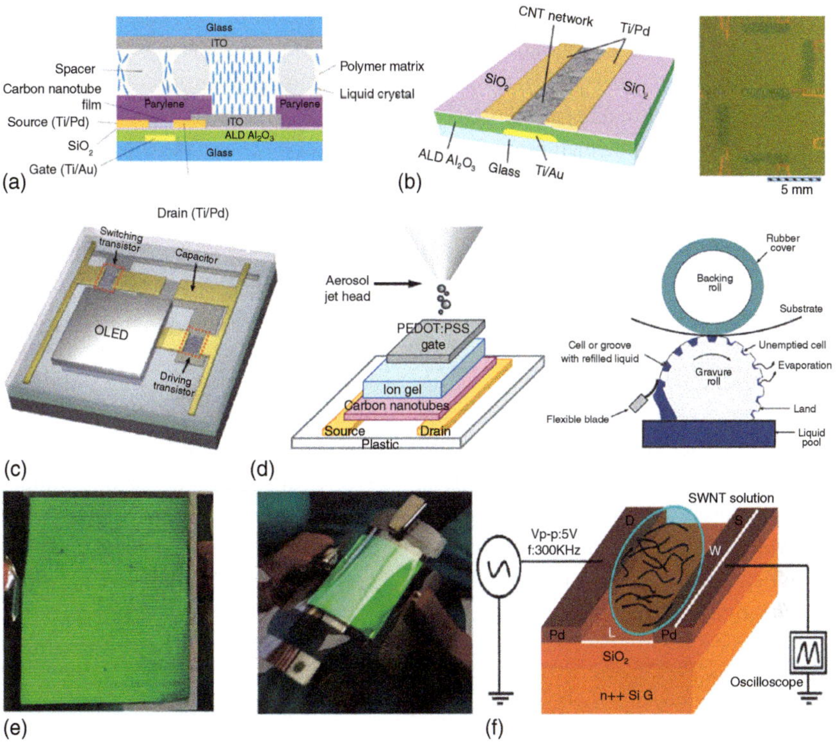


**Figure 9.7** CNT–TFT application in LCD and OLED. (a) Schematic of LCD with CNT–TFT as a driving device. Source: Cong et al. [116]/with permission of American Chemical Society. (b) The schematic of CNT–TFT and optical performance image of the LCD. Source: Cong et al. [116]/with permission of American Chemical Society. (c) Pixel structure of CNT–TFT OLED with one switching transistor and driving transistor. Source: Zhang et al. [118]/with permission of American Chemical Society. (d) Schematic of inkjet and gravure printing for the fabrication of CNT–TFT. Source: Refs. [119, 120]/with permission of American Chemical Society, with permission of Elsevier. (e) Flexible 64 × 64 pixels OLED panel with CNT–TFT. Source: Zhao et al. [121]. Reproduced with permission of American Chemical Society. (f) Schematic of dielectrophoresis alignment of CNT. Source: Adapted from Joo et al. [122].

In order to avoid it, CNT film is prepared at low temperature in the air by direct solution deposition and even be printed on a large scale [115, 123, 126]. The main printing techniques shown in Figure 9.7d include inkjet printing and roll-to-roll gravure printing [119, 120].

Although CNT–TFT can be fabricated by solution deposition or directly printed at a low cost, there are some challenges that remain for practical applications. Because of the different electronic properties and chirality indices, CNTs are a mixture of semiconductor and metallic tubes [1]. The mixed CNTs are used in two strategies in the display: separation of semiconducting CNTs and controlling the percolation [127]. Although the purity of separated semiconducting CNTs has reached 99.9%, further improvement is still required. The CNT–TFT exhibited device mobility ranging from tens to hundreds of $cm^2\ (V\,s)^{-1}$, which is much higher than that of amorphous Si (~1 $cm^2$ $(V\,s)^{-1}$) [4, 116, 128]. The high mobility of carriers provides a high current for the driving of OLED. Meanwhile, the on/off ratio of CNT–TFT can be modified up to $10^7$, which is much higher than that of amorphous Si (~$10^6$) [121, 126, 129]. Based on the high-performance CNT–TFT, a flexible OLED 64 × 64 pixels matrix is fabricated, as shown in Figure 9.7e [121]. Up to now, the separation process and percolation control are still very complex and expensive, which will hinder the mass production of CNT–TFT [1].

Besides the impurity of metallic CNTs, uniformity is another challenge that CNT–TFT faced because of the solution deposition and printing fabrication. The uniformity problem comes from the CNT network and the drying process, such as the coffee-ring effect. As for the liquid film related to roughness, suitable liquid viscosity, surface tension, and coating speed are required. In order to improve the mobility and uniformity of the CNTs network, various alignment methods are adopted, including self-assembly, dielectrophoresis, and vacuum filtration [122, 130, 131]. However, it is difficult to fabricate large-scale CNTs film using the self-assembly method [122]. For the dielectrophoresis method, large size and highly aligned CNT film can be prepared. At the same time, the alternating current electric field is more effective for metallic CNTs, as shown in Figure 9.7f [130]. For the vacuum filtration method, it is efficient and simple, but the yield is very low. Therefore, more efforts are needed for the alignment of CNTs and direct printing methods as they can achieve roll-to-roll mass production.

### 9.3.4 Transparent Electrode and Touch Panel in the Display

For the current LCD and OLED, the typical transparent electrode material is still indium tin oxide (ITO). However, with the increasing demand for flexible and transparent displays, the limitations of ITO, particularly high fabrication cost, fragile mechanical property, and migration to the active emission layer, hinder the development of future displays [1, 127, 132]. Because of the transparency, flexibility, and high conductivity, it is promising to use CNTs film as a transparent electrode, the touch panel of the display.

For application as transparent electrode, the most important parameters are resistance and transparency. The sheet resistance of CNT is around 1000 $\Omega\,sq^{-1}$, and the optical transparency in the visible range is around 90–95% for the thickness of 5–10 nm [133]. In order to improve the electrical conductivity of CNTs film without decreasing the transparency, modification methods, such as various doping of CNTs and coating of metal particles, are applied [1, 127, 134]. Based on the improvement mentioned above, the CNTs film shows comparable conductivity and transparency (24 $\Omega\,sq^{-1}$ and 83.4%) with ITO and exhibits much better flexibility [134, 135]. Figure 9.8a shows the principal schematic and optical image of CNT touch screen in LCD. The touch position can be easily detected by measuring the pulse voltage of the top and bottom layer of CNT film separated by polymer dots [134]. CNT film has also been applied as hole-injection anode material for OLED typically [5]. Figure 9.8b shows the structure of OLED with CNT film as the anode [136]. The CNT film deposited on the substrate is connected with hole transportation layer,

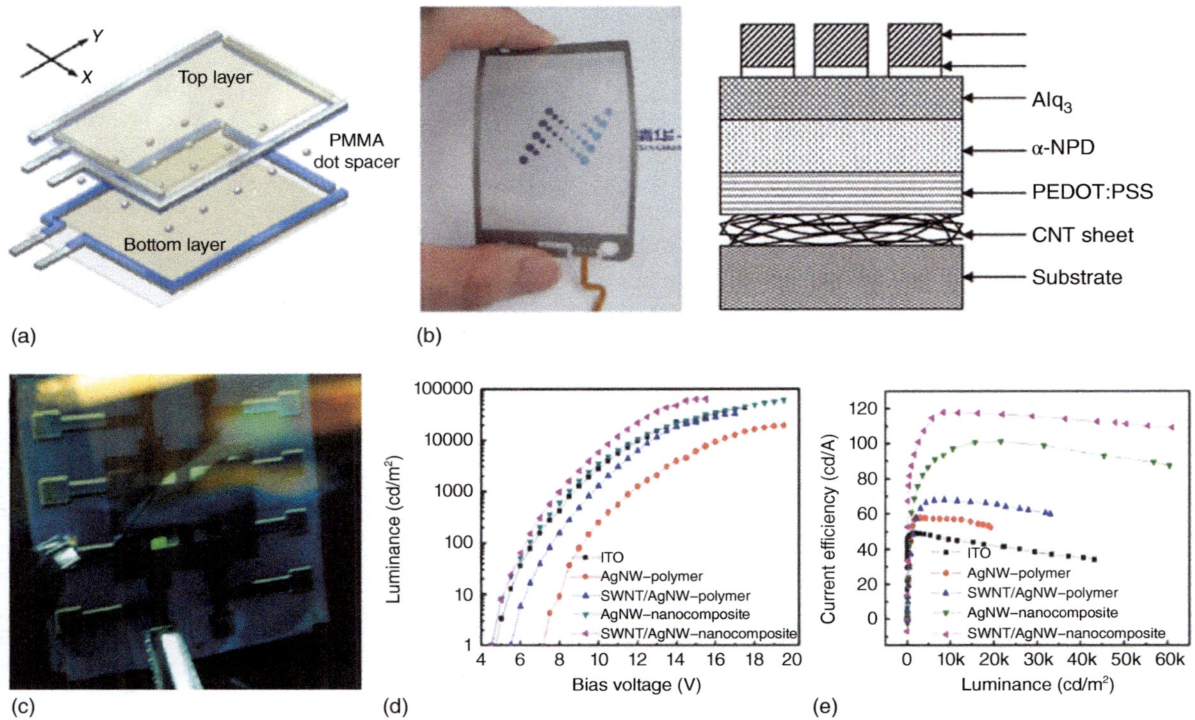


**Figure 9.8** CNT film as transparent electrode and touch panel of the display. (a) Schematic and image of CNT film as touch panel of the display. Source: Feng et al. [134]/with permission of John Wiley and Sons. (b) Schematic of OLED with CNT as the anode. Source: Williams et al. [136]/with permission of AIP Publishing. (c) Optical image of OLED with CNT film as a transparent electrode. Source: Ou et al. [137]. Reproduced with permission of American Chemical Society. (d) and (e) Improved luminance and current efficiency of CNT–OLED by introduction of silver nanowire and barium strontium titanate nanoparticles. Source: Li et al. [138]/with permission of Springer Nature.

poly(3,4-ethylenedioxythiophene):poly(styrenesulfonate) (PEDOT:PSS), and the emission layer. The luminance of the OLED is around several thousand cd $m^{-2}$, which is high enough to display applications [134, 136, 139]. Figure 9.8c shows the image of OLED with CNT film as the transparent electrode [137]. The current efficiency is always lower than that of ITO-based OLED because of the lower electric conductivity, and the luminance of CNT OLED will be much higher than that of ITO–OLED when the conductivity of CNT film is improved. Therefore, ITO-modification, silver nanowires, graphene, doping, and CNT-polymer composite were applied to improve the luminance to tens of thousands of cd $m^{-2}$ at a current efficiency of over ten or even 100 cd $A^{-1}$ [138, 140]. For example, after the modification of doping, polymer encapsulation, and PEDOT:PSS coating on the top, the CNT–OLED achieved luminance of 9000 cd $m^{-2}$ with an efficiency of 10 cd $A^{-1}$ [137]. Figure 9.8d shows that the OLED luminance was increased to 100 000 cd $m^{-2}$, and current efficiency was increased to 120 cd $A^{-1}$ with silver nanowire and barium strontium titanate modification [138]. What is more, the CNT film can also be used as an electron transport layer and cathode/anode in OLED.

## 9.4 Display Based on Graphene and Graphene Oxide

Graphene and GO have many distinguished properties, such as anisotropic optical properties, low sheet resistance, high-optical transmission, excellent mechanical strength, and flexibility, making them competitive candidates for the transparent electrode and liquid-crystal material of display [3, 141]. The anisotropic reflective and transmissive optical properties of graphene and GO can be controlled by applying an electric field, magnetic field, and mechanical shear stress, thus generating patterns on the graphene and GO flakes suspension [142–144]. The high conductivity and transparency of graphene make it one of the excellent candidates for driving transparent electrodes and touch screens [3, 145, 146]. With the fabrication technology development of graphene, large-scale and high-quality graphene film can be prepared at a relatively low price and widely used in practical display devices.

### 9.4.1 Graphene and Graphene Oxide as Liquid-Crystal Materials

As the classical two-dimensional carbon-based nanomaterials, the graphene and GO exhibit a very large aspect ratio. The unique geometry will generate optical reflection and transmission anisotropy. The transmission and reflection properties of graphene and GO planes are quite different from those of the edge. For the graphene and GO solvent dispersion or solution, when the orientation of the flakes is controlled or with the increasing solution concentration, the phase of solution will transfer from isotropic to nematic phase and exhibits anisotropic optical properties [147]. Therefore, graphene and GO flakes can be applied as liquid-crystal materials in reflective and transmissive LCD.

The orientation control or alignment methods of graphene and GO dispersion include mechanical shearing, electric field, and magnetic field [142–144, 147, 148].

For GO dispersion, Shen et al. demonstrated that the GO LCs could be transferred to a nematic phase when applied with a 10 kHz high-frequency alternating electric field [143]. The GO LCs dispersion with the nematic phase can transmit light, but the other part of the solution with isotropic phase will block the back illumination, thus displaying a bright and dark pattern. Figure 9.9a shows the electric-field-controlled GO LCs display and the on/off state [143]. The on/off response time of the GO LC display can be further improved to about 0.2 second by optimizing the GO flakes size [150]. By applying mechanical shearing flow, the GO flakes can orient along with the flow direction and exhibit different transmission and reflection performances. Figure 9.9b shows the schematic of mechanical alignment of GO dispersion by sweeping a stylus parallel and perpendicular to the surface to align the GO flakes horizontally and vertically. The GO liquid can be

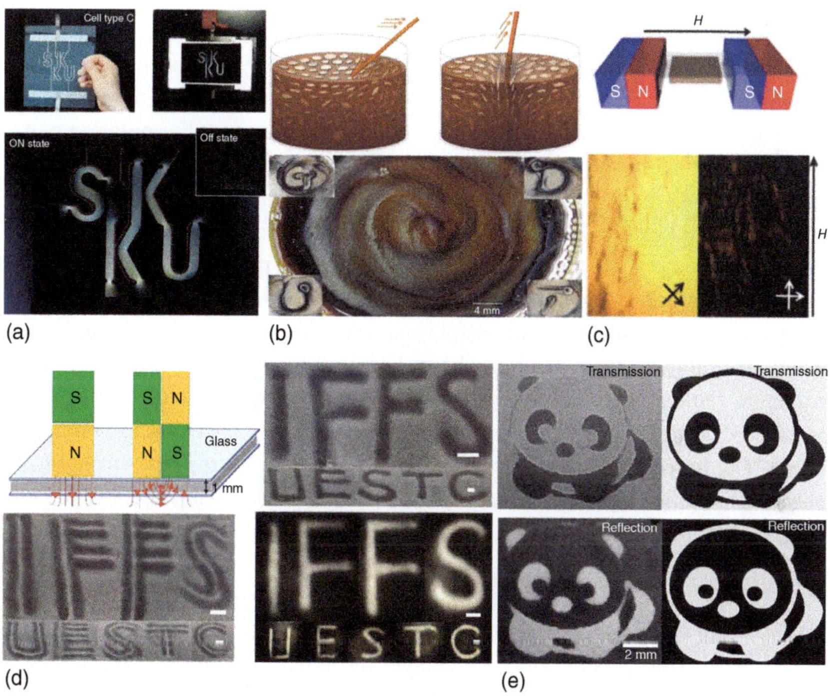


**Figure 9.9** Graphene and graphene oxide as the liquid-crystal materials of LCD. (a) High-frequency alternating electric-field-controlled GO LC transmission display. Source: Shen et al. [143]. Reproduced with permission of Springer Nature. (b) Schematic and patterns of GO handwriting reflective display. Source: He et al. [144]. Reproduced with permission of Royal Society of Chemistry. (c) Magnetic field alignment of GO solution and the bright/dark transmission images. Source: Kim et al. [148]. Reproduced with permission of John Wiley and Sons, Inc. (d) The magnetic field graphene display with one or two magnets to write onside. Letters are shown in a reflective and transmissive mode. Source: Lin et al. [142]. Reproduced with permission of John Wiley and Sons, Inc. (e) Patterned graphene thin-film display in complementary reflective and transmissive mode with planarly aligned graphene. Source: Lin et al. [149]. Reproduced with permission of John Wiley and Sons, Inc.

written with different letters or patterns in this way and works as a handwriting reflective display [144]. The GO solution can also be aligned by a static magnetic field and display a bright and dark transmission pattern, as shown in Figure 9.9c [3]. However, because of the extremely low diamagnetism, this alignment takes several hours, and iron oxide is required for fast switching, thus hindering the display development.

Compared with GO, graphene flakes have the highest diamagnetism at room temperature except for the superconductor. The orientation of few layers and multilayers graphene flakes can be easily aligned by common commercial NdFeB magnets [142]. The orientation of graphene flakes is parallel to the magnetic field, i.e. the graphene flakes orient vertically and horizontally when applying the magnets pole and middle part of two magnets. The vertically aligned graphene flakes have a weak reflection to ambient light but strong transmission to back illumination, thus displaying dark reflection and a bright transmission pattern. On the contrary, the horizontally aligned graphene flakes exhibit strong reflection but weak transmission, thus having bright reflection and dark transmission patterns. Figure 9.9d is the schematic of the graphene magnetic display, the reflection, and transmission display images [142]. For the static magnetic field, it can only achieve one-dimensional orientation control of graphene. There are two orientation freedoms for graphene flakes. By using a rotating magnetic field, Lin et al. demonstrated the planar alignment of graphene with all flakes oriented in one direction and paralleled to each other [149]. The planarly aligned graphene has almost perfect alignment order parameters and enhances the display contrast dramatically. Figure 9.9e shows the high contrast and complementary transmission and reflection display pattern with planarly aligned graphene.

Although graphene and GO have been successfully applied as LC materials in a display, the display contrast and switching response time need to be further improved. With the fast development of OLED, conventional LCD has less competitiveness. The reflective and handwriting display may find more breakthroughs in displays that require less power consumption, flexible handwriting, and low cost.

### 9.4.2 Graphene Transparent Electrode in the Display

The transparent electrode is widely used in display devices, such as LCD and OLED. With the increasing production of displays, the current dominant transparent electrode material ITO is hard to meet the high amount, low cost, and flexibility requirements. The advantageous properties of graphene can meet these requirements; thus, graphene film finds many applications in LCD electrodes, touch screens, and anode of OLED.

The graphene flakes or graphene film can be transferred to a glass substrate and applied as a driving electrode of liquid crystals instead of ITO film. The typical structure of graphene LCD electrodes is shown in Figure 9.10a [146]. The optical absorption of single-layer graphene is only 2%, and it is still much lower for multilayers graphene than conventionally used ITO film (15–18%). The sheet resistance of graphene can be reduced to tens of Ohm by chemical doping. However, for ITO film, the reduction of resistance is related to increased thickness and affects

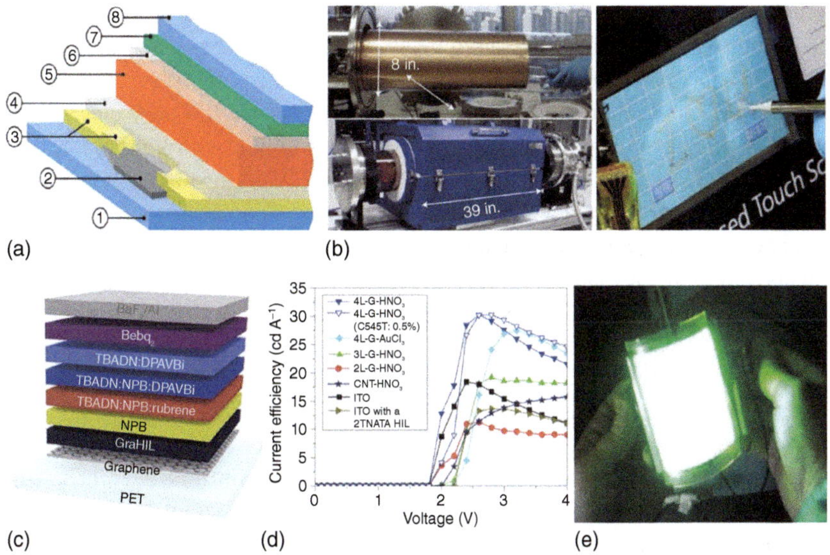


**Figure 9.10** Graphene transparent electrode in the display. (a) Graphene film as the alignment driving electrode of LCD. Source: Blake et al. [146]. Reproduced with permission of American Chemical Society. (b) Graphene electrode as a touch screen of a display. Source: Bae et al. [145]. Reproduced with permission of Springer Nature. (c) Schematic of graphene OLED structure. Source: Han et al. [151]/with permission of Springer Nature. (d) Luminous efficiency of OLED with various layers of graphene and ITO anode. Source: Han et al. [151]/with permission of Springer Nature. (e) Four-inch OLED with graphene anode. Source: Zhang et al. [152]. Reproduced with permission of Springer Nature.

the optical transparency. With the development of the fabrication technology of graphene, graphene film can be fabricated into tens of inches with the roll-to-roll CVD method and applied as a touch screen of a display [3, 145]. Figure 9.10b shows the preparation process with copper foil and the graphene touch-screen display [145]. The successful fabrication of high-quality and large-scale graphene film greatly promotes the replacement of ITO film, especially in the flexible display. Besides the CVD growth of graphene film, magnetic alignment of graphene flakes in polymer and reduction of GO sheets film on the substrate can also produce scalable and transparent graphene electrodes [153, 154].

The brittle and increasingly expensive properties of ITO also require graphene as a transparent electrode of OLED. The graphene layer is always used as the anode to drive the injection of holes into the luminous organic layer [155]. The graphene layer can be deposited by solution or grown by CVD, which will provide high-quality graphene film [151, 156]. In OLED graphene anode, the main challenges are reducing sheet resistance and improving the work function. The higher sheet resistance of graphene (about hundreds of $\Omega\,sq^{-1}$) than ITO (tens of $\Omega\,sq^{-1}$) will increase the operating voltage and limit the luminous efficiency. The lower work function (about 4.4 eV) than ITO (4.7–4.9 eV) will hinder the hole injection between the electrode and organic layer *N*,*N*′-bis(naphthalene-1-yl)-*N*,*N*′-bis(phenyl) benzidine (work function about 5.4 eV), because of the high injection barrier [151]. The structure of graphene OLED is shown in Figure 9.10c. In order to solve these problems, various methods have been applied. Kim et al. increased the work function of graphene anode to 5.4 eV and reduced the sheet resistance to 90 Ω $sq^{-1}$ by doing with bis(trifluoromethanesulfonyl)amide (TFSA, $[CF_3SO_2]_2NH$) [157]. Kwon et al. prepared graphene with the work function of 4.7 eV by doping with a fluoropolymer and realized a 104% luminance ratio to ITO-based OLED. Seo et al. demonstrated the improved work function of 4.9 eV by inserting gold nanoparticles into graphene layers and reducing the sheet resistance to 110 Ω $sq^{-1}$ [158]. Han et al. improved the work function to 5.95 eV by coating the graphene with a high work function conducting polymer layer and reduced the sheet resistance to 30 Ω $sq^{-1}$ by chemical doping [151]. The achieved current efficiency for fluorescent OLED is 30.2 cd $A^{-1}$, much higher than that of ITO (13.7 cd $A^{-1}$), as shown in Figure 9.10d [151]. The luminance quantum efficiency of graphene-based OLED can be further improved by optimizing the device structure [159, 160]. In the work of Wu et al., the hole conductivity, the work function of graphene, and quantum efficiency of green phosphorescence were increased by doping of boron [161]. Liu et al. inserted the poly(9,9-di-*n*-octylfluorene-*alt*-(1,4-phenylene-(4-*sec*-butylphenyl)imino)-1,4-phenylene) between the graphene and the transfer material poly(methyl methacrylate) to protect the graphene during fabrication and work as the hole-injection layer, realizing current efficiency of 74.5 cd $A^{-1}$ [162]. They also modified the work function to 4.94 eV with high dielectric polymer poly(vinylidene chloride) and improved the current efficiency to 80 cd $A^{-1}$ [163]. Besides the high-efficiency luminance of OLED, large-scale, stretchable, and flexible electrodes are required in a display application. Zhang et al. demonstrated a clean and damage-free transfer method with rosin as the support layer during graphene transfer, thus achieving four-inch graphene-based OLED [152]. Figure 9.10e shows the large size of flexible graphene OLED [152]. Shin et al. then demonstrated high-quality flexible OLED by combining graphene and silver nanowires in PEDOT:PSS film, achieving stretching, bending, rolling, and twisting of the device [164].

## 9.5 Summary and Outlook

The excellent photoluminescent and electroluminescent properties of carbon dots, electron emission of CNTs, high electrical conductivity, flexibility, and optical transparency of CNT and graphene make carbon-based nanomaterials distinguished component materials in display devices. The luminance and contrast of the display have been greatly improved by carbon dots and CNT-based display. The CNT and graphene electrode decreased the operation voltage and power consumption for the electrical properties. What is more, CNT and graphene film promote the development of flexible and transparent displays. With the development of display technologies, high resolution, low cost, low-power consumption, high contrast, transparent, wearable, and flexible displays are expected. Carbon-based nanomaterials meet these requirements very well, especially in the flexible and folding display. Up to now, the quality and cost of mass production of carbon-based nanomaterials are still needed for further improvement. Therefore, the quality and uniformity must be tackled with more research devotion for practical applications in the display. With the production of commercial carbon-based displays, the cost will be gradually reduced to an acceptable level.

## References


1 Jiang, K. (2017). Industrial applications of carbon nanotubes, Chap. 4. In: *Carbon Nanotubes for Displaying* (ed. H. Peng, Q. Li and T. Chen), 101–127. Elsevier.

2 Sun, Y.-P., Zhou, B., Lin, Y. et al. (2006). Quantum-sized carbon dots for bright and colorful photoluminescence. *Journal of the American Chemical Society* 128: 7756–7757, https://doi.org/10.1021/ja062677d.

3 Jo, G., Choe, M., Lee, S. et al. (2012). The application of graphene as electrodes in electrical and optical devices. *Nanotechnology* 23: 112001, https://doi.org/10.1088/0957-4484/23/11/112001.

4 De Volder, M.F.L., Tawfick, S.H., Baughman, R.H., and Hart, A.J. (2013). Carbon nanotubes: present and future commercial applications. *Science* 339: 535, https://doi.org/10.1126/science.1222453.

5 Opatkiewicz, J., LeMieux, M.C., and Bao, Z. (2010). Nanotubes on display: how carbon nanotubes can be integrated into electronic displays. *ACS Nano* 4: 2975–2978, https://doi.org/10.1021/nn101092d.

6 Yuan, F.L., Li, S.H., Fan, Z.T. et al. (2016). Shining carbon dots: synthesis and biomedical and optoelectronic applications. *Nano Today* 11: 565–586, https://doi.org/10.1016/j.nantod.2016.08.006.

7 Xu, X., Ray, R., Gu, Y. et al. (2004). Electrophoretic analysis and purification of fluorescent single-walled carbon nanotube fragments. *Journal of the American Chemical Society* 126: 12736–12737, https://doi.org/10.1021/ja040082h.

8 Bourlinos, A.B., Stassinopoulos, A., Anglos, D. et al. (2008). Photoluminescent carbogenic dots. *Chemistry of Materials* 20: 4539–4541, https://doi.org/10.1021/cm800506r.

9 Wei, Y., Liu, P., Jiang, K., and Fan, S. (2012). A display module implemented by the fast high-temperatue response of carbon nanotube thin yarns. *Nano Letters* 12: 2548–2553, https://doi.org/10.1021/nl3007769.

10 Baker, S.N. and Baker, G.A. (2010). Luminescent carbon nanodots: emergent nanolights. *Angewandte Chemie (International Ed. in English)* 49: 6726–6744, https://doi.org/10.1002/anie.200906623.

11 Fan, Z., Li, Y., Li, X. et al. (2014). Surrounding media sensitive photoluminescence of boron-doped graphene quantum dots for highly fluorescent dyed crystals, chemical sensing and bioimaging. *Carbon* 70: 149–156, https://doi.org/10.1016/j.carbon.2013.12.085.

12 Wang, Z., Yuan, F., Li, X. et al. (2017). 53% efficient red emissive carbon quantum dots for high color rendering and stable warm white-light-emitting diodes. *Advanced Materials* 29: 1702910, https://doi.org/10.1002/adma.201702910.

**13** Tepliakov, N.V., Kundelev, E.V., Khavlyuk, P.D. et al. (2019). sp(2)-sp(3)-hybridized atomic domains determine optical features of carbon dots. *ACS Nano* 13: 10737–10744, https://doi.org/10.1021/acsnano.9b05444.

**14** Ding, H., Yu, S.B., Wei, J.S., and Xiong, H.M. (2016). Full-color light-emitting carbon dots with a surface-state-controlled luminescence mechanism. *ACS Nano* 10: 484–491, https://doi.org/10.1021/acsnano.5b05406.

**15** Yang, S.-T., Cao, L., Luo, P.G. et al. (2009). Carbon dots for optical imaging *in vivo*. *Journal of the American Chemical Society* 131: 11308–11309, https://doi.org/10.1021/ja904843x.

**16** Zhang, X., Zhang, Y., Wang, Y. et al. (2013). Color-switchable electroluminescence of carbon dot light-emitting diodes. *ACS Nano* 7: 11234–11241, https://doi.org/10.1021/nn405017q.

**17** Lan, X., Ren, H., Yang, X., et al. (2020). A facile microwave-assisted synthesis of highly crystalline red carbon dots by adjusting the reaction solvent for white light-emitting diodes. *Nanotechnology* 31: 215704, https://doi.org/10.1088/1361-6528/ab71b6.

**18** Yuan, F., He, P., Xi, Z. et al. (2019). Highly efficient and stable white LEDs based on pure red narrow bandwidth emission triangular carbon quantum dots for wide-color gamut backlight displays. *Nano Research* 12: 1669–1674, https://doi.org/10.1007/s12274-019-2420-x.

**19** Zhao, H., Liu, L., Liu, Z. et al. (2011). Highly selective detection of phosphate in very complicated matrixes with an off–on fluorescent probe of europium-adjusted carbon dots. *Chemical Communications* 47: 2604–2606, https://doi.org/10.1039/C0CC04399K.

**20** Chung, Y.J., Kim, J., and Park, C.B. (2020). Photonic carbon dots as an emerging nanoagent for biomedical and healthcare applications. *ACS Nano* 14: 6470–6497, https://doi.org/10.1021/acsnano.0c02114.

**21** Jiang, K., Sun, S., Zhang, L. et al. (2015). Red, green, and blue luminescence by carbon dots: full-color emission tuning and multicolor cellular imaging. *Angewandte Chemie (International Ed. in English)* 54: 5360–5363, https://doi.org/10.1002/anie.201501193.

**22** Zheng, M., Ruan, S., Liu, S. et al. (2015). Self-targeting fluorescent carbon dots for diagnosis of brain cancer cells. *ACS Nano* 9: 11455–11461, https://doi.org/10.1021/acsnano.5b05575.

**23** Cheng, S., Ye, T., Mao, H. et al. (2020). Electrostatically assembled carbon dots/boron nitride nanosheet hybrid nanostructures for thermal quenching-resistant white phosphors. *Nanoscale* 12: 524–529, https://doi.org/10.1039/c9nr07785e.

**24** Yuan, F., Yuan, T., Sui, L. et al. (2018). Engineering triangular carbon quantum dots with unprecedented narrow bandwidth emission for multicolored LEDs. *Nature Communications* 9: 2249, https://doi.org/10.1038/s41467-018-04635-5.

**25** Zhang, H., Huang, H., Ming, H. et al. (2012). Carbon quantum dots/$Ag_3PO_4$ complex photocatalysts with enhanced photocatalytic activity and stability under visible light. *Journal of Materials Chemistry* 22: 10501–10506, https://doi.org/10.1039/c2jm30703k.

**26** Shen, J., Zhu, Y., Yang, X., and Li, C. (2012). Graphene quantum dots: emergent nanolights for bioimaging, sensors, catalysis and photovoltaic devices. *Chemical Communications* 48: 3686–3699, https://doi.org/10.1039/C2CC00110A.

**27** Yuan, F., Wang, Z., Li, X. et al. (2017). Bright multicolor bandgap fluorescent carbon quantum dots for electroluminescent light-emitting diodes. *Advanced Materials* 29: 1604436, https://doi.org/10.1002/adma.201604436.

**28** Tian, Z., Zhang, X., Li, D. et al. (2017). Full-color inorganic carbon dot phosphors for white-light-emitting diodes. *Advanced Optical Materials* 5: 1700416, https://doi.org/10.1002/adom.201700416.

**29** Yuan, T., Yuan, F., Li, X. et al. (2019). Fluorescence-phosphorescence dual emissive carbon nitride quantum dots show 25% white emission efficiency enabling single-component WLEDs. *Chemical Science* 10: 9801–9806, https://doi.org/10.1039/c9sc03492g.

**30** Namdari, P., Negahdari, B., and Eatemadi, A. (2017). Synthesis, properties and biomedical applications of carbon-based quantum dots: an updated review. *Biomedicine & Pharmacotherapy* 87: 209–222, https://doi.org/10.1016/j.biopha.2016.12.108.

**31** Yu, S.-J., Kang, M.-W., Chang, H.-C. et al. (2005). Bright fluorescent nanodiamonds: no photobleaching and low cytotoxicity. *Journal of the American Chemical Society* 127: 17604–17605, https://doi.org/10.1021/ja0567081.

**32** Wang, Q., Zheng, H., Long, Y. et al. (2011). Microwave–hydrothermal synthesis of fluorescent carbon dots from graphite oxide. *Carbon* 49: 3134–3140, https://doi.org/10.1016/j.carbon.2011.03.041.

**33** Zhou, J., Booker, C., Li, R. et al. (2007). An electrochemical avenue to blue luminescent nanocrystals from multiwalled carbon nanotubes (MWCNTs). *Journal of the American Chemical Society* 129: 744–745, https://doi.org/10.1021/ja0669070.

**34** Liu, H., Ye, T., and Mao, C. (2007). Fluorescent carbon nanoparticles derived from candle soot. *Angewandte Chemie (International Ed. in English)* 46: 6473–6475, https://doi.org/10.1002/anie.200701271.

**35** Dong, Y., Zhou, N., Lin, X. et al. (2010). Extraction of electrochemiluminescent oxidized carbon quantum dots from activated carbon. *Chemistry of Materials* 22: 5895–5899, https://doi.org/10.1021/cm1018844.

**36** Hu, S.-L., Niu, K.-Y., Sun, J. et al. (2009). One-step synthesis of fluorescent carbon nanoparticles by laser irradiation. *Journal of Materials Chemistry* 19: 484–488, https://doi.org/10.1039/b812943f.

**37** Yang, F., LeCroy, G.E., Wang, P. et al. (2016). Functionalization of carbon nanoparticles and defunctionalization—toward structural and mechanistic elucidation of carbon "quantum" dots. *Journal of Physical Chemistry C* 120: 25604–25611, https://doi.org/10.1021/acs.jpcc.6b08171.

**38** Tarasenka, N., Stupak, A., Tarasenko, N. et al. (2017). Structure and optical properties of carbon nanoparticles generated by laser treatment of graphite in liquids. *ChemPhysChem* 18: 1074–1083, https://doi.org/10.1002/cphc.201601182.

**39** Li, H., He, X., Kang, Z. et al. (2010). Water-soluble fluorescent carbon quantum dots and photocatalyst design. *Angewandte Chemie (International Ed. in English)* 49: 4430–4434, https://doi.org/10.1002/anie.200906154.

**40** Yan, Y., Li, H., Wang, Q. et al. (2017). Controllable ionic liquid-assisted electrochemical exfoliation of carbon fibers for the green and large-scale preparation of functionalized graphene quantum dots endowed with multicolor emission and size tunability. *Journal of Materials Chemistry C* 5: 6092–6100, https://doi.org/10.1039/c7tc01342f.

**41** Tan, X., Li, Y., Li, X. et al. (2015). Electrochemical synthesis of small-sized red fluorescent graphene quantum dots as a bioimaging platform. *Chemical Communications* 51: 2544–2546, https://doi.org/10.1039/c4cc09332a.

**42** Bao, L., Zhang, Z.L., Tian, Z.Q. et al. (2011). Electrochemical tuning of luminescent carbon nanodots: from preparation to luminescence mechanism. *Advanced Materials* 23: 5801–5806, https://doi.org/10.1002/adma.201102866.

**43** Ponomarenko, L.A., Schedin, F., Katsnelson, M.I. et al. (2008). Chaotic Dirac billiard in graphene quantum dots. *Science* 320: 356–358, https://doi.org/10.1126/science.1154663.

**44** Lee, J., Kim, K., Park, W.I. et al. (2012). Uniform graphene quantum dots patterned from self-assembled silica nanodots. *Nano Letters* 12: 6078–6083, https://doi.org/10.1021/nl302520m.

**45** Tao, H., Yang, K., Ma, Z. et al. (2012). *In vivo* NIR fluorescence imaging, biodistribution, and toxicology of photoluminescent carbon dots produced from carbon nanotubes and graphite. *Small* 8: 281–290, https://doi.org/10.1002/smll.201101706.

**46** Pan, D., Zhang, J., Li, Z., and Wu, M. (2010). Hydrothermal route for cutting graphene sheets into blue-luminescent graphene quantum dots. *Advanced Materials* 22: 734–738, https://doi.org/10.1002/adma.200902825.

**47** Peng, J., Gao, W., Gupta, B.K. et al. (2012). Graphene quantum dots derived from carbon fibers. *Nano Letters* 12: 844–849, https://doi.org/10.1021/nl2038979.

**48** Bourlinos, A.B., Stassinopoulos, A., Anglos, D. et al. (2008). Surface functionalized carbogenic quantum dots. *Small* 4: 455–458, https://doi.org/10.1002/smll.200700578.

**49** Bhunia, S.K., Saha, A., Maity, A.R. et al. (2013). Carbon nanoparticle-based fluorescent bioimaging probes. *Scientific Reports* 3: 1473, https://doi.org/10.1038/srep01473.

**50** Li, H., He, X., Liu, Y. et al. (2011). One-step ultrasonic synthesis of water-soluble carbon nanoparticles with excellent photoluminescent properties. *Carbon* 49: 605–609, https://doi.org/10.1016/j.carbon.2010.10.004.

**51** Liu, R., Wu, D., Liu, S. et al. (2009). An aqueous route to multicolor photoluminescent carbon dots using silica spheres as carriers. *Angewandte Chemie (International Ed. in English)* 48: 4598–4601, https://doi.org/10.1002/anie.200900652.

**52** Tian, L., Ghosh, D., Chen, W. et al. (2009). Nanosized carbon particles from natural gas soot. *Chemistry of Materials* 21: 2803–2809, https://doi.org/10.1021/cm900709w.

**53** Hsu, P.C., Chen, P.C., Ou, C.M. et al. (2013). Extremely high inhibition activity of photoluminescent carbon nanodots toward cancer cells. *Journal of Materials Chemistry B* 1: 1774–1781, https://doi.org/10.1039/c3tb00545c.

**54** Zhu, S., Meng, Q., Wang, L. et al. (2013). Highly photoluminescent carbon dots for multicolor patterning, sensors, and bioimaging. *Angewandte Chemie (International Ed. in English)* 52: 3953–3957, https://doi.org/10.1002/anie.201300519.

**55** Xia, C., Tao, S., Zhu, S. et al. (2018). Hydrothermal addition polymerization for ultrahigh-yield carbonized polymer dots with room temperature phosphorescence via nanocomposite. *Chemistry - A European Journal* 24: 11303–11308, https://doi.org/10.1002/chem.201802712.

**56** Miao, X., Qu, D., Yang, D. et al. (2018). Synthesis of carbon dots with multiple color emission by controlled graphitization and surface functionalization. *Advanced Materials* 30: 1704740, https://doi.org/10.1002/adma.201704740.

**57** Pan, L., Sun, S., Zhang, A. et al. (2015). Truly fluorescent excitation-dependent carbon dots and their applications in multicolor cellular imaging and multidimensional sensing. *Advanced Materials* 27: 7782–7787, https://doi.org/10.1002/adma.201503821.

**58** Shao, J., Zhu, S., Liu, H. et al. (2017). Full-color emission polymer carbon dots with quench-resistant solid-state fluorescence. *Advancement of Science* 4: 1700395, https://doi.org/10.1002/advs.201700395.

**59** Qu, D., Zheng, M., Du, P. et al. (2013). Highly luminescent S,N co-doped graphene quantum dots with broad visible absorption bands for visible light photocatalysts. *Nanoscale* 5: 12272–12277, https://doi.org/10.1039/c3nr04402e.

**60** Liu, J., Li, D., Zhang, K. et al. (2018). One-step hydrothermal synthesis of nitrogen-doped conjugated carbonized polymer dots with 31% efficient red emission for *in vivo* imaging. *Small* 14: 1613–6810, https://doi.org/10.1002/smll.201703919.

**61** Jia, H., Wang, Z., Yuan, T. et al. (2019). Electroluminescent warm white light-emitting diodes based on passivation enabled bright red bandgap emission carbon quantum dots. *Advancement of Science* 6: 1900397, https://doi.org/10.1002/advs.201900397.

**62** Meng, T., Yuan, T., Li, X. et al. (2019). Ultrabroad-band, red sufficient, solid white emission from carbon quantum dot aggregation for single component warm white light emitting diodes with a 91 high color rendering index. *Chemical Communications* 55: 6531–6534, https://doi.org/10.1039/c9cc01794a.

**63** Tian, Z., Tian, P., Zhou, X. et al. (2019). Ultraviolet-pumped white light emissive carbon dot based phosphors for light-emitting devices and visible light communication. *Nanoscale* 11: 3489–3494, https://doi.org/10.1039/c9nr00224c.

**64** Lim, S.Y., Shen, W., and Gao, Z. (2015). Carbon quantum dots and their applications. *Chemical Society Reviews* 44: 362–381, https://doi.org/10.1039/C4CS00269E.

**65** Hummers, W.S. and Offeman, R.E. (1958). Preparation of graphitic oxide. *Journal of the American Chemical Society* 80: 1339, https://doi.org/10.1021/ja01539a017–1339.

**66** Eda, G., Lin, Y.Y., Mattevi, C. et al. (2010). Blue photoluminescence from chemically derived graphene oxide. *Advanced Materials* 22: 505–509, https://doi.org/10.1002/adma.200901996.

**67** Fu, M., Ehrat, F., Wang, Y. et al. (2015). Carbon dots: a unique fluorescent cocktail of polycyclic aromatic hydrocarbons. *Nano Letters* 15: 6030–6035, https://doi.org/10.1021/acs.nanolett.5b02215.

**68** Jang, M.H., Ha, H.D., Lee, E.S. et al. (2015). Is the chain of oxidation and reduction process reversible in luminescent graphene quantum dots? *Small* 11: 3773–3781, https://doi.org/10.1002/smll.201500206.

**69** Nie, H., Li, M., Li, Q. et al. (2014). Carbon dots with continuously tunable full-color emission and their application in ratiometric pH sensing. *Chemistry of Materials* 26: 3104–3112, https://doi.org/10.1021/cm5003669.

**70** Yang, Y., Zheng, Y., Cao, W. et al. (2015). High-efficiency light-emitting devices based on quantum dots with tailored nanostructures. *Nature Photonics* 9: 259–266, https://doi.org/10.1038/nphoton.2015.36.

**71** Karatum, O., Jalali, H.B., Sadeghi, S. et al. (2019). Light-emitting devices based on type-II InP/ZnO quantum dots. *ACS Photonics* 6: 939–946, https://doi.org/10.1021/acsphotonics.8b01618.

**72** Tang, L., Ji, R., Li, X. et al. (2014). Deep ultraviolet to near-infrared emission and photoresponse in layered N-doped graphene quantum dots. *ACS Nano* 8: 6312–6320, https://doi.org/10.1021/nn501796r.

**73** Sun, C., Zhang, Y., Wang, Y. et al. (2014). High color rendering index white light emitting diodes fabricated from a combination of carbon dots and zinc copper indium sulfide quantum dots. *Applied Physics Letters* 104: 261106, https://doi.org/10.1063/1.4886415.

**74** Sun, C., Zhang, Y., Sun, K. et al. (2015). Combination of carbon dot and polymer dot phosphors for white light-emitting diodes. *Nanoscale* 7: 12045–12050, https://doi.org/10.1039/C5NR03014E.

**75** Li, X., Liu, Y., Song, X. et al. (2015). Intercrossed carbon nanorings with pure surface states as low-cost and environment-friendly phosphors for white-light-emitting diodes. *Angewandte Chemie (International Ed. in English)* 54: 1759–1764, https://doi.org/10.1002/anie.201406836.

**76** Chen, Y., Zheng, M., Xiao, Y. et al. (2016). A self-quenching-resistant carbon-dot powder with tunable solid-state fluorescence and construction of dual-fluorescence morphologies for white light-emission. *Advanced Materials* 28: 312–318, https://doi.org/10.1002/adma.201503380.

**77** He, J., He, Y., Chen, Y. et al. (2017). Solid-state carbon dots with red fluorescence and efficient construction of dual-fluorescence morphologies. *Small* 13: 1700075, https://doi.org/10.1002/smll.201700075.

**78** Hu, S., Ding, Y., Chang, Q. et al. (2015). Self-assembly of fluorescent carbon dots in a *N*,*N*-dimethylmethanamide solution via Schiff base reaction. *Nanoscale* 7: 4372–4376, https://doi.org/10.1039/C4NR07119K.

**79** Shen, C., Wang, J., Cao, Y., and Lu, Y. (2015). Facile access to B-doped solid-state fluorescent carbon dots toward light emitting devices and cell imaging agents. *Journal of Materials Chemistry C* 3: 6668–6675, https://doi.org/10.1039/c5tc01156f.

**80** Kwon, W., Kim, Y.H., Lee, C.L. et al. (2014). Electroluminescence from graphene quantum dots prepared by amidative cutting of tattered graphite. *Nano Letters* 14: 1306–1311, https://doi.org/10.1021/nl404281h.

**81** Gupta, V., Chaudhary, N., Srivastava, R. et al. (2011). Luminscent graphene quantum dots for organic photovoltaic devices. *Journal of the American Chemical Society* 133: 9960–9963, https://doi.org/10.1021/ja2036749.

**82** Iijima, S. (1991). Helical microtubules of graphitic carbon. *Nature* 354: 56–58, https://doi.org/10.1038/354056a0.

**83** Waldrop, M.M. (2016). The chips are down for Moore's law. *Nature* 530: 144–147, https://doi.org/10.1038/530144a.

**84** Amara, H. and Bichara, C. (2017). Single-walled carbon nanotubes. *Topics in Current Chemistry* 375: 55, https://doi.org/10.1007/s41061-017-0141-8.

**85** Bati, A.S.R., Yu, L., Batmunkh, M., and Shapter, J.G. (2019). Recent advances in applications of sorted single-walled carbon nanotubes. *Advanced Functional Materials* 29: 1902273, https://doi.org/10.1002/adfm.201902273.

**86** Choi, W.B., Jin, Y.W., Kim, H.Y. et al. (2001). Electrophoresis deposition of carbon nanotubes for triode-type field emission display. *Applied Physics Letters* 78: 1547–1549, https://doi.org/10.1063/1.1349870.

**87** Zhang, Y., Deng, S.Z., Xu, N.S., and Chen, J. (2008). Fully sealed carbon nanotube flat-panel light source and its application as thin film transistor–liquid-crystal display backlight. *Journal of Vacuum Science and Technology B* 26: 1033, https://doi.org/10.1116/1.2927920.

**88** Liu, P., Liu, L., Wei, Y. et al. (2009). Fast high-temperature response of carbon nanotube film and its application as an incandescent display. *Advanced Materials* 21: 3563–3566, https://doi.org/10.1002/adma.200900473.

**89** Liu, P., Zhou, D., Wei, Y. et al. (2015). Load characteristics of a suspended carbon nanotube film heater and the fabrication of a fast-response thermochromic display prototype. *ACS Nano* 9: 3753–3759, https://doi.org/10.1021/nn506410y.

**90** Liu, P., Liu, L., Jiang, K., and Fan, S. (2011). Carbon-nanotube-film microheater on a polyethylene terephthalate substrate and its application in thermochromic displays. *Small* 7: 732–736, https://doi.org/10.1002/smll.201001662.

**91** Fan, S., Chapline, M.G., Franklin, N.R. et al. (1999). Self-oriented regular arrays of carbon nanotubes and their field emission properties. *Science* 283: 512–514, https://doi.org/10.1126/science.283.5401.512.

**92** Li, Y.K., Zhu, C.C., and Liu, X.H. (2002). Field emission display with carbon nanotubes cathode: prepared by a screen-printing process. *Diamond and Related Materials* 11: 1845–1847, https://doi.org/10.1016/S0925-9635(02)00171-1.

**93** Tabatabaei, M.K., Ghafourifard, H., Koohsorkhi, J. et al. (2011). Remote and direct plasma regions for low-temperature growth of carbon nanotubes on glass substrates for display applications. *Journal of Physics D: Applied Physics* 44: 115401, https://doi.org/10.1088/0022-3727/44/11/115401.

**94** Liu, Q.F., Ding, G.F., Yang, Z.Q. et al. (2012). Development of an ac plasma display panel with a low discharge voltage utilizing an electrophoretic carbon nanotube as a field emission layer. *Journal of Micromechanics and Microengineering* 22: 075014, https://doi.org/10.1088/0960-1317/22/7/075014.

**95** Liu, Q.F., Yang, Z.Q., Wang, Y., and Ding, G.F. (2012). Electrophoretic carbon nanotube field emission layer for plasma display panels. *Nano-Micro Letters* 4: 247–252, https://doi.org/10.1007/Bf03353722.

**96** Choi, Y.C., Lee, J.W., Lee, S.K. et al. (2008). The high contrast ratio and fast response time of a liquid crystal display lit by a carbon nanotube field emission backlight unit. *Nanotechnology* 19: 235306, https://doi.org/10.1088/0957-4484/19/23/235306.

**97** Kim, Y.C., Park, S.H., Lee, C.S. et al. (2015). A 46-inch diagonal carbon nanotube field emission backlight for liquid crystal display. *Carbon* 91: 304–310, https://doi.org/10.1016/j.carbon.2015.04.093.

**98** Lee, S.Y. (2009). *LCD Backlights* (ed. C.L. Anthony), 1–20. Wiley.

**99** Fu, W., Liu, P., Tang, J. et al. (2008). Spherical field emission cathode based on carbon nanotube paste and its application in luminescent bulbs. *Journal of Vacuum Science and Technology B* 26: 1404, https://doi.org/10.1116/1.2958243.

**100** Kim, Y.C., Kang, H.S., Cho, E. et al. (2009). Building a backlight unit with lateral gate structure based on carbon nanotube field emitters. *Nanotechnology* 20: 095204, https://doi.org/10.1088/0957-4484/20/9/095204.

**101** Delepierre, G., Mahfouz, R., Aires, F.J.C.S., and Dijon, J. (2010). Green backlighting for TV liquid crystal display using carbon nanotubes. *Journal of Applied Physics* 108: 044308, https://doi.org/10.1063/1.3448021.

**102** Hochberg, F., Seitz, H.K., and Brown, A.V. (1973). A thin-film integrated incandescent display. *IEEE Transactions on Electron Devices* 20: 1002–1005, https://doi.org/10.1109/t-ed.1973.17781.

**103** Song, W., Kinloch, I.A., and Windle, A.H. (2003). Nematic liquid crystallinity of multiwall carbon nanotubes. *Science* 302: 1363, https://doi.org/10.1126/science.1089764.

**104** Malik, A., Prakash, J., Kumar, A. et al. (2012). Copper oxide decorated multi-walled carbon nanotubes/ferroelectric liquid crystal composites for faster display devices. *Journal of Applied Physics* 112: 054309, https://doi.org/10.1063/1.4748958.

**105** Lim, Y.J., Bhattacharyya, S.S., Tie, W. et al. (2013). Effects of carbon nanotubes on electro-optic characteristics in vertically aligned liquid crystal display. *Liquid Crystals* 40: 1202–1208, https://doi.org/10.1080/02678292.2013.806997.

**106** Lagerwall, J., Scalia, G., Haluska, M. et al. (2007). Nanotube alignment using lyotropic liquid crystals. *Advanced Materials* 19: 359–364, https://doi.org/10.1002/adma.200600889.

**107** Fu, W., Liu, L., Jiang, K. et al. (2010). Super-aligned carbon nanotube films as aligning layers and transparent electrodes for liquid crystal displays. *Carbon* 48: 1876–1879, https://doi.org/10.1016/j.carbon.2010.01.026.

**108** Lee, W.-K., Choi, Y.S., Kang, Y.-G. et al. (2011). Super-fast switching of twisted nematic liquid crystals on 2D single wall carbon nanotube networks. *Advanced Functional Materials* 21: 3843–3850, https://doi.org/10.1002/adfm.201101345.

**109** Shin, H.S., Manda, R., Kim, T.H. et al. (2020). Multi-layered carbon nanotube UV polariser for photo-alignment of liquid crystals. *Liquid Crystals* 47: 1604–1611, https://doi.org/10.1080/02678292.2020.1751318.

**110** Jung, M., Noh, Y., Suh, D., and Ahn, S.E. (2018). Flexible and thermally stable optical polarizers based on highly aligned carbon nanotube sheets for the visible spectral range. *Advanced Materials Technologies* 3, https://doi.org/10.1002/admt.201800203.

**111** Ren, L., Pint, C.L., Arikawa, T. et al. (2012). Broadband terahertz polarizers with ideal performance based on aligned carbon nanotube stacks. *Nano Letters* 12: 787–790, https://doi.org/10.1021/nl203783q.

**112** Zubair, A., Tsentalovich, D.E., Young, C.C. et al. (2016). Carbon nanotube fiber terahertz polarizer. *Applied Physics Letters* 108: 141107, https://doi.org/10.1063/1.4945708.

**113** Arutyunyan, N.R., Kanygin, M.A., Pozharov, A.S. et al. (2016). Light polarizer in visible and THz range based on single-wall carbon nanotubes embedded into poly(methyl methacrylate) film. *Laser Physics Letters* 13: 065901, https://doi.org/10.1088/1612-2011/13/6/065901.

**114** Cole, M.T., Doherty, M., Parmee, R. et al. (2014). Ultra-broadband polarisers based on metastable free-standing aligned carbon nanotube membranes. *Advanced Optical Materials* 2: 929–937, https://doi.org/10.1002/adom.201400238.

**115** Chen, K., Gao, W., Emaminejad, S. et al. (2016). Printed carbon nanotube electronics and sensor systems. *Advanced Materials* 28: 4397–4414, https://doi.org/10.1002/adma.201504958.

**116** Zhang, J., Fu, Y., Wang, C. et al. (2016). Carbon nanotube macroelectronics for active matrix polymer-dispersed liquid crystal displays. *ACS Nano* 10: 10068–10074, https://doi.org/10.1021/acsnano.6b04951.

**117** Liang, X., Xia, J., Dong, G. et al. (2016). Carbon nanotube thin film transistors for flat panel display application. *Topics in Current Chemistry* 374: 80, https://doi.org/10.1007/s41061-016-0083-6.

**118** Zhang, J., Fu, Y., Wang, C. et al. (2011). Separated carbon nanotube macroelectronics for active matrix organic light-emitting diode displays. *Nano Letters* 11: 4852–4858, https://doi.org/10.1021/nl202695v.

**119** Ha, M., Xia, Y., Green, A.A. et al. (2010). Printed, sub-3V digital circuits on plastic from aqueous carbon nanotube inks. *ACS Nano* 4: 4388–4395, https://doi.org/10.1021/nn100966s.

**120** Yin, X. and Kumar, S. (2006). Flow visualization of the liquid emptying process in scaled-up gravure grooves and cells. *Chemical Engineering Science* 61: 1146–1156, https://doi.org/10.1016/j.ces.2005.07.039.

**121** Zhao, T.Y., Zhang, D.D., Qu, T.Y. et al. (2019). Flexible 64 × 64 pixel AMOLED displays driven by uniform carbon nanotube thin-film transistors. *ACS Applied Materials & Interfaces* 11: 11699–11705, https://doi.org/10.1021/acsami.8b17909.

**122** Joo, Y., Brady, G.J., Arnold, M.S., and Gopalan, P. (2014). Dose-controlled, floating evaporative self-assembly and alignment of semiconducting carbon nanotubes from organic solvents. *Langmuir* 30: 3460–3466, https://doi.org/10.1021/la500162x.

**123** Cao, X., Lau, C., Liu, Y. et al. (2016). Fully screen-printed, large-area, and flexible active-matrix electrochromic displays using carbon nanotube thin-film transistors. *ACS Nano* 10: 9816–9822, https://doi.org/10.1021/acsnano.6b05368.

**124** Zhang, J., Wang, C., and Zhou, C. (2012). Rigid/flexible transparent electronics based on separated carbon nanotube thin-film transistors and their application in display electronics. *ACS Nano* 6: 7412–7419, https://doi.org/10.1021/nn3026172.

**125** McCarthy, M.A., Liu, B., Donoghue, E.P. et al. (2011). Low-voltage, low-power, organic light-emitting transistors for active matrix displays. *Science* 332: 570, https://doi.org/10.1126/science.1203052.

**126** Chen, P., Fu, Y., Aminirad, R. et al. (2011). Fully printed separated carbon nanotube thin film transistor circuits and its application in organic light emitting diode control. *Nano Letters* 11: 5301–5308, https://doi.org/10.1021/nl202765b.

**127** Hu, L., Hecht, D.S., and Gruner, G. (2010). Carbon nanotube thin films: fabrication, properties, and applications. *Chemical Reviews* 110: 5790–5844, https://doi.org/10.1021/cr9002962.

**128** Sangwan, V.K., Ortiz, R.P., Alaboson, J.M. et al. (2012). Fundamental performance limits of carbon nanotube thin-film transistors achieved using hybrid molecular dielectrics. *ACS Nano* 6: 7480–7488, https://doi.org/10.1021/nn302768h.

**129** Xia, J., Zhao, J., Meng, H. et al. (2018). Performance enhancement of carbon nanotube thin film transistor by yttrium oxide capping. *Nanoscale* 10: 4202–4208, https://doi.org/10.1039/c7nr08676h.

**130** Shekhar, S., Stokes, P., and Khondaker, S.I. (2011). Ultrahigh density alignment of carbon nanotube arrays by dielectrophoresis. *ACS Nano* 5: 1739–1746, https://doi.org/10.1021/nn102305z.

**131** He, X., Gao, W., Xie, L. et al. (2016). Wafer-scale monodomain films of spontaneously aligned single-walled carbon nanotubes. *Nature Nanotechnology* 11: 633–638, https://doi.org/10.1038/nnano.2016.44.

**132** Lo, M.-F., Ng, T.-W., Mo, H.-W., and Lee, C.-S. (2013). Direct threat of a UV-ozone treated indium-tin-oxide substrate to the stabilities of common organic semiconductors. *Advanced Functional Materials* 23: 1718–1723, https://doi.org/10.1002/adfm.201202120.

**133** Hecht, D.S., Hu, L., and Irvin, G. (2011). Emerging transparent electrodes based on thin films of carbon nanotubes, graphene, and metallic nanostructures. *Advanced Materials* 23: 1482–1513, https://doi.org/10.1002/adma.201003188.

**134** Feng, C., Liu, K., Wu, J.-S. et al. (2010). Flexible, stretchable, transparent conducting films made from superaligned carbon nanotubes. *Advanced Functional Materials* 20: 885–891, https://doi.org/10.1002/adfm.200901960.

**135** Park, S., Vosguerichian, M., and Bao, Z. (2013). A review of fabrication and applications of carbon nanotube film-based flexible electronics. *Nanoscale* 5: 1727–1752, https://doi.org/10.1039/c3nr33560g.

**136** Williams, C.D., Robles, R.O., Zhang, M. et al. (2008). Multiwalled carbon nanotube sheets as transparent electrodes in high brightness organic light-emitting diodes. *Applied Physics Letters* 93: 183506, https://doi.org/10.1063/1.3006436.

**137** Ou, E.C., Hu, L., Raymond, G.C. et al. (2009). Surface-modified nanotube anodes for high performance organic light-emitting diode. *ACS Nano* 3: 2258–2264, https://doi.org/10.1021/nn900406n.

**138** Li, L., Liang, J., Chou, S.Y. et al. (2014). A solution processed flexible nanocomposite electrode with efficient light extraction for organic light emitting diodes. *Scientific Reports* 4: 4307, https://doi.org/10.1038/srep04307.

**139** Gu, Z.Z., Tian, Y., Geng, H.Z. et al. (2019). Highly conductive sandwich-structured CNT/PEDOT:PSS/CNT transparent conductive films for OLED electrodes. *Applied Nanoscience* 9: 1971–1979, https://doi.org/10.1007/s13204-019-01006-4.

**140** Liu, Y., Jung, E., Wang, Y. et al. (2014). "Quasi-freestanding" graphene-on-single walled carbon nanotube electrode for applications in organic light-emitting diode. *Small* 10: 944–949, https://doi.org/10.1002/smll.201301829.

**141** Chen, X., Shehzad, K., Gao, L. et al. (2020). Graphene hybrid structures for integrated and flexible optoelectronics. *Advanced Materials* 32: 1902039, https://doi.org/10.1002/adma.201902039.

**142** Lin, F., Zhu, Z., Zhou, X. et al. (2017). Orientation control of graphene flakes by magnetic field: broad device applications of macroscopically aligned graphene. *Advanced Materials* 29: 1604453, https://doi.org/10.1002/adma.201604453.

**143** Shen, T.Z., Hong, S.H., and Song, J.K. (2014). Electro-optical switching of graphene oxide liquid crystals with an extremely large Kerr coefficient. *Nature Materials* 13: 394–399, https://doi.org/10.1038/nmat3888.

**144** He, L., Ye, J., Shuai, M. et al. (2015). Graphene oxide liquid crystals for reflective displays without polarizing optics. *Nanoscale* 7: 1616–1622, https://doi.org/10.1039/c4nr06008c.

**145** Bae, S., Kim, H., Lee, Y. et al. (2010). Roll-to-roll production of 30-inch graphene films for transparent electrodes. *Nature Nanotechnology* 5: 574–578, https://doi.org/10.1038/nnano.2010.132.

**146** Blake, P., Brimicombe, P.D., Nair, R.R. et al. (2008). Graphene-based liquid crystal device. *Nano Letters* 8: 1704–1708, https://doi.org/10.1021/nl080649i.

**147** Lin, F., Tong, X., Wang, Y. et al. (2015). Graphene oxide liquid crystals: synthesis, phase transition, rheological property, and applications in optoelectronics and display. *Nanoscale Research Letters* 10: 435, https://doi.org/10.1186/s11671-015-1139-1.

**148** Kim, J.E., Han, T.H., Lee, S.H. et al. (2011). Graphene oxide liquid crystals. *Angewandte Chemie (International Ed. in English)* 50: 3043–3047, https://doi.org/10.1002/anie.201004692.

**149** Lin, F., Yang, G., Niu, C. et al. (2018). Planar alignment of graphene sheets by a rotating magnetic field for full exploitation of graphene as a 2D material. *Advanced Functional Materials* 28, https://doi.org/10.1002/adfm.201805255.

**150** Ahmad, R.T., Hong, S.H., Shen, T.Z., and Song, J.K. (2015). Optimization of particle size for high birefringence and fast switching time in electro-optical switching of graphene oxide dispersions. *Optics Express* 23: 4435–4440, https://doi.org/10.1364/OE.23.004435.

**151** Han, T.-H., Lee, Y., Choi, M.-R. et al. (2012). Extremely efficient flexible organic light-emitting diodes with modified graphene anode. *Nature Photonics* 6: 105–110, https://doi.org/10.1038/nphoton.2011.318.

**152** Zhang, Z., Du, J., Zhang, D. et al. (2017). Rosin-enabled ultraclean and damage-free transfer of graphene for large-area flexible organic light-emitting diodes. *Nature Communications* 8: 14560, https://doi.org/10.1038/ncomms14560.

**153** Le Ferrand, H., Bolisetty, S., Demirors, A.F. et al. (2016). Magnetic assembly of transparent and conducting graphene-based functional composites. *Nature Communications* 7: 12078, https://doi.org/10.1038/ncomms12078.

**154** Savchak, M., Borodinov, N., Burtovyy, R. et al. (2018). Highly conductive and transparent reduced graphene oxide nanoscale films via thermal conversion of polymer-encapsulated graphene oxide sheets. *ACS Applied Materials & Interfaces* 10: 3975–3985, https://doi.org/10.1021/acsami.7b16500.

**155** Sun, T., Wang, Z.L., Shi, Z.J. et al. (2010). Multilayered graphene used as anode of organic light emitting devices. *Applied Physics Letters* 96: 133301, https://doi .org/10.1063/1.3373855.

**156** Wu, J., Agrawal, M., Becerril, H.A. et al. (2010). Organic light-emitting diodes on solution-processed graphene transparent electrodes. *ACS Nano* 4: 43–48, https://doi.org/10.1021/nn900728d.

**157** Kim, D., Lee, D., Lee, Y., and Jeon, D.Y. (2013). Work-function engineering of graphene anode by bis(trifluoromethanesulfonyl)amide doping for efficient polymer light-emitting diodes. *Advanced Functional Materials* 23: 5049–5055, https://doi.org/10.1002/adfm201301386.

**158** Jung-Tak Seo, J.H., Lim, T., Lee, K.-H. et al. (2014). Fully transparent quantum dot light-emitting diode integrated with graphene anode and cathode. *ACS Nano* 8: 12476–12482.

**159** Lee, J., Han, T.H., Park, M.H. et al. (2016). Synergetic electrode architecture for efficient graphene-based flexible organic light-emitting diodes. *Nature Communications* 7: 11791, https://doi.org/10.1038/ncomms11791.

**160** Li, N., Oida, S., Tulevski, G.S. et al. (2013). Efficient and bright organic light-emitting diodes on single-layer graphene electrodes. *Nature Communications* 4: 2294, https://doi.org/10.1038/ncomms3294.

**161** Wu, T.L., Yeh, C.H., Hsiao, W.T. et al. (2017). High-performance organic light-emitting diode with substitutionally boron-doped graphene anode. *ACS Applied Materials & Interfaces* 9: 14998–15004, https://doi.org/10.1021/acsami .7b03597.

**162** Liu, L., Shang, W., Han, C. et al. (2018). Two-in-one method for graphene transfer: simplified fabrication process for organic light-emitting diodes. *ACS Applied Materials & Interfaces* 10: 7289–7295, https://doi.org/10.1021/acsami .7b19039.

**163** Liu, L., Ye, D., Dong, R. et al. (2020). Work function-tunable graphene-polymer composite electrodes for organic light-emitting diodes. *ACS Applied Energy Materials* 3: 4068–4077, https://doi.org/10.1021/acsaem.0c00580.

**164** Shin, H., Sharma, B.K., Lee, S.W. et al. (2019). Stretchable electroluminescent display enabled by graphene-based hybrid electrode. *ACS Applied Materials & Interfaces* 11: 14222–14228, https://doi.org/10.1021/acsami.8b22135.

# 10

# Enhanced Carbon-Based Materials and Their Applications

*Poh Choon Ooi*[1], *Mengying Xie*[2], *and Chang Fu Dee*[1]

[1] *Institute of Microengineering and Nanoelectronics (IMEN), Universiti Kebangsaan Malaysia, 43600 Bangi, Selangor, Malaysia*
[2] *State Key Laboratory of Precision Measuring Technology and Instrument, College of Precision Instruments and Opto-electronics Engineering, Tianjin University, Tianjin, 300072, China*

This book covers an overview of enhanced carbon-based nanomaterials, presenting essential aspects, especially properties, synthesis methods, and functionalization of carbon-based nanomaterials. Subsequently, in the following applications, Chapters 3–9, it is apparent through the wide-ranging literature coverage that all forms of carbon nanomaterials have significant promise for diverse electronics applications to enhance the device's performance, including sensors, resistive switching memories, optoelectronic devices, thermoelectrics, piezoelectric energy harvesters, actuators, and displays. The carbon-based nanomaterials enhanced the device's performance attributed to their fascinating unique structural dimensions, and in addition to excellent mechanical, electrical, thermal, optical, and chemical properties.

Although several promising signs of progress have been accomplished, however, some challenges must be addressed before the complete commercial potential of carbon nanomaterials rivals their semiconductor counterpart. For example, in the field of nonvolatile memory (NVM) application, despite many research groups having proposed the potential switching behavior of graphene oxide (GO)-based NVM devices, nevertheless, there is still lacking definite evidence to support the claim. The switching behavior is highly associated with the random distribution of the GO flakes during the device fabrication, deposition process, and intrinsic variability, especially the occurrence and organization of oxygen functionalities [1]. Besides, it is also crucial to ensure the reliable performance of all mentioned devices (sensors, resistive switching memories, optoelectronics, thermoelectrics, piezoelectric energy harvesters, actuators, and displays) to meet the commercial application demands. Additionally, global researchers still need to put further effort into attaining mature and large-scale production with good uniformity graphene-based material synthesis technologies for the commercial market [1, 2].

Besides the previously mentioned applications, in the path toward the development of carbon-based integrated circuits, some progress has been achieved where

*Enhanced Carbon-Based Materials and Their Applications*, First Edition.
Edited by Poh Choon Ooi, Mengying Xie, and Chang Fu Dee.

researchers are able to align carbon nanotube (CNT) arrays with a misalignment of ±9° and tunable density from 100 to 200 CNTs per micrometer on a 10-cm silicon wafer [3]. Researchers from Stanford University attempt to use metallic CNT and graphene as interconnectors to link conventional silicon transistors to form an integrated circuit with a speed of 1 GHz and above [4, 5]. Those are the attempt to transition hybrid devices and circuits as a preparation for future development toward fully carbon-based integrated circuits. Most carbon-based nanostructures are only used as interconnection, which will be discussed separately in this book.

On the other hand, currently, various green synthesis methods of carbon-based nanomaterials are attracting significant attention to produce carbon-based nanomaterials, namely, CNTs and graphene, despite a wide range of conventional synthesis methods that have been developed. Generally, the conventional synthesis methods, for instance, arc discharge, laser ablation, mechanical exfoliation, chemically derived, epitaxial growth on SiC, and chemical vapor deposition to synthesize the CNTs and graphene, demonstrated the disadvantages of expensive structural defects due to high operation temperature, high-energy consumption, low yield, small-sized graphene, time-consuming, impurities, hazardous reducing agents, and challenging to obtain monolayer graphene. Conversely, the green synthesis process shows the advantages of energy-saving reactions or inexpensive starting materials. More importantly, the techniques are capable of converting low-value waste materials into high-value carbon-based nanomaterials. Hence, transforming the waste materials, for example, plastics, biomass, industrial, vegetable, and animal oil, could contribute to sustainable materials processing and recycling technology in the near future [6, 7].

So far, this book mainly focuses on carbon nanomaterials that are utilized to enhance electronics applications. Nonetheless, thanks to the unique properties of carbon at the nanoscale, carbon-based nanomaterials also demonstrated promising potential to be used in other applications. For instance, graphene oxide, graphene, CNTs, graphene, and carbon quantum dots have attracted the interest of researchers to explore their possible contribution in the fields of biomedical (biosensing, drug delivery, and cancer therapy), automotive (brake pad), aerospace, defense, and wastewater treatment applications. Although some satisfactory results have been obtained at the initial stage of the aforementioned studies, the challenges pertaining to sustainable carbon-based products can only enter the consumer market after one-to-two decades. Therefore, academia and industry's intense research of high-quality carbon-based final products is crucial during the devices development stage [8-13].

## References

**1** Shen, Z., Zhao, C., Qi, Y. et al. (2020). Memristive non-volatile memory based on graphene materials. *Micromachines* 11: 341.

**2** Rao, N., Singh, R., and Bashambu, L. (2021). Carbon-based nanomaterials: synthesis and prospective applications. *Materials Today: Proceedings* 44: 608–614.

**3** Liu, L., Han, J., Xu, L. et al. (2020). Aligned, high-density semiconducting carbon nanotube arrays for high-performance electronics. *Science* 368: 850–856.

**4** Close, G.F., Yasuda, S., Paul, B. et al. (2008). A 1 GHz integrated circuit with carbon nanotube interconnects and silicon transistors. *Nano Letters* 8: 706–709.

**5** Chen, X., Akinwande, D., Lee, K. et al. (2010). Fully integrated graphene and carbon nanotube interconnects for gigahertz high-speed CMOS electronics. *IEEE Transactions on Electron Devices* 57: 3137–3143.

**6** Deng, J., You, Y., Sahajwalla, V., and Joshi, R.K. (2016). Transforming waste into carbon-based nanomaterials. *Carbon* 96: 105–115.

**7** Porto, L.S., Silva, D.N., de Oliveira, A.E.F. et al. (2019). Carbon nanomaterials: Synthesis and applications to development of electrochemical sensors in determination of drugs and compounds of clinical interest. *Reviews in Analytical Chemistry* 38: 20190017.

**8** Bertolazzi, S., Bondavalli, P., Roche, S. et al. (2019). Nonvolatile memories based on graphene and related 2D materials. *Advanced Materials* 31: 1806663.

**9** Selvaraj, S.K., Ramesh, R., Narendhra, T. et al. (2021). New developments in carbon-based nanomaterials for automotive brake pad applications and future challenges. *Journal of Nanomaterials* 2021: 6787435.

**10** Nasir, S., Hussein, M.Z., Zainal, Z., and Yusof, N.A. (2018). Carbon-based nanomaterials/allotropes: a glimpse of their synthesis, properties and some applications. *Materials* 11: 295.

**11** Nasrollahzadeh, M., Sajjadi, M., Iravani, S., and Varma, R.S. (2021). Carbon-based sustainable nanomaterials for water treatment: state-of-art and future perspectives. *Chemosphere* 263: 128005.

**12** Sajjadi, M., Nasrollahzadeh, M., Jaleh, B. et al. (2021). Carbon-based nanomaterials for targeted cancer nanotherapy: recent trends and future prospects. *Journal of Drug Targeting* 29: 716–741.

**13** Maiti, D., Tong, X., Mou, X., and Yang, K. (2019). Carbon-based nanomaterials for biomedical applications: a recent study. *Frontiers in Pharmacology* 9: 1401.

# Index



*Enhanced Carbon-Based Materials and Their Applications*, First Edition.
Edited by Poh Choon Ooi, Mengying Xie, and Chang Fu Dee.